MTP International Review of Science

Volume 12

Analytical Chemistry—Part 1

Edited by **T. S. West**
Imperial College, University of London

Butterworths · London
University Park Press · Baltimore

THE BUTTERWORTH GROUP

ENGLAND
Butterworth & Co (Publishers) Ltd
London: 88 Kingsway, WC2B 6AB

AUSTRALIA
Butterworths Pty Ltd
Sydney: 586 Pacific Highway 2067
Melbourne: 343 Little Collins Street, 3000
Brisbane: 240 Queen Street, 4000

NEW ZEALAND
Butterworths of New Zealand Ltd
Wellington: 26–28 Waring Taylor Street, 1

SOUTH AFRICA
Butterworth & Co (South Africa) (Pty) Ltd
Durban: 152–154 Gale Street

ISBN 0 408 70273 7

UNIVERSITY PARK PRESS

U.S.A. and CANADA
University Park Press Inc
Chamber of Commerce Building
Baltimore, Maryland, 21202

Library of Congress Cataloging in Publication Data

West, Thomas Summers.
  Analytical chemistry.

  (Physical chemistry, series one, v. 12–13)
  (MTP international review of science)
  1. Chemistry, Analytic. I. Title.
QD453.2.P58 Vol. 12–13 [QD75.2] 541'.3'08s [543]
ISBN 0–8391–1026–X (pt. 1)   72–8644

First Published 1973 and © 1973
MTP MEDICAL AND TECHNICAL PUBLISHING CO. LTD.
Seacourt Tower
West Way
Oxford, OX2 OJW
and
BUTTERWORTH & CO. (PUBLISHERS) LTD.

Filmset by Photoprint Plates Ltd., Rayleigh, Essex
Printed in England by Redwood Press Ltd., Trowbridge, Wilts
and bound by R. J. Acford Ltd., Chichester, Sussex

Consultant Editor's Note

The MTP International Review of Science is designed to provide a comprehensive, critical and continuing survey of progress in research. The difficult problem of keeping up with advances on a reasonably broad front makes the idea of the Review especially appealing, and I was grateful to be given the opportunity of helping to plan it.

This particular 13-volume section is concerned with Physical Chemistry, Chemical Crystallography and Analytical Chemistry. The subdivision of Physical Chemistry adopted is not completely conventional, but it has been designed to reflect current research trends and it is hoped that it will appeal to the reader. Each volume has been edited by a distinguished chemist and has been written by a team of authoritative scientists. Each author has assessed and interpreted research progress in a specialised topic in terms of his own experience. I believe that their efforts have produced very useful and timely accounts of progress in these branches of chemistry, and that the volumes will make a valuable contribution towards the solution of our problem of keeping abreast of progress in research.

It is my pleasure to thank all those who have collaborated in making this venture possible – the volume editors, the chapter authors and the publishers.

Cambridge   A. D. Buckingham

MTP International Review of Science

Analytical Chemistry—Part 1

MTP International Review of Science

Publisher's Note

The MTP International Review of Science is an important new venture in scientific publishing, which we present in association with MTP Medical and Technical Publishing Co. Ltd. and University Park Press, Baltimore. The basic concept of the Review is to provide regular authoritative reviews of entire disciplines. We are starting with chemistry because the problems of literature survey are probably more acute in this subject than in any other. As a matter of policy, the authorship of the MTP Review of Chemistry is international and distinguished; the subject coverage is extensive, systematic and critical; and most important of all, new issues of the Review will be published every two years.

In the MTP Review of Chemistry (Series One), Inorganic, Physical and Organic Chemistry are comprehensively reviewed in 33 text volumes and 3 index volumes, details of which are shown opposite. In general, the reviews cover the period 1967 to 1971. In 1974, it is planned to issue the MTP Review of Chemistry (Series Two), consisting of a similar set of volumes covering the period 1971 to 1973. Series Three is planned for 1976, and so on.

The MTP Review of Chemistry has been conceived within a carefully organised editorial framework. The over-all plan was drawn up, and the volume editors were appointed, by three consultant editors. In turn, each volume editor planned the coverage of his field and appointed authors to write on subjects which were within the area of their own research experience. No geographical restriction was imposed. Hence, the 300 or so contributions to the MTP Review of Chemistry come from many countries of the world and provide an authoritative account of progress in chemistry.

To facilitate rapid production, individual volumes do not have an index. Instead, each chapter has been prefaced with a detailed list of contents, and an index to the 13 volumes of the MTP Review of Physical Chemistry (Series One) will appear, as a separate volume, after publication of the final volume. Similar arrangements will apply to the MTP Review of Organic Chemistry (Series One) and to subsequent series.

Butterworth & Co. (Publishers) Ltd.

**Physical Chemistry
Series One**
Consultant Editor
A. D. Buckingham
*Department of Chemistry
University of Cambridge*

Volume titles and Editors

1 THEORETICAL CHEMISTRY
 Professor W. Byers Brown, *University of Manchester*

2 MOLECULAR STRUCTURE AND PROPERTIES
Professor G. Allen, *University of Manchester*

3 SPECTROSCOPY
Dr. D. A. Ramsay, F.R.S.C.,
National Research Council of Canada

4 MAGNETIC RESONANCE
Professor C. A. McDowell, F.R.S.C.,
University of British Columbia

5 MASS SPECTROMETRY
Professor A. Maccoll, *University College, University of London*

6 ELECTROCHEMISTRY
Professor J. O'M Bockris, *University of Pennsylvania*

7 SURFACE CHEMISTRY AND COLLOIDS
Professor M. Kerker, *Clarkson College of Technology, New York*

8 MACROMOLECULAR SCIENCE
Professor C. E. H. Bawn, F.R.S.,
University of Liverpool

9 CHEMICAL KINETICS
Professor J. C. Polanyi, F.R.S.,
University of Toronto

10 THERMOCHEMISTRY AND THERMODYNAMICS
Dr. H. A. Skinner, *University of Manchester*

11 CHEMICAL CRYSTALLOGRAPHY
Professor J. Monteath Robertson, F.R.S.,
University of Glasgow

12 ANALYTICAL CHEMISTRY —PART 1
Professor T. S. West, *Imperial College, University of London*

13 ANALYTICAL CHEMISTRY — PART 2
Professor T. S. West, *Imperial College, University of London*

INDEX VOLUME

Physical Chemistry
Series One

Consultant Editor
A. D. Buckingham

Preface

During recent years analytical chemistry has probably undergone more radical change than most other branches of chemistry. The increased sophistication of instrumentation has of course been an outstanding feature, but more fundamental changes are also evident in the type of information which is now being provided. On the elemental side, techniques such as atomic absorption and atomic fluorescence spectroscopy have provided completely unequivocal direct analysis for virtually every metal in the periodic table. Selective-ion electrodes take over for anions and non-metals where flame spectroscopy leaves off and immobilised enzyme and antibiotic membranes offer particularly exciting new vistas of development. Not only do analytical techniques now offer increased sensitivity and selectivity of analysis, they also provide information on structure, conformation, location on surfaces and many other aspects.

In common with the other volumes of this series, these on analytical chemistry present reviews of selected areas by authors who are readily recognised authorities on their topics. Each contributor has been asked, wherever practical, to make a selective assessment of the present status of the subject rather than write a comprehensive review. Some areas are 'new', e.g. electron spectroscopy for chemical analysis and selective-ion electrodes, whilst others are well tried and fully recognised, but have been subject to recent developments, e.g. organic microanalysis and liquid–liquid distribution (solvent extraction). In all cases, the authors have been given complete freedom to review their topics individually within a minimum framework of size and scope. The authors have been drawn deliberately from a wide range of countries in an attempt to present as geographically balanced a view of analytical chemistry as possible. The editor has made a conscious effort to avoid duplication, but obviously some repetition is inevitable in topics which are inter-related. It is probable, however, that such duplication as exists may be beneficial rather than otherwise since two ways of expressing a common theme, particularly from different individual and national backgrounds, are more likely to be complementary and enlightening than purely repetitive or confusing.

Many topics that could or should have been reviewed are obviously missing; the selection of topics is, arbitrarily, that of the editor.

London   T. S. West

Contents

1
Electron Spectroscopy for Chemical Analysis

C. J. ALLAN and K. SIEGBAHN
University of Uppsala, Sweden

1.1 INTRODUCTION

1.1.1 Brief historical introduction

In the past, the study of matter and its interaction with radiation has been confined almost exclusively to the measurement and detection of electromagnetic radiation. The excitation and de-excitation of atoms and molecules have principally been studied by means of the photons which are absorbed or emitted when an electron makes a transition from one quantised state to another. Such investigations have been continuously refined so that today the techniques of spectroscopy in the visible, infrared, ultraviolet, and x-ray regions of the electromagnetic field are well developed and of wide application. These methods, together with those of n.m.r., e.s.r., Mössbauer spectroscopy and microwave spectroscopy, have been chiefly responsible for our understanding of atomic and molecular structure.

An alternate method of studying the structure of atoms and molecules is to study the electrons emitted when a sample is irradiated. It has long been known that, when a metal or other material is irradiated with ultraviolet light, electrons are emitted. As early as 1905 Einstein postulated his law of photoelectricity

$$E = h\nu - w \qquad (1.1)$$

where E is the energy of the emitted electron, ν is the frequency of the incident radiation, h is Planck's constant, and w is the minimum energy required to remove an electron from the irradiated sample and is a constant characteristic of the material. The enunciation of this law was, in fact, one of the early

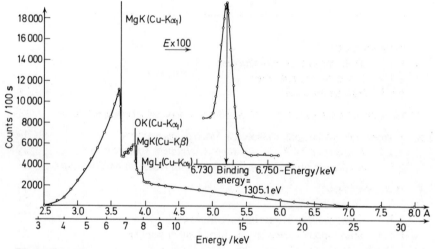

Figure 1.1 Electron spectrum obtained from magnesium oxide with copper x-radiation. Edges are found at energies corresponding to atomic levels of magnesium and oxygen. A sharp electron line can be resolved from each edge as shown in the insert where the energy scale has been expanded by a factor of 100

steps in the development of the quantum theory. Although the early work on the photoelectric effect involved the emission of conduction electrons from metals, equation (1.1) may be applied to photo-electrons ejected from deeper lying atomic (molecular) orbitals. Then $E = h\nu - E_B$ where E_B is the binding energy of the electron.

Despite its rather long history, the study of photo-electrons as a tool in the investigation of matter met with limited success until the middle of the 1950s. Although the photo-electrons have a well-defined energy on leaving the atom (molecule) they are energy-degraded as they pass through the sample and the energy spectrum of the emitted electrons does not show a well-defined peak, but is characterised by a long tail with an edge at the high-energy side. The position of this high energy edge is a measure of the energy of the electrons before degradation but the accuracy with which it could be determined was rather poor, which limited electron spectroscopy as a competitor to other spectroscopic techniques.

The development of high-resolution, double focusing electron spectrometers as a tool in nuclear physics[1] initiated a new approach to the old problem and high-resolution studies of photo-electrons induced by soft x-rays were begun. Some difficulties were encountered in designing and developing the appropriate equipment for this purpose but, after a few years development, the first successful attempts were made to record an x-ray induced electron spectrum with adequate resolution. The resulting spectrum, Figure 1.1, showed the general features described above but it was found that sharp lines could be resolved from the edges of the energy veils. These lines correspond to electrons which have not undergone any energy losses. Thus they provide direct measures of the binding energies of the inner shells from which the electrons were ejected. The accuracy with which an electron peak could be located, and hence the accuracy with which the binding energy could be measured, was comparable to or better than that of x-ray spectroscopy and was essentially set by the inherent width of the atomic levels via the uncertainty principle.

Since that time, electron spectroscopy has developed rapidly until, today, it is an established branch of spectroscopy. In this article we shall present a brief review of some of the principles and applications of this technique. For the most part we shall confine ourselves to a discussion of electrons excited from free molecules, i.e. gases, using soft x-rays. The study of solids by means of ESCA and the use of ultraviolet radiation to induce spectra are also important areas of research but in this review we shall deal with these topics only briefly. In addition to discussing well-developed principles[2, 3] we shall also mention a few areas which have not been fully examined but which show signs of promise.

1.1.2 Some basic principles

If a photon of energy hv, is incident on a free molecule, the molecule may absorb the photon and eject an electron. The energy of the emitted photoelectron is then given by $E = hv - E_B$ where E_B is the ionisation energy of the electron in the molecule. The electrons of a molecule occupy discrete energy levels and the resulting electron spectrum shows distinct and separate peaks corresponding to these levels. Figure 1.2, for example, shows the ESCA spectra of the simplest molecules, the mono-atomic noble gases, induced by Mg-Kα radiation, 1253.6 eV. In the orbital picture of atomic (molecular) structure we think of the different electronic levels as corresponding to different atomic (molecular) orbitals. The different peaks observed in an ESCA spectrum then correspond to electrons being ejected from the different occupied orbitals of the molecule. The ionisation energy, E_B, introduced above is then given by $E_B = E - E^+$ where E is the total electrostatic binding energy of the neutral species, i.e. the total energy required to remove all the electrons and nuclei to infinity, and E^+ is the corresponding binding energy of the resulting ion.

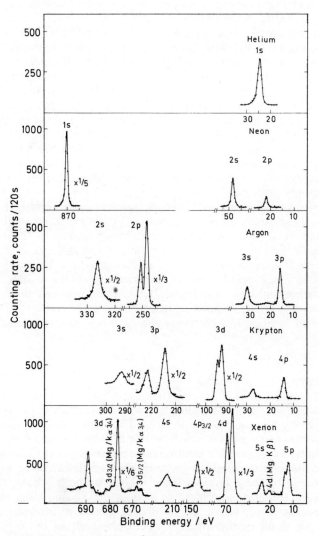

Figure 1.2 ESCA spectra from the noble gases He, Ne, Ar, Kr and Xe excited by Mg-Kα radiation

The width of the observed photo-electron peak depends on three factors, (i) the inherent width of the exciting radiation, (ii) the inherent width of the level under study, and (iii) the resolution of the analyser, typically $c.\ 5 \times 10^{-4}$. When using x-rays to excite spectra, the first of these factors usually predominates. For example, the observed full width at half maximum intensity of the neon 1s line, Figure 1.2, is 0.80 eV while the natural width of the Mg-K$\alpha_{1,2}$ radiation used to excite the spectrum is $c.0.7$ eV. The most commonly used x-rays are the Mg-Kα line, 1253.6 eV, and the Al-Kα line, 1486.6 eV. It turns out that for most applications these two lines are adequate. If it is necessary to excite electrons from core levels which cannot be reached by these x-rays then characteristic x-rays from a higher Z material must be used, but one then suffers a corresponding loss in resolution because of an increase in the natural line width of the x-rays. Using Mg-Kα, line widths of 1.0 eV for the core levels of gaseous compounds are typically obtained.

A great deal of work has been done using ultraviolet radiation to excite the electrons. The He(I) resonance line, 21.21 eV, has principally been used,

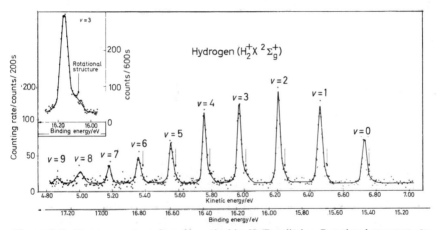

Figure 1.3 Electron spectrum from H_2 excited by He(I) radiation. Rotational structure can be resolved from the vibrational peaks as indicated by the insert

but He(II), 40.81 eV, and Ne(I), 16.67 and 16.85 eV, have also been employed. Using He(I) only those orbitals with a binding energy $\lesssim 20$ eV can be reached and, therefore, only the outer orbitals can be studied. On the other hand, the much lower kinetic energy of the electrons and the small inherent width of the radiation results in line widths $c.\ 10$–20 meV. Under these conditions the vibronic structure of the lines can be resolved. Figure 1.3, for example, shows the spectrum excited from H_2, demonstrating the vibration of the H_2^+ ion excited during the photo-ionisation process. Here, the edge on the high energy side of the vibration peaks is rotational structure. In Figure 1.2 the argon 3p doublet is not resolved. Figure 1.4 shows the same doublet excited by He(I) radiation. Here the two lines are completely resolved.

For completeness it should be pointed out that a complementary method of examining the electronic structure of matter is to study the Auger and auto-ionisation spectra of atoms and molecules. If, for example, an ion has a vacancy in the 1s orbital, K shell, then the ion may de-excite by an electron transferring from the $2p_{\frac{1}{2}}$ orbital (L_{II} shell). When this occurs, a photon with energy $hv = E_K - E_{L_{II}}$ may be emitted *or* a secondary electron from another shell, for example, the L_{III} shell, may be emitted. When the original

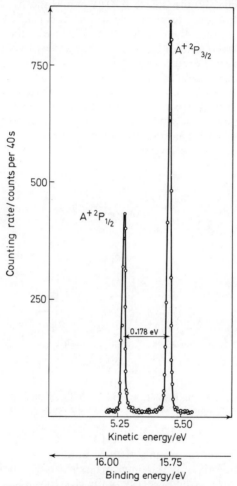

Figure 1.4 The Ar-3p doublet excited by He(I) radiation

species is an ion, the process is due to the Auger effect and when the original species is an excited, but neutral species, the process is due to auto-ionisation. In either case the emitted electron has a line width which is independent of the exciting radiation and a unique energy related to the orbital binding energies of the electrons involved. Thus for the $KL_{II}L_{III}$ Auger transition discussed above we have

$$E_{kin} = E_K^* - E_{L_{II}, L_{III}}^{**}$$

where E_K^* is the electrostatic binding energy of the ion with a K shell vacancy and $E_{L_{II}, L_{III}}^{**}$ is the binding energy of the doubly charged ion, having vacancies in the L_{II} and L_{III} shell. As might be expected, the Auger and auto-ionisation processes give rise to complicated spectra, the interpretation of which is not straightforward. Auger and auto-ionisation spectra can be excited by x-rays, but the yield obtained using electron bombardment is much higher and this is the normal method of exciting such spectra. Figure 1.5 shows part of the carbon Auger and auto-ionisation spectrum from CO excited by electron impact. Here, the observed widths are dependent only on the natural widths of the levels involved and the resolution of the spectrometer. Again, it is possible to resolve vibrational structure, as indicated in the figure. Of particular interest in this figure is the intense line at 250.5 eV, which has been

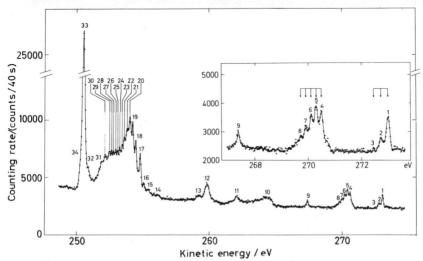

Figure 1.5 Part of the carbon Auger and auto-ionisation electron spectrum from CO excited by electron impact. The insert figure shows the vibrational structure in some of the auto-ionisation lines

identified as involving the C-1s level[4]. The width of this peak, including the resolution of the spectrometer, is 0.18 eV and this implies that the natural line width of the C-1s level must be less than 0.18 eV.

When one excites spectra from solid samples rather than gaseous ones, the energy of the photo-electron is given by

$$E = h\nu - E_B - e\Phi$$

where Φ is the work function. For metal samples Φ is a constant of the spectrometer. For an insulator the situation is not so clear because of the possibility of a charging up of the sample. Hnatowich et al.[5], for example, have studied the charging of $BaSO_4$ and have observed shifts of c. 2 eV, by depositing thin layers of gold or palladium on to the sample and observing the apparent change in the binding energies of core lines from the metals. By biasing the sample and observing that the signals from the $BaSO_4$ and from the metal, deposited on to the salt, shifted by equal amounts they also demonstrated that the metal was in electrical equilibrium with the salt.

Thus they were able to correct for the charge-up of the sample within an accuracy of 0.1 eV.

For solid samples an important property of the ESCA technique is the small amount of material required to obtain a signal. By measuring the intensity of the Au-4f$_{7/2}$ peak, ($E_B \approx 83$ eV), relative to that of the Cr-3p line, ($E_B \approx 43$ eV), for different thicknesses of Au deposited on to a Cr backing, Baer[6] et al. have shown that, for Mg-Kα, the mean escape depth of these photo-electrons is c. 22 Å. These results imply that one can obtain measurable signals from even a fraction of a molecular monolayer of material. However, great care must be taken in the preparation of solid samples so that surface contamination is reduced as much as possible. Often this can be done by the simple expedient of heating the source to a few hundred degrees, either in

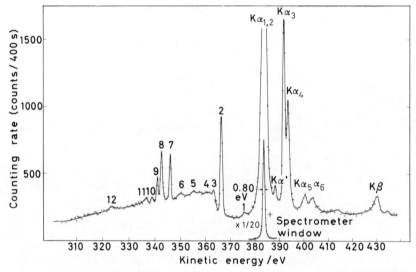

Figure 1.6 Ne-1s electron spectrum excited by Mg-Kα radiation. The main peak (0) is the Ne 1s line excited by Mg-K$\alpha_{1,2}$. To the higher-kinetic-energy side of this peak are lines excited by Mg-K x-ray satellites. The peaks to the lower kinetic energy side are due to shake-up, shake-off and inelastic scattering as explained in the text

the vacuum of the spectrometer, typically c. 10^{-5}–10^{-6} Torr, or in a hydrogen atmosphere of c. 10^{-2}–10^{-1} Torr. When this is not possible, as for example in the investigation of low-temperature effects, ultra-high vacuum conditions are required.

Whereas for solids, the basic requirement is surface purity, for gases the main point of interest is the pressure in the collision chamber. One, of course, wants to obtain as high a signal intensity as possible and since the density of target molecules is directly proportional to the gas pressure one wants to operate at as high a pressure as possible. There are, however, two considerations which limit the pressure. Firstly, the pressure in the energy analyser must be kept low so that the mean free path of the electron is at least as long as the distance the electron must travel through the analyser. This usually means a pressure in the analyser of less than 10^{-4} Torr. By using differential pumping one can conveniently maintain a pressure gradient difference between the analyser and the source of about a factor of 10^3–10^4. This then limits the source pressure to *c*. 0.1–1 Torr. Secondly, as one increases the pressure in the collision chamber, the mean path of the electrons decreases. If the mean free path becomes smaller than the dimensions of the collision chamber, the intensity of the main peak then falls because electrons are lost through inelastic collisions with the molecules. Such inelastically scattered electrons produce line spectra to the low-kinetic-energy side of the main ESCA peak. Figure 1.6, for example, shows the electron spectrum of Ne, excited by Mg-Kα, in the region of the 1s peak. The peaks labelled 2, 3 and 4 have been identified as arising from inelastically scattered electrons. Although such peaks can provide information on excited states of the neutral molecule they are potentially troublesome when comparing intensities and when studying 'shake-up' processes. This later process will be dealt with later in this article. The effect of these considerations is to limit the useful pressure range to the region from *c*. 10^{-2} to *c*. 10^{-1} Torr.

1.1.3   Instrumentation

Figure 1.7 illustrates the basic principles of electron spectroscopy, showing sample, excitation source (x-rays, u.v.-radiation or electrons), electron-energy analyser, and a detector for registering the analysed electrons. The electrons excited from the sample are analysed by either an electrostatic or

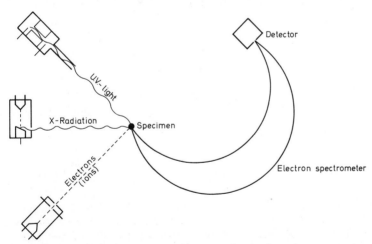

Figure 1.7   Principle of electron spectroscopy showing its main components: specimen, radiation source, electron-energy analyser and detector

a magnetic spectrometer. External magnetic fields including that of the earth must be eliminated and this is done by means of a Helmholtz coil system or, in the case of an electrostatic instrument, by means of mu-metal shields. Gases can be studied directly by using differential pumping to maintain a high vacuum in the spectrometer, typically *c*. 10^{-6} Torr, or by condensing the gas on to a cold finger in the sample chamber. Solid samples, preferably thin, should have surfaces free of impurities and adsorbed gases. Besides heating the sample, this requirement is often met by using ion bombardment to clean the sample, *in situ*, or by producing the sample, *in situ*, by sublimation. The sample handling technique is presently subject to careful studies and so far no generally applicable treatment can be recommended for routine work. Various commercial instruments, therefore, have different methods for sample preparation and treatment. Details can be found in the brochures of the manufacturers, such as Varian Associates, A.E.I., Vacuum Generators, McPherson, Hewlett-Packard, and, for u.v.-excitation, Perkin Elmer.

When using x-ray excitation, the electron line widths obtainable are essentially set by the inherent width of the exciting x-ray line. One can

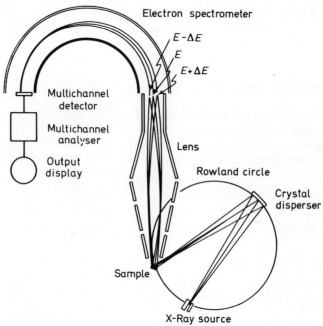

Figure 1.8 An ESCA spectrometer using x-ray monochromation and dispersion compensation

overcome this limitation by monochromating the x-rays by means of x-ray diffraction. This technique has been incorporated in the recent design of Hewlett-Packard. By means of 'dispersion compensation' the total x-ray line is used for excitation, but the inherent width of the x-ray does not contribute to the electron line width. Figure 1.8 shows the principle of this design. A spherically-bent quartz crystal diffracts the Al-Kα line along the Rowland circle. The crystal has to be of such a quality that the various wavelengths contributing to the finite width of the x-ray are correctly dispersed. The photo-electrons from a given shell are then expelled from the sample surface with slightly varying energies, depending on the geometrical position of the sample surface element. The next step is to project—at the

Si 2p's: Graphical deconvolution   [silicon-wafer]

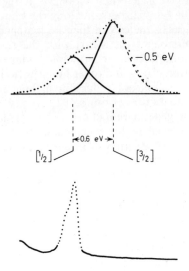

Figure 1.9   The Si 2p$_\frac{1}{2}$, 2p$_\frac{3}{2}$ doublet obtained monochromatised Al-Kα x-rays

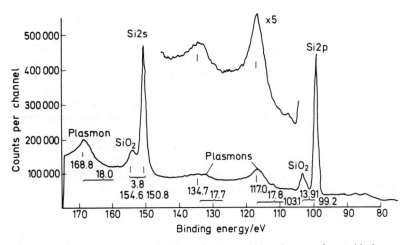

Figure 1.10 ESCA spectrum of silicon showing evidence for a surface oxide layer and showing first and second order plasmons

entrance of the analyser – an electron image of the irradiated section of the sample, using the expelled photo-electrons, where the electrons of different energy are arranged in a radial sequence at the analyser entrance which exactly reflects the x-ray dispersion along the sample surface. The dispersion of the electron spectrometer can then be matched to compensate precisely the crystal dispersion of the x-ray, with the result that all electrons are brought to a common focus, in principle, independently of their spread in energies. Aberration errors and other imperfections result in a finite electron line width, but a significant reduction in line width is, in fact, obtained. Furthermore, contributions from satellite x-ray lines in the spectrum are eliminated and a general reduction in background is achieved. A multi-detector system placed in the focal plane of the spectrometer more than compensates for the reduction in intensity caused by the x-ray diffraction part of the arrangement.

Figures 1.9 and 1.10 are illustrative spectra, taken at high resolution, with this instrument. Figure 1.9 shows a detailed study of the Si-$2p_{1/2,\,3/2}$ doublet. These levels are separated by 0.6 eV and a deconvolution of the spectrum yields the two resolved lines, each having a line width of 0.5 eV. Figure 1.10 is a wide scan from a Si crystal which has been briefly exposed to air. One can observe low intensity lines to the low kinetic energy side of each of the Si-2p and -2s lines. These are characteristic of SiO_2 and demonstrate that a few angstroms thick layer of SiO_2 was formed on the surface. Peaks attributable to plasmon excitations can also be seen in the Si spectrum. Furthermore, it can be observed that the 2s and 2p lines are of different widths.

This brief discussion of ESCA instrumentation can only serve as an introduction to the topic. For a more complete treatment the reader may consult Refs. 2, 3, 7 and 8.

1.2   ESCA SPECTRA FROM THE VALENCE REGION OF FREE MOLECULES

Historically, chemists have viewed the binding between atoms of a molecule as a property of the valence electrons and have neglected the inner electrons, regarding them as forming an inert core. This distinction between core electrons and valence electrons is maintained in ESCA. The core electrons are effected by the inter-atomic binding, as is clearly illustrated by chemical shifts, but this effect is secondary, reflecting the distribution of valence electrons among the atoms of the molecule. The separation of core and valence electrons is well demonstrated by the electron spectrum obtained from a simple molecule like CO (Figure 1.11). At low kinetic energy (high binding energy), we can see two peaks corresponding to ionisation from the O-1s and C-1s core orbitals. There is then a wide region which is free of lines and then in the range of ionisation energies from *c.* 10 to 40 eV one can see a number of peaks corresponding to ionisation of valence electrons. It is, therefore, natural to distinguish between the spectra of core electrons and those of valence electrons and we begin by considering the valence region.

In a molecule, as in an atom, we think of the electrons as occupying definite orbitals, with different orbitals corresponding to different binding energies. The innermost core orbitals are thought to be essentially atomic in character and are highly localised on their respective atoms, in keeping with our concept of an inert core. The outermost, valence orbitals, on the other hand, extend throughout the molecule and are responsible for the binding of the atoms in the molecule. This qualitative picture of molecular orbitals is conveniently described quantitatively in the Hartree–Fock approximation.

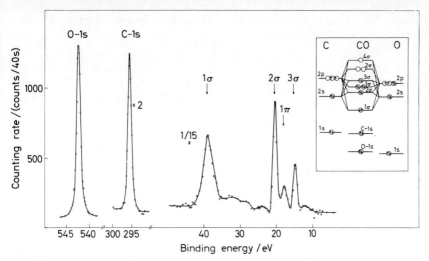

Figure 1.11 Electron spectrum of carbon monoxide excited by Mg-Kα radiation. Both the core and valence orbital regions of the spectrum are shown

The Hartree–Fock approach to molecular orbital theory has been dealt with extensively in a number of texts and articles and we shall only briefly discuss it in so far as it applies to electron spectroscopy.

For a closed shell system of $2n$ electrons we assume that the electrons move in spin orbitals, $\psi = R(r)\chi(s)$, and that the total wave function is given by

$$\Psi = \frac{1}{2\sqrt{n}} \sum_P (-1)^P P[R_1(1)\alpha(1)R_1(2)\beta(2) \cdots R_N(2N)\beta(2N)]$$

where the sum is over all permutations P of the electron coordinates. The Hamiltonian, in atomic units, is given by

$$H = \sum_i \left[-\tfrac{1}{2}\nabla_i^2 - \sum_A \frac{Z_A}{r_{iA}} \right] + \tfrac{1}{2}\sum_{i,j}\frac{1}{r_{ij}} = \sum_i H(i) + \tfrac{1}{2}\sum_{i,j}\frac{1}{r_{ij}}$$

Then the total energy of the system, assuming orthogonality of the molecular orbitals is

$$E = 2\sum_i^n H_{ii} + \sum_{i,j}(2J_{ij} - K_{ij}) \tag{1.2}$$

where

$$H_{ii} = \langle R_i(1)|H(1)|R_i(1)\rangle \tag{1.3}$$

$$J_{ij} = \langle R_i(1)R_j(2)|1/r_{ij}|R_i(1)R_j(2)\rangle \tag{1.4}$$

$$K_{ij} = \langle R_i(1)R_j(2)|1/r_{ij}|R_i(2)R_j(1)\rangle \tag{1.5}$$

The molecular orbitals are usually approximated by a linear combination of atomic orbitals, LCAO, i.e. we set

$$R_i = \sum_\mu C_{\mu i} \Phi_\mu \tag{1.6}$$

where the Φ_μ are atomic functions. In order for the R_i to be orthonormal the number of atomic orbitals in the expansion must be greater than, or equal to, the number of occupied molecular orbitals and the expansion coefficients must satisfy the constraint

$$\sum_{\mu\nu} C_{\mu i}^* C_{\nu j} S_{\mu\nu} = \delta_{ij}$$

where $S_{\mu\nu}$ is the overlap integral $\langle \Phi_\mu(1) | \Phi_\nu(1) \rangle$.

Substituting equation (1.6) into equations (1.3), (1.4) and (1.5) we obtain:

$$H_{ii} = \sum_{\mu\nu} C_{\mu i}^* C_{\nu i} \langle \Phi_\mu(1) | H(1) | \Phi_\nu(1) \rangle = \sum_{\mu\nu} C_{\mu i}^* C_{\nu i} H_{\mu\nu} \tag{1.7}$$

$$J_{ij} = \sum_{\mu\lambda\nu\sigma} C_{\mu i}^* C_{\lambda j}^* C_{\nu i} C_{\sigma j} \langle \mu\nu | \lambda\sigma \rangle \tag{1.8}$$

$$K_{ij} = \sum_{\mu\lambda\nu\sigma} C_{\mu i}^* C_{\lambda j}^* C_{\nu i} C_{\sigma j} \langle \mu\lambda | \nu\sigma \rangle \tag{1.9}$$

where $\langle \mu\nu | \lambda\sigma \rangle = \langle \Phi_\mu(1)\Phi_\lambda(2) | r_{12}^{-1} | \Phi_\nu(1)\Phi_\sigma(2) \rangle$

Setting $P_{\mu\nu} = 2 \sum_i C_{\mu i}^* C_{\nu i}$ we then obtain for the total energy

$$E = \sum_{\mu\nu} P_{\mu\nu} H_{\mu\nu} + \tfrac{1}{2} \sum_{\mu\nu\lambda\sigma} P_{\mu\nu} P_{\lambda\sigma} [\langle \mu\nu | \lambda\sigma \rangle - \tfrac{1}{2} \langle \mu\lambda | \nu\sigma \rangle] \tag{1.10}$$

The Hartree–Fock procedure consists of finding a minimum value of E by varying the coefficients $C_{\mu i}$ subject to the constraint

$$\sum_{\mu\nu} C_{\mu i}^* C_{\nu j} S_{\mu\nu} = \delta_{ij}$$

We then obtain the Roothaan equations

$$\sum_\nu (F_{\mu\nu} - E_i S_{\mu\nu}) C_{\nu i} = 0 \tag{1.11}$$

where $F_{\mu\nu} = H_{\mu\nu} + \sum_{\lambda\sigma} P_{\lambda\sigma} [\langle \mu\nu | \lambda\sigma \rangle - \tfrac{1}{2} \langle \mu\lambda | \nu\sigma \rangle]$.

Here the eigenvalues, E_i, can be shown to be

$$E_i = H_{ii} + \sum_{j=1}^n (2J_{ij} - K_{ij}) \tag{1.12}$$

and are called the one-electron orbital energies. The ith orbital energy, E_i, is essentially the energy of the electron in R_i interacting with the nuclei and the other $2n - 1$ electrons.

Since the Fock matrix $F_{\mu\nu}$ is a quadratic function of the coefficients, $C_{\mu i}$, the Roothaan equations of the LCAO coefficients are cubic and must be solved by a self-consistent iterative procedure, and the calculations are referred to as self-consistent field calculations, SCF. The Roothaan equations may be solved *ab initio*, where all electrons are included and where all integrals are evaluated. Using present day computers the one-centre integrals can be evaluated in a matter of seconds, or at most minutes, but the two-centre integrals can require several hours. Considerable reduction in computation time can be achieved by using the symmetry of the molecule. Examples of two programmes presently used for *ab initio* calculations which include this feature are IBMOL[9] and REFLECT[10]. By introducing various approximations an appreciable saving in computer time can be obtained but with a reduction in the significance of the results. The most commonly-used approximation is the CNDO method introduced by Pople, Santry and Segal[11]. The basic approximation is that of zero differential overlap for all products of different atomic orbitals $\Phi_\mu \Phi_\nu$. Then

$$\langle \mu\nu | \lambda\sigma \rangle = \delta_{\mu\nu} \delta_{\lambda\sigma} \langle \mu\mu | \lambda\lambda \rangle \qquad (1.13)$$

Further, only valence electrons are treated explicitly, the inner shells being treated as part of a rigid core.

For a more complete discussion of the derivation of the Roothaan equations and their solution the reader may refer to texts on the subject, such as those by McWeeny and Sutcliffe[12] and Pople and Beveridge[13], among others.

One of the applications of such calculations to electron spectroscopy arises because the orbital energies, E_i, approximate the ionisation energy of an electron in the molecular orbital R_i. This approximation is based on the assumption that there is no re-organisation of the other $2n-1$ electrons on ionisation and is known as Koopmans' theorem, or Koopmans' approximation. The calculated orbital energies thus provide a valuable guide for the interpretation of valence-orbital specta. Let us consider, for example, the case of CO. The experimental ionisation energies obtained from the spectrum shown in Figure 1.11 are compared with SCF orbital energies taken from a calculation by Neumann and Moskowitz[14] in Table 1.1. We have included the core levels for completeness. The labelling of the orbitals is based on the irreducible representation of the symmetry group of the molecule, $C_{\infty v}$. See, for example, Schonland[15].

Table 1.1 Comparison of the observed ionisation energies of CO and those obtained from SCF calculations using Koopmans' theorem

Orbital	Ionisation energy	Calculated orbital energy
3σ	14.5	15.08
1π	17.2	17.35
2σ	20.1	21.85
1σ	38.3	41.40
C-1s	295.9	309.13
O-1s	542.1	562.21

Here we can see a very good correlation between the orbital energies and the ionisation potentials and one can be quite confident about the assignment of the peaks in the valence orbital region of the electron spectrum. Such good agreement between calculated orbital energies and ionisation energies is not always the case, but one can frequently determine the origin of the various lines simply by comparing the two. It should also be noted that quite often the peaks one obtains when using x-rays for excitation are often composites containing contributions from two or more closely spaced levels. In such cases the higher resolution obtained using u.v. radiation is usually sufficient to separate the lines, provided of course, that the ionisation energy is lower than the u.v. cut-off. Deep lying molecular orbitals, however, can only be reached using x-ray excitation. In many cases then, the information obtained from the two types of excitation is complementary and may be combined to give a better picture of the molecular structure.

The spectrum of CO shows another interesting feature. The intensities of the valence peaks, i.e. the area under the curves, varies considerably from peak to peak. Hence the peak intensity does not reflect, directly, the number of electrons in a particular orbital. If one looks at the spectra from the noble gases, Figure 1.2, one can see this more directly. Thus for Ne, the 2s line is *c.* three times as intense as the 2p peak even though there are three times as many 2p electrons as 2s electrons. This then means that the photo-ionisation

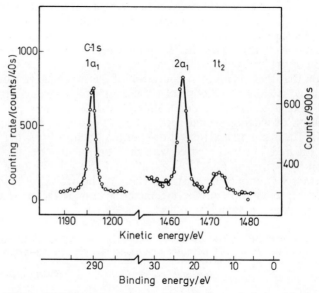

Figure 1.12 Electron spectrum of methane excited by Mg-Kα radiation

cross-section for a Ne-2s electron is *c.* 9 times greater than for a Ne-2p electron. Similarly for Ar the 3s photo-ionisation cross-section is *c.* 1.4 times the 3p cross-section, while for Kr the 4s and 4p cross-sections are equal. These numbers, of course, hold only for Mg-Kα radiation. It appears, from a large body of experimental data, that the photo-ionisation cross-section for valence electrons in a molecule shows a similar dependence. This dependence is best described in terms of the LCAO approximation. Thus for a given molecular orbital we can define a fractional atomic parentage, $P_i(A\lambda)$, which gives the relative strength of the λ type atomic orbital, centred on atom A, in the LCAO expansion of the *i*th molecular orbital. Using CO as an example $P_{1\sigma}(C\text{-}2s)$ is the relative carbon 2s atomic character of the 1σ molecular orbital. The photo-ionisation cross-section for the *i*th molecular orbital, I_i, is then related to the atomic parentage of that orbital by an expression of the form

$$I_i = \sum_{A\lambda} P_i(A\lambda)I^i_{A\lambda} \qquad (1.14)$$

where $I^i_{A\lambda}$ is the relative photo-ionisation cross-section for an $A\lambda$ electron in the ith orbital. If the relative photo-ionisation cross-sections are known, then one can predict quantitatively peak intensities using the atomic parentage obtained from LCAO–SCF calculations. To obtain the relative photo-ionisation cross-sections for a particular x-ray such as Mg-Kα, is not easy, however. As a first approximation we can assume $I^i_{A\lambda}$ is independent of the molecular orbital. Further, since no electron peak is observed for H_2 using Mg-Kα we conclude that the photo-ionisation cross-section for hydrogen is negligible for this radiation. Qualitatively, it appears that for the second-row elements C, N, O, and F, 2s character is seen approximately ten times as strongly as 2p character, in agreement with the relative intensities of the Ne-2s and Ne-2p lines. Although the photo-ionisation cross-section is expected to vary from element to element, qualitatively, at least, the major factor determining the valence peak intensities is the relative amount of total s and total p character. Thus the low intensity of the 1π orbital of CO is a consequence of its having no s character whatsoever. Here it should be remembered that the π orbital is doubly degenerate and, therefore, contains twice as many electrons as the σ orbitals.

As a second example, we can consider the valence orbital spectrum of CH_4 induced by Mg-Kα radiation, Figure 1.12. This spectrum consists of only two peaks, one corresponding to the triply degenerate $1t_2$ orbital and one corresponding to the single $2a_1$ orbital. From an *ab initio* SCF calculation we expect the binding energy of the $1t_2$ orbital to be lowest[16]. In the LCAO picture the $1t_2$ orbital consists of carbon 2p and hydrogen 1s atomic orbitals and the $2a_1$ is made up of carbon 2s and hydrogen 1s orbitals. Thus the line from the $1t_2$ orbital is expected to be considerably weaker than that from the

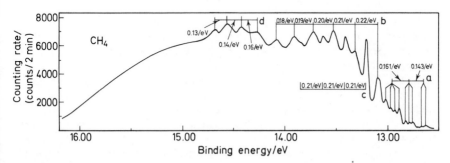

Figure 1.13   The electron spectrum from the $1t_2$ orbital of CH_4 excited by He(I) radiation

$2a_1$ orbital. Hence the origins of the two peaks in the CH_4 valence spectrum are clearly indicated. Here it may be noted that the $1t_2$ orbital has recently been studied with He(I) radiation and the spectrum, Figure 1.13, has been discussed in terms of the Jahn–Teller effect[17].

Although the variation of peak intensity with atomic parentage has been

used quite successful, qualitatively, little quantitative work has been done. Recently, however, a quantitative prediction of peak intensities for the valence orbital spectra of C_4H_4O, C_4H_5N, C_4H_4S and C_6H_6 excited by Mg-Kα radiation has been made using the atomic populations of the orbitals obtained from *ab initio* SCF calculations[18]. For this work the carbon s photo-ionisation cross-section relative to that of carbon p, the oxygen s cross-section relative to oxygen p and the sulphur s cross-section relative to sulphur p were determined from the observed intensities of the valence lines in the photo-electron spectra of CH_4, H_2O and H_2S and the atomic populations of the orbitals determined from *ab initio* calculations. The nitrogen s cross-section relative to nitrogen p was taken to be the average of the values for carbon and oxygen. The hydrogen s cross-section was assumed to be negligible. The oxygen, nitrogen and sulphur s cross-sections relative to carbon s were treated as parameters. The results obtained agreed reasonably well with experiment. A second attempt at calculating relative peak intensities has been made for the linear molecule C_3O_2[19]. Here the same values determined above for the carbon s to p and oxygen s to p cross-sections were used for σ orbitals but to obtain a good fit it was necessary to reduce the ratios by a factor of 2 for π orbitals. Again the oxygen s cross-section to carbon s was treated as a parameter. The fit obtained to the experimental spectrum was excellent. However, it may be noted that the carbon s photo-ionisation cross-section relative to that of oxygen s differed by a factor of 2 between C_4H_4O and C_3O_2. It is believed that this difference resulted, at least in part, from difficulties in background for the C_4H_4O spectrum. The C_4H_4O spectrum has been re-calculated using the carbon-to-oxygen ratio obtained from C_3O_2 and reasonable agreement with experiment has been obtained, although the agreement is not as good as that obtained using the ratio determined from the C_4H_4O spectrum[20]. The technique has also been applied to the molecules SF_6 and CF_4 with satisfactory results[20]. It is felt that these first attempts at quantitatively predicting peak intensities have been moderately successful and that this is an area of electron spectroscopy where much profitable work might be done.

The relative dependence of valence-orbital peak intensities on atomic parentage is expected to vary with the energy of the exciting radiation. Thus, for example, the Ne-2s line excited by Mg-Kα radiation is three times as intense as the 2p line whereas using 100 eV photons it is only *c.* 0.15 times as intense[21]. This suggests that additional information on the atomic parentage of valence orbitals might be gained by exciting spectra with x-rays of considerably different energies. One of these should be either the Mg-Kα or Al-Kα x-ray and, because of resolution considerations, the second one should be lower in energy. One possibility is to use the M_VN_{III} line from Y at 132.3 eV. This line has been studied by Krause[22] and has an inherent width *c.* 0.5 eV. It is sufficiently energetic to reach all valence orbitals and is sufficiently different in energy from the Mg-Kα (Al-Kα) x-ray that any change in dependence of peak intensity on atomic parentage should be apparent. It is perhaps worth pointing out that in going to a lower energy x-ray, the photo-ionisation cross-section for hydrogen 1s may become sufficiently large that H contributions cannot be neglected as is the case with Mg-Kα radiation.

1.3 A MORE DETAILED DISCUSSION OF PHOTO-IONISATION

1.3.1 The use of SCF calculations in electron spectroscopy

In the previous section we have seen how one can use Koopmans' approximation as an aid in interpreting the electron spectra of molecules. The use of Koopmans' theorem implies a number of approximations. Firstly, the ionisation energy for removing an electron from an orbital, R_k, is given by the orbital energy, E_k, only if all the other orbitals are left unchanged. This is the so-called frozen-orbital approximation and implies that there is no rearrangement of the $2N - 1$ electrons during the photo-ionisation process. In general this, of course, is not true. Secondly, the Hartree–Fock operator does not include any relativistic effects. Particularly for inner shells these effects can be large. Hartmann and Clementi[23] have computed relativistic corrections to the Hartree–Fock orbital energies for a number of atoms. For argon they obtained the following corrections for the various sub-shells: 1s, 33.3; 2s, 6.4; 2p, 7.0; 3s, 0.7; and 3p, 0.6 eV. If the relativistic effects are the same for both ion and molecule, they cancel and the use of Koopmans' theorem introduces no relativistic errors. This may well be the case for ionisation of outer electrons. The relativistic corrections to the orbital energies of the neutral Ar atom and of the Ar^{6+} ion formed by removing all six 3p electrons are the same to within one-tenth of an eV[23]. However, this is probably not the case for ionisation from an inner orbital. Thirdly, the Hartree–Fock approximation does not include any correlation effects. Again, if the correlation energies were the same for the ion and the molecule, the two would cancel. However, the correlation energy is, to a large extent, a result of pair interactions between electrons in the same orbital, and will certainly be different in the ion and the molecule. In addition to these approximations, the methods used to solve the Roothaan equations often involve further approximations so that the orbital energies are themselves inexact.

In view of the preceding it is obvious that care must be taken in using Koopmans' theorem to interpret the electron spectra of molecules. By performing separate Hartree–Fock calculations for molecule and ion one might expect to obtain better theoretical estimates of the ionisation potentials. This appears to be the case for the core levels of atoms[24, 25]. Few such calculations have been carried out for molecules, however. Schwartz[26] has calculated the C-1s binding energy of CH_4, the N-1s energy of NH_3 and the O-1s energy of H_2O in this way and has obtained values of 291.0, 405.7, and 539.4 eV respectively. These values are in excellent agreement with the experimental binding energies 290.7, 405.6, and 539.7 eV. This good agreement may be, in part, fortuitous since Gianturco and Guidotti have carried out similar calculations with a larger basis set than that employed by Schwartz and obtained values of 283.8, 400.9 and 512.3 eV[27]. However, for both calculations, the C-1s and N-1s binding energies estimated from the difference in total energies of the ion and molecule are closer to experiment than the estimates obtained from Koopmans' theorem. For the O-1s level of H_2O, Gianturco and Guidotti obtain somewhat better agreement using Koopmans' approximation, but Schwartz obtains better agreement using the difference in total energies of the ion and molecule. P. Siegbahn[10] has calculated the O-1s binding energy of C_4H_4O from the total energies of the molecule and ion and has also obtained better agreement with experiment in this way than that obtained from Koopmans' theorem.

For core electrons the rearrangement energy is the main source of error in applying Koopmans' theorem. For valence electrons, however, correlation

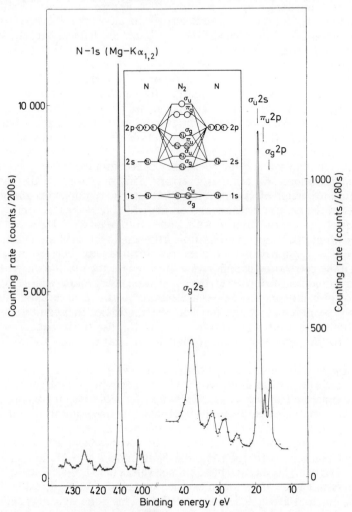

Figure 1.14 The electron spectrum of nitrogen excited by Mg-Kα radiation. Both the core and valence orbital regions of the spectrum are shown

effects can be more important. Since the correlation energy is generally not the same for all states of the ion, care must be exercised in interpreting the ionisation potentials of valence electrons obtained by taking the difference of total energies of the molecule and the corresponding ion. A good illustration of the hazards is provided by the valence spectrum of N_2, Figure 1.14.

The three outermost peaks have also been observed with He(I) excitation and have been assigned to the $3\sigma_g$, $1\pi_u$, and $2\sigma_u$ molecular orbitals respectively[28]. Cade et al.[29] have calculated the ionisation potentials for the $3\sigma_g$, $1\pi_u$ and $2\sigma_u$ orbitals from the energies of the molecule and the appropriate ion. They found that the predicted ionisation potential of the $1\pi_u$ orbital was smaller than that of the $3\sigma_g$. Thus the ordering of the valence peaks is incorrectly predicted.

Because of these difficulties a certain amount of caution must be exercised in using Koopmans' theorem and SCF calculations in the interpretation of electron specta. Nevertheless, such calculations do provide a valuable aid in elucidating electron specta.

1.3.2 Shake-up and shake-off processes in electron spectroscopy

There exists an experimental quantity which, within the Hartree–Fock approximation, is more directly comparable with the orbital energies obtained from SCF calculations using Koopmans' theorem, than the ionisation potential. So far in our discussion of photo-ionisation we have implicitly assumed that the ion formed on removing an electron from a given orbital is in its 'ground state'. This is not always the case. During the ionisation process a second electron may be promoted from a filled orbital to an unfilled but bound orbital or to the continuum. In the former case we observe a sharp line to the low-kinetic-energy side of the main photo-electron peak at an energy $E = hv - E_B - E^*$ where E^* is the additional excitation energy. In the latter case two electrons are emitted and we observe a continuum to the low kinetic energy side of the main peak. The energies of the two emitted electrons are related through the equation $E_1 + E_2 = hv - E_{B_1}^* - E_{B_2}^*$ where $E_{B_1}^* + E_{B_2}^*$ is the energy required to remove the two electrons. We shall refer to these two processes as shake-up and shake-off. The terms monopole-excitation and monopole-ionisation have also been used[30].

Experimentally, shake-up and shake-off have been studied in some detail for the noble gases and more recently in some simple linear molecules[19, 30–33]. For the sake of argument let us consider the Ne-2s spectrum. A sample spectrum, excited by Mg-Kα, is shown in Figure 1.6[33]. To the low energy side of the main Ne-2s line can be seen a number of additional peaks. These peaks have three different origins: namely, discrete energy loss peaks caused by photo-electrons being inelastically scattered from neutral Ne atoms (peaks 2, 3 and 4), shake-up peaks corresponding to simultaneous ionisation of the 1s orbital and the excitation of an outer electron (peaks 7, 8, 9, 10, 11 and 12) and finally lines due to K$\alpha_{3, 4}$ satellites with the same origin as the types just mentioned. Although the discrete energy-loss lines provide information on excited states of the neutral molecule their presence complicates the identification of the shake-up peaks and indicates the necessity of running gaseous samples at low pressure. The Ne-1s shake-up data taken from Siegbahn et al.[33] and from Carlson, Krause, and Moddeman[30] is summarised in Table 1.2. There is substantial agreement between the two sets of experimental data. Krause, Carlson, and Dismuskes[32] have also studied the Ne-1s shake-off spectrum and found its total intensity to be 16.5% of the main peak.

Table 1.2 Summary of Ne-1s shake-up data

Final-state Configuration	Shake-up energy/eV			Relative line intensity		
	Siegbahn et al.	Carlson et al.	Theoretical	Siegbahn et al.	Carlson et al.	Theoretical
$1s\,2s^2\,2p^6\ ^2S$	0	0	0	100	100	100
$1s\,2s^2\,2p^5\,3p\ ^2S$	37.3	37.2	35.6	2.4 ± 0.4	3.1 ± 0.2	2.3
$1s\,2s^2\,2p^5\,3p\ ^2S$	40.7	40.6	39.5	2.6 ± 0.6	3.1 ± 0.3	2.9
$1s\,2s^2\,2p^5\,4p\ ^2S$	42.3	42.2	40.5	1.5 ± 0.4	1.8 ± 0.3	—
$1s\,2s^2\,2p^5\,5p\ ^2S$	44.2	44.4	42.4	0.5 ± 0.2	0.6 ± 0.2	—
$1s\,2s^2\,2p^5\,4p\ ^2S$	46.4	≈46	44.6	0.6 ± 0.2	0.8 ± 0.4	—
$1s\,2s\,2p^6\,3s\ ^2S$	60	59.3	61	—	≈1	—
$1s\,2s\,2p^6\,4s\ ^2S$	68				≈0.2	—

Theoretically, shake-up arising from ionisation of an inner orbital can be adequately described in terms of the sudden approximation[33-35]. Following this approach, the excitation of the valence electron is treated independently of the core ionisation and occurs because of the sudden perturbation in the potential experienced by the outer electrons following the removal of the core electron. The transition is then between two different states of the ion and is of the monopole type. For Ne-1s, the initial state of the ion has the configuration $1s\ 2s^2\ 2p^6\ {}^2S$. Applying the monopole selection rules shake-up can only occur to states of the ion with the configuration $1s\ 2s^2\ 2p^5\ np\ {}^2S$ or $1s\ 2s\ 2p^6\ ns\ {}^2S$. The shake-up states have three unpaired electrons and, therefore, there are two independent doublet spin functions. Bagus and Gelius[33] have calculated the excitation energy for the various final-state configurations and the assignment of the shake-up lines follows from the calculation as indicated in Table 1.2. Carlson et al.[30] have calculated the transition probabilities for the various shake-up states using the sudden approximation and the agreement with the experimental intensities is good. See Table 1.2.

Manne and Åberg[37] and Meldner and Perez[36] have argued that the single-particle orbital energies obtained from Hartree–Fock SCF calculations using Koopmans' approximation are more directly comparable to the centroid of the ionisation spectrum, including shake-up and shake-off, than to the normal or lowest ionisation energy which corresponds to the 'ground state' of the ion. The rearrangement energy is included in the mean ionisation potential and this error is minimised. Meldner and Perez[36] have used the Ne-1s shake-up spectrum of Figure 1.6 and the shake-off spectrum obtained by Krause, Carlson and Dismukes[32] to determine the experimental mean ionisation energy for Ne-1s. They obtained a value of 892 eV. Manne and Åberg have used the shake-up and shake-off data of Krause et al.[32] and have determined a value of 886 eV for the experimental mean ionisation potential. Theoretically, the Ne-1s orbital energy is 892 eV, much closer to the experimental mean ionisation potential than to the normal experimental ionisation energy, 870 eV. The good agreement, in part, results from the fact that, for Ne, correlation and relativistic effects are small compared to the rearrangement energy[36]. It should, however, be emphasised that, experimentally, it is not as easy to determine the mean ionisation energy as it is to determine the ionisation energy peak.

For molecules the shake-up associated with photo-ionisation of the core levels of N_2, O_2, CO and CO_2 has been studied[30, 38]. Gelius et al.[19] have recently observed this phenomenon in C_3O_2 (Figure 1.15). For molecules the shake-up intensity may be relatively high, some 15–20% of that of the main line or higher, and is consequently not a negligible effect. The relative energies and intensities of the shake-up lines from the different core levels show some similarities, but there are also numerous differences. The shake-up transitions which occur thus depend on the atom from which the core electron is removed. This dependence may, in turn, yield information about the charge densities of the molecular orbitals involved in the shake-up transition. Thus, for example, Gelius et al.[19] have concluded that there is no C-1s shake-up transition associated with the central carbon corresponding to the first O-1s shake-up line of C_3O_2, but that there is such a transition associated with the other carbons. This implies that the valence orbitals involved have a much smaller charge density at the central carbon than at the other two carbons. The analysis of shake-up in molecules is much more complicated than for atoms since it involves the overlap of the valence orbitals of molecular ions with core vacancies. At the present time, wave-functions for such states

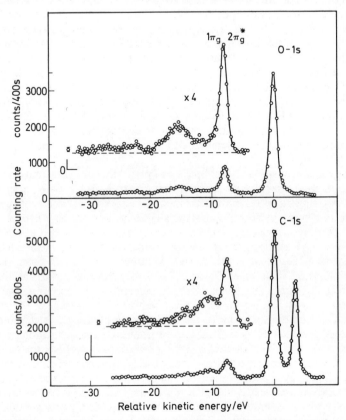

Figure 1.15 The electron spectrum of C_3O_2 excited by Mg-Kα radiation in the region of the O-1s and C-1s core orbitals. The peaks to the low-kinetic-energy side of the main lines are due to shake-up

are almost totally non-existent. In fact one way of testing such calculations, when they become available, is by comparing the predicted shake-up intensities with experiment.

Shake-up and shake-off accompanying ionisation of core levels involves transitions between outer orbitals. However, if photo-ionisation takes place from an outer orbital it is possible to promote another electron from the same orbital into a continuum or discrete state. This effect has been observed in He, Ne, and Ar[30] and in the valence spectrum of N_2 (Figure 1.14) the fourth innermost peak may also be a shake-up line[38]. Theoretically double-ionisation and shake-up cross-sections have been calculated for He. Since both electrons originate from the same orbital correlation effects might be expected to be important. This is, in fact, the case. Using Hartree–Fock single-electron wave-functions the cross-sections are considerably under-estimated but by including correlation explicitly good agreement with experiment is obtained[30, 39].

For completeness it may also be pointed out that similar effects have also been observed in the solid state. Novakov[40], Rosencwaig, Wertheim and Guggenheim[41] and Novakov and Prins[42] have observed shake-up satellites in association with ionisation of the 2p levels of a number of different transition metals in various compounds. The shake-up transitions have been attributed to transitions between the valence (d) band and the unfilled conduction (s) band. For a given ion the satellites are found to depend on the ligand and may, therefore, provide information on the crystal and ligand fields to which the ion is exposed. A similar dependence on ligand was observed by Barber et al.[43] for the O-1s and C-1s satellites of a number of transition metal carbonyls. Novakov and Prins[42] have recently shown that the satellite lines observed in the Cu-2p spectra of Cu_2O, CuCl and CuBr depend on the presence of adsorbed oxygen (or water) but that they are nevertheless attributable to the compound itself. Wertheim and Rosencwaig[44] have also observed satellites in the spectra of a number of alkali halides.

1.4   CORE SPECTRA

1.4.1   Chemical shifts

Although the core levels of an atom are essentially inert as far as chemical bonding is concerned they are still affected by the chemical environment. The core binding energies of a given atom are different for the different molecules of which the atom is a part. Thus, for example, a comparison of the C-1s and O-1s binding energies for CO and CO_2 (Figure 1.16) shows that, in CO_2, the C-1s electron is more tightly bound than in CO, but that the O-1s binding energy is larger for CO than for CO_2. Such changes in binding energy with chemical environment are called chemical shifts. Qualitatively, the shifts can be simply explained using the concept of electronegativity. Oxygen is more electronegative than carbon and, therefore, reduces the electron charge density on the carbon atoms. This, in turn, increases the binding energy of the carbon core electrons and reduces that of the oxygen. Since CO_2 contains twice the number of oxygens that CO does, the C-1s binding energy of CO_2 is increased correspondingly more than that of CO. This is indicated schematically in Figure 1.16. Based on this simple picture one would expect the core binding energies of a given atom to be different even within a given molecule if the atom occupies two or more inequivalent positions. This is in fact the case as demonstrated by the C-1s spectrum of Me_2CO, Figure 1.17. Here the C-1s binding energy of the carbonyl carbon is greater than that of the methyl carbons because of the greater electro-negativity of the oxygen compared to hydrogen. It is interesting to note that the intensities of the two peaks reflect the number of the different carbons and can in fact be used to identify the lines. Thus from a consideration of electro-negativity one is led to consider the chemical shifts as being electrostatic in

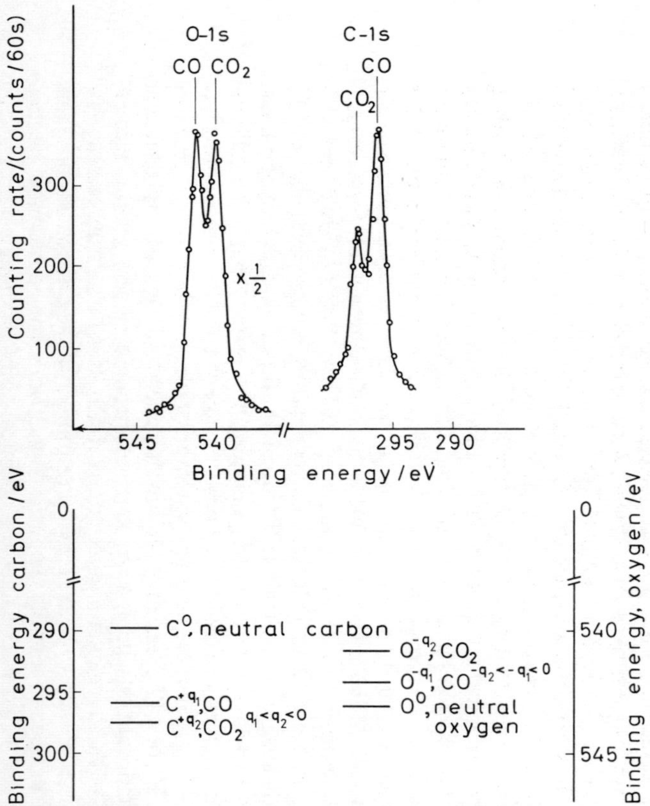

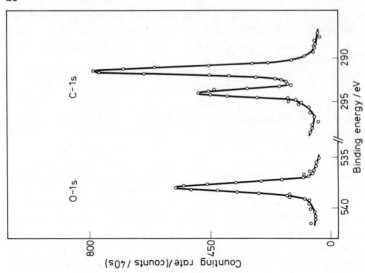

Figure 1.17 The electron spectrum from the O-1s and C-1s levels of acetone, excited by Mg-Kα radiation. Two carbon lines are obtained corresponding to the different valence states of the carbon atoms in the molecule

Figure 1.16 The electron spectrum from the O-1s and C-1s levels of CO and CO_2 excited by Mg-Kα radiation. The schematic indicates the effect of charge transfer on the core levels of the atoms when they are bound in a molecule. The binding energies for the neutral atoms have been estimated and have not been obtained directly from experiment

origin and reflecting the distribution of electronic charge in the molecule. This, in turn, has led to the development of the electrostatic or potential model of chemical shifts.

In the potential model the chemical shifts are seen as a change in the electrostatic potential energy of the given core electron which results when the atom becomes part of a molecule. Since the core electrons remain localised, it can be shown that the change in potential results primarily from the change in distribution of the valence electrons. In the LCAO molecular orbital picture the charge distribution of the valence electrons can be divided among

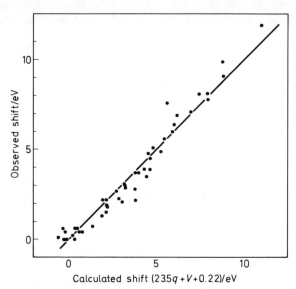

Figure 1.18 Comparison between measured shifts and shifts calculated with the potential model using charges obtained from CNDO/2 calculations

the atoms using the concept of gross atomic charge. The electrostatic interaction of a core electron of atom A with the valence electrons can then be written

$$E_A = k_A q_A + \sum_{B \neq A} q_B / R_{AB} \qquad (1.15)$$

where k_A is a constant which represents the average interaction between the core electron and a valence electron on atom A. The chemical shift of the core electron is then given by

$$\Delta E_A = k_A q_A + V_A + l \qquad (1.16)$$

where $V_A = \sum_{B \neq A} q_B / R_{AB}$ is the molecular potential at atom A and l is a constant for a given type of atom and is used to define the reference level to which the shifts are referred. This equation has been used to correlate observed chemical shifts of a number of atoms with gross atomic charges obtained from SCF calculations and the results are generally good. Figure 1.18 shows the correlation obtained for the C-1s chemical shifts of a large series of compounds[45]. Here k was treated as a parameter and was determined by a least squares fit. The agreement between predicted and measured shifts is generally very good. In using equation (1.16) one attributes the chemical shifts entirely to the ground-state properties of the molecule. Since we think of the core levels as being essentially atomic in nature, one might well expect relativistic and correlation effects to change little from molecule to molecule and hence to have a small effect on chemical shifts. On the other hand, the relaxation energy is not necessarily the same for all molecules. Since experimentally, one obtains the difference in total energy of the neutral molecule and the ion, the relaxation energy is included in the experimental chemical shifts. The good agreement obtained between experimental and predicted C-1s chemical shifts indicates that, at least for carbon, the relaxation energy is, to a good approximation, constant from molecule to molecule. This is not universally true, however. For example, the C-1s binding energy of C_6H_6 is 0.5 eV less than that of CH_4. Using equation (1.16) and CNDO charges the benzene C-1s binding energy is predicted to be 0.8 eV greater. Not only is the predicted shift incorrect quantitatively, it is in the opposite direction from the measured shift. Using simple electronegativities one would also predict a higher binding energy for the benzene C-1s electron. This discrepancy probably results from C_6H_6 having a larger relaxation energy than CH_4 presumably because of the greater mobility of the benzene π-electrons. Such an effect has been discussed by Thomas[46]. In general, however, the potential model predicts the C-1s shifts quite well.

Similar correlations have been obtained for S-2p, O-1s, and N-1s chemical shifts using charges obtained from CNDO calculations[47, 48]. Having determined the constants k and l for different atoms from such correlations one can then do the reverse and use equation (1.16) to determine the gross atomic charges for a given molecule from the measured shifts.

For large molecules it is often impractical to determine gross atomic charges from calculations. Therefore, it is desirable to use a simpler method of systematising chemical shifts. Since the molecular potential, V_A, can be expected to be approximately linearly related to q_A, one might expect a linear relation between the core binding energies and the charge of the atom, i.e.

$$E_A = kq_A + l \tag{1.17}$$

One could, of course, use charges obtained from MO calculations as in the potential model, but a less involved technique is to estimate the atomic charge using Pauling's valence-bond model. In this model the charge on an atom in a molecule is obtained as the sum of the partial ionic character of the bonds between the atom and its neighbours. Thus for the ith atom

$$q_i = Q_i \pm \sum_{i \neq j} nI_{ij} \tag{1.18}$$

where Q_i is the formal charge on the ith atom, I_{ij} is the partial ionic character of the bond between atoms i and j, and n is the bond number, $n = 1$ for a single bond, 2 for a double bond etc. The partial ionic character is obtained from the electronegativity difference between the bonding partners and is given by

$$I_{ij} = 1 - \exp(-0.25\Delta\chi_{ij}^2) \tag{1.19}$$

where $\Delta\chi_{ij}$ is the electronegativity difference between atoms i and j. The charges calculated in this way only account for the influence of nearest neighbours but it would appear that this is, at least to a first approximation, sufficient. Thus Figure 1.19 shows the correlation between C-1s chemical shifts and the Pauling charge[45]. A good linear correlation is obtained, but there is some scatter which is interpreted as being a result of secondary substituent effects. Good correlations have also been obtained for S-2p binding energies[48]. N-1s binding energies have also been correlated with Pauling charges[49]. The

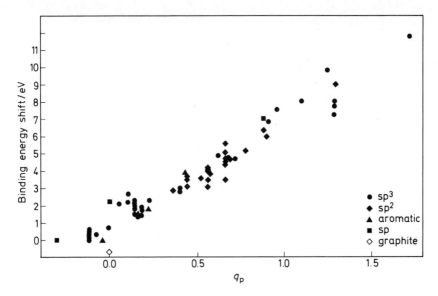

Figure 1.19 Correlation between C-1s chemical shifts and the Pauling charge parameter, q_p

results are not as good as those obtained for carbon and sulphur, but they are satisfactory. Somewhat poorer results were first obtained for P-2p binding energies[50]. For P-2p, however, Hedman et al.[51] have recently shown that by excluding ionic compounds, and compounds for which the phosphorus is bound to two or more sulphur atoms, a very good correlation could be regained, Figure 1.20. The poorer correlation for ionic compounds suggests that the influence of the molecular potential may be important. Here it is worth noting that, for ionic crystals, the molecular potential should also include a sum over all the ions in the crystal in order to account for the Madelung potential. On the other hand, the correlation of C-1s and S-2p experimental shifts for ionic compounds is of the same quality as for covalent compounds indicating that, in these cases, the effects of the crystal potential tend to average out. Correlations between chemical shifts and atomic charges determined from various types of molecular calculation have now been obtained for C-1s, N-1s, S-2p, P-2p, B-1s, Cr-3p, Si-2p, Se-3p and Se-3d and As-3d electrons[47, 48, 49, 52—57].

Another method of determining chemical shifts is by using additive

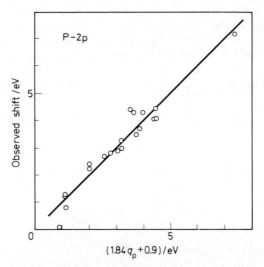

Figure 1.20 Correlation between P-2p chemical shifts and the Pauling charge parameter, q_p

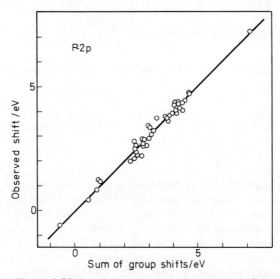

Figure 1.21 Correlation between the observed P-2p chemical shifts and the sum of the group shifts

group shifts. The shift, ΔE, for a given atom is given by the sum of shifts ΔE_{gr} caused by each of the groups attached to the atom under question. Thus

$$\Delta E = \sum_{gr} \Delta E_{gr} + l \tag{1.20}$$

The use of group shifts follows from equation (1.15) and the requirement of molecular electroneutrality. Thus

$$q_A = -\sum_{B \neq A} q_B \tag{1.21}$$

and

$$\Delta E = kq_A + \sum_{B \neq A} \frac{q_B}{R_{AB}} + l = \sum_{B \neq A} q_B \frac{1}{R_{AB}} - k \times l \tag{1.22}$$

By separating the sum over the atoms B into partial sums over each group we obtain equation (1.20). Group shifts for carbon and phosphorus have been obtained by a least squares fit of equation (1.20) to experimental data[45, 51]. One can then obtain good correlations between the sum of the group shifts and the experimental shifts. Figure 1.21 shows the results obtained for P-2p. The use of additive group shifts implies that the shifts caused by groups attached to the same atom are independent of one another. This is probably not entirely correct since highly-electronegative substituents induce a large positive charge on the atom which will, in turn, affect the influence of the other substituents. However, the good correlations obtained for carbon and phosphorus indicate that such effects are of second order.

Group electronegativities have been derived from the group shifts. These should be valid for all substituted elements whereas the group shifts derived from C-1s and P-2p chemical shifts are only applicable to these elements. Where comparison is possible, the group electronegativities derived from phosphorus and carbon are in good agreement except for $-S$[51].

Jolly and Hendrickson have developed a method of correlating chemical shifts with thermochemical data[58]. The method is based on the assumption that atomic cores with the same charge are chemically equivalent. Then the increase in an atom's core charge by one unit caused by photo-ionisation corresponds to replacing it by the atomic nucleus of the next element in the Periodic Table. By way of illustration we can consider the difference in N-1s binding energies of NH_3 and N_2. An asterisk * indicates a 1s vacancy. Then the N-1s photo-ionisation of NH_3 and N_2 can be written

$$NH_3 \longrightarrow NH_3^+* + e^-$$
$$N_2 \longrightarrow N_2^+* + e^-$$

Hence the difference in binding energies is the energy of the reaction

$$NH_3 + N_2^+* \longrightarrow N_2 + NH_3^+*$$
$$\Delta E = E_B(NH_3) - E_B(N_2) \tag{1.23}$$

Then for any process in which an N^{6+} core is replaced by an O^{6+} core we assume $\Delta E = 0$. Thus for the reaction

$$NH_3^+* + NO^+ \longrightarrow OH_3^+ + N_2^+* \tag{1.24}$$

$\Delta E = 0$.

By combining the two reactions we obtain

$$NH_3 + NO^+ \longrightarrow OH_3^+ + N_2$$
$$\Delta E = E_B(NH_3) - E_B(N_2)$$

(1.25)

Good correlation has been obtained in this way for N-1s chemical shifts[58, 59]. Using such correlations unknown heats of formation can be estimated from core binding energies.

Barbet *et al.*[60] have demonstrated a linear correlation between Sn-4d binding energies measured by ESCA and Mössbauer isomeric shifts for a series of Sn^{IV} octahedral complexes. For a different stereochemistry and valence state of the central atom one might expect similar linear correlations between Mössbauer chemical shifts and core binding energies and hence one expects to find families of such linear correlation. Mateescu and Riemenschneider have recently shown that within classes of closely related compounds good correlations can be obtained between C-1s binding energies and the ^{13}C n.m.r. chemical shifts[61]. It seems that within such classes the

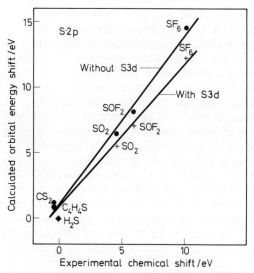

Figure 1.22 Correlation between the observed S-2p chemical shifts and the shifts in S-2p orbital energies from *ab initio* calculations with and without sulphur 3d polarisation functions

variation of the diamagnetic shielding is the main factor determining the n.m.r. chemical shifts and that other factors such as paramagnetic contributions remain relatively constant so that it is the diamagnetic term which correlates with the C-1s binding energies.

Another method of predicting chemical shifts is by comparing calculated orbital energies obtained from *ab initio* SCF calculations. Figure 1.22 shows the correlation obtained for S-2p chemical shifts for calculations with and without sulphur 3d functions in the basis set. The effect of including the 3d functions is to reduce the slope of the correlation line from 1.29 to 1.09.

Perfect agreement between experiment and theory corresponds to a slope of 1.00. Good agreement between shifts predicted from orbital energies and experiment has also been obtained for carbon. Since the orbital energies are calculated only for the neutral molecule, this agreement reinforces the conclusion that the chemical shifts are primarily a property of the neutral molecule. Further, it supports the assumption that the rearrangement energy does not vary greatly from molecule to molecule.

1.4.2   Some further remarks on core lines

If the initial state of a molecule has non-zero angular momentum, J, then, when a core electron is ionised, the hole state can couple to J in more than one way to form two or more final states. This effect can be termed spin, multiplet ψ, or exchange splitting. It should not be confused with normal spin–orbit splitting. The simplest system to exhibit spin splitting would be atomic lithium. The Li^+ ion could be in either the $1s2s$ $^3S_1$ configuration or in the $1s2s$ $^1S_0$ configuration. The ESCA spectrum would then exhibit two peaks whose relative intensities would be given by the multiplicities of the states $1:3$ and whose separation is $2H$ where H is the $1s2s$ exchange integral[62].

Such splitting was first observed in the ESCA spectra of the paramagnetic gases O_2, and NO, Figure 1.23[63]. The ground state configuration of O_2 is $^3\Sigma_g^-$. Emission of a 1s electron leads to states of $^2\Sigma^-$ or $^4\Sigma^-$ symmetry. The O-1s spectrum thus shows two peaks whose intensities are in the ratio $2:1$. The splitting can be estimated theoretically by evaluating the appropriate exchange integral. Using neutral-molecule wave-functions, it is predicted to be 1.2 eV, in good agreement with the experimental value of 1.1 eV. For NO the ground state is $2\pi_{\frac{1}{2}}$ which couples with the unpaired 1s electron to give the $^1\pi$ and $^3\pi$ final states. The intensities are expected to be in the ratio $1:3$. For N-1s the doublet is resolved and the separation is found to be 1.5 eV. For O-1s the splitting is smaller and is observed as a broadening of the O-1s line. A deconvolution, assuming the intensities are in the ratio $1:3$ then gives a splitting of 0.7 eV. Theoretically the N-1s and O-1s splittings are estimated to be 0.9 and 0.7 respectively. Although the O-1s estimate is in good agreement with experiment the N-1s splitting is poorly predicted. However, the theory is qualitatively correct in predicting a larger N-1s splitting.

Such splitting has also been observed in the transition elements with unfilled d shells. The lines from the 3s levels of Mn and Fe in MnF_2, MnO, MnO_2 and FeF_3 are split by as much as 6–7 eV[64, 65]. Free-ion theoretical calculations overestimate the splitting by about a factor of two, but splittings taken from unrestricted Hartree–Fock calculations for $(MnF_6)^{4-}$ are in reasonable agreement with the Mn-3s splitting in MnF_2 and MnO. The 3s lines of Fe, Co and Ni metal are also split. For Fe, the splitting has been observed both above and below the Curie temperature and the effect is the same for both the paramagnetic and ferromagnetic states. The 3p levels of Mn and Fe and the 2p level of Mn also showed multiplet structure. For Eu gas the 3d electron spectrum is anomalous compared to the 3d spectra of Xe and Yb and this has been attributed to spin coupling of the 3d hole state with the half-filled 4f shell[64]. Clark and Adams have also recently observed such multiplet splitting in the Cr-3s lines of chromocene and tris-hexafluoracetonyl acetonate chromium(III)[66], and Hedén *et al.*[67a] and Wertheim and Cohen[67b] have observed these effects in several rare earth metals and compounds.

The multiplet splitting reported above is primarily dependent on the coupling of a single-particle hole state with unpaired electrons. Novakov

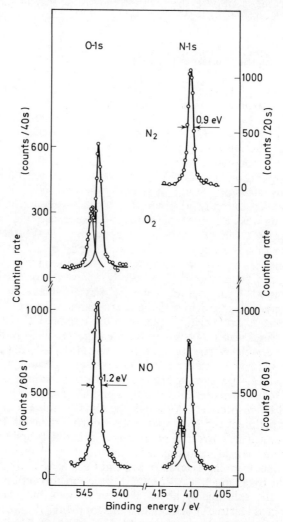

Figure 1.23 ESCA spectra from N_2, NO, and O_2 showing spin splitting of the 1s levels in the paramagnetic molecules

and Hollander have also observed a splitting of the $p_{\frac{3}{2}}$ levels of thorium and uranium in several compounds[62, 68]. It has been suggested that this splitting is caused by an effective electric-field gradient in the interior of the atom and the term electrostatic splitting has been used to describe this effect.

Another property of core lines which is of interest is the line width for which the full width at half-maximum, FWHM, is normally used as a measure. This width includes a contribution from the natural width of the ionic state under excitation and this in turn is related to the lifetime of the state. It should not be too surprising, therefore, to find that the width of a given core line varies somewhat from compound to compound. In some instances, however, the FWHM of a given type of core line is not constant even within a given compound. Thus, for example, the observed widths of the two N-1s lines of N_2O differ by some 0.1 eV[69]. Since the various factors contributing to the line-width do not combine linearly, but approximately as a sum of squares this rather moderate difference in widths represents a considerable difference in the widths of the ionic levels. A similar effect has been observed for the two C-1s lines of C_3O_2 [19]. The explanation of this effect is still forthcoming. The lines obtained from solid samples generally have a larger FWHM than the corresponding lines from gaseous samples. Again this effect is not completely understood. Here it may be noted that the lines obtained from the innermost valence orbitals are frequently very broad. This broadening has been attributed to a shortened lifetime of the ionic state because of Coster–Kronig transitions. For completeness we may note that an additional source of line broadening is vibrational excitation but that normally, for x-ray excitation, such effects are small.

Besides the inherent interest, a knowledge of line-widths is important in electron spectroscopy if one wishes to analyse a given peak into components. Such a situation might arise, for example, when an expected chemical shift is too small to produce two resolved lines.

1.4.3   Some applications of core level spectroscopy

Since the core levels are localised on the atoms of a molecule each atom gives characteristic lines in an ESCA spectrum which may be used to identify the presence of the atom. Thus, electron spectroscopy can be used as a tool in qualitative analysis. Figure 1.24 shows the ESCA spectrum obtained from a sample of air run at a partial pressure of c. 0.1 Torr. The 1s lines of O_2 and N_2 are easily recorded. The 2p line of Ar, whose abundance in air is c. 1%, was also recorded, indicating the sensitivity of the technique. As a further indication of the sensitivity of the method, Figure 1.25 shows part of the ESCA spectrum for the vitamin B_{12}, Figure 1.26. The single Co atom, which is only one atom among 180 others, can be readily distinguished. Not only is it possible to distinguish between the different atoms of a sample but, because of the chemical shift effect, one can frequently distinguish between atoms of a given type in different molecules. Thus, for example, Figure 1.27 shows the O-1s and C-1s lines recorded for a mixture of CO, CO_2 and CH_4. Here one can see three carbon lines corresponding to the three different gases. From a consideration of electronegativities, one would expect the C-1s binding energy to be greatest for CO_2 and least for CH_4. Hence it is possible to determine the origin of the lines. By varying the partial pressure of the gases this determination has been confirmed. As mentioned in Section 1.1.2, the technique is very sensitive to surface contamination when using solid samples. Although this can introduce certain difficulties in using solid

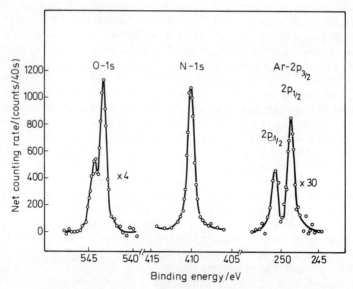

Figure 1.24 ESCA spectrum from air excited by Mg-Kα radiation

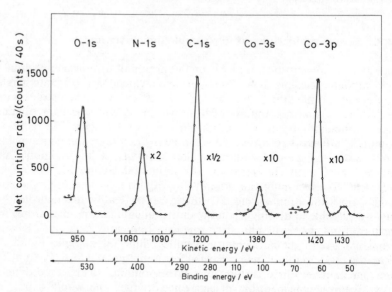

Figure 1.25 Electron spectrum from vitamin B_{12} showing the lines from oxygen, nitrogen, carbon and cobalt

samples, this sensitivity provides a very convenient method for monitoring surface contamination and for studying surface processes.

In addition to being useful in qualitative analysis, ESCA can, in principle, be useful for quantitative analysis. Thus in Figure 1.27 the intensity of the C-1s lines, as determined by the area of the peaks, is a measure of the relative amounts of CO_2, CO and CH_4. To do elemental analysis one can, in principle, calibrate the spectrometer using compounds of known constituency. At present little work has been done in this area but, at least for relatively simple systems, one can obtain reasonably accurate quantitative results. For example,

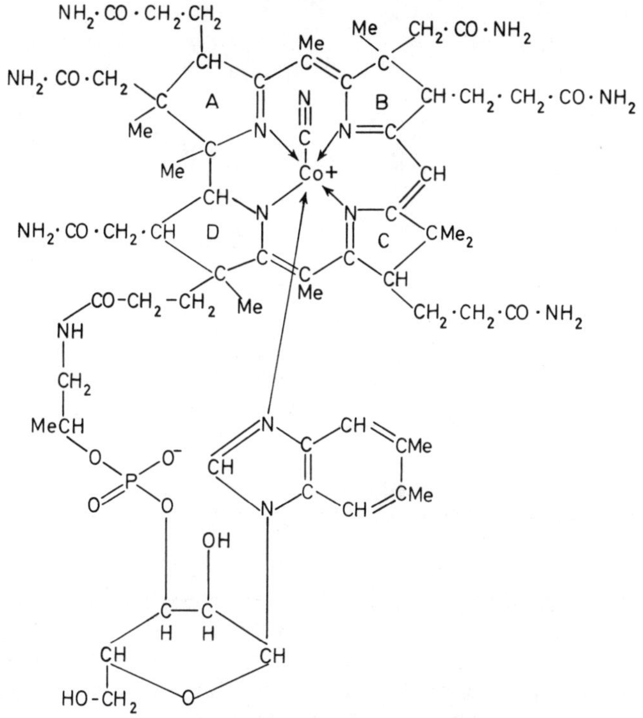

Figure 1.26   The vitamin B_{12} molecule

the relative amounts of zinc and copper in brass alloys have been determined to within a few per cent[70]. In another study the relative number of carbon and chlorine atoms in a number of organic compounds was determined with a reasonable accuracy[71]. Progress in quantitative analysis is not, however, simply a matter of accurate calibration. For gases, discrete energy losses through inelastic scattering and shake-up and shake-off all reduce the intensity of the main photo-peak. The first effect can often be eliminated or reduced to negligible levels by using a sufficiently low gas pressure, but shake-up and shake-off effects are always present. Further as we have seen they can represent a non-negligible fraction of the peak intensity. For solids, shake-up effects are also present. To date little work has been done in this area, but it is clear that for accurate elemental analysis this area must be investigated in more detail.

Another factor which must be taken into account in determining any such calibration is the geometry of the experiment. Only electrons emitted into a small solid angle are accepted. Usually electrons emitted c. 90 degrees to the photon direction are accepted. If for some reason electrons emitted in some other direction are accepted, then the relative photo-ionisation cross-section can be changed. Krause has examined the angular dependence of photo-electrons from the 1s shell of Ne at various energies[72]. For Mg-Kα radiation

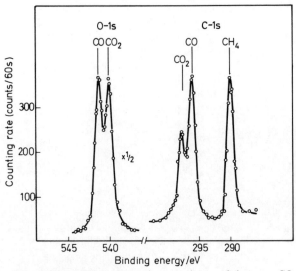

Figure 1.27 ESCA spectrum from a mixture of the gases CO, CO_2 and CH_4

he found that the differential photo-ionisation cross-section decreased by c. 50 % in going from 90 to 45 degrees. This effect is thus relatively important. It is worth mentioning that in the valence region of the spectrum the angular dependence of the differential photo-ionisation cross-section may be useful in determining the origin of a given line or lines[73].

ESCA can also be a useful tool in structure analysis. Lindberg[74] has recently reviewed this application in organic chemistry so we shall limit ourselves to a few illustrative examples. We can take as a first example the dioxide of cystine for which two possible structures exist.

$$
\begin{array}{cc}
\begin{array}{c}
\text{O} \\
| \\
\text{R—S—S—R} \\
| \\
\text{O}
\end{array}
&
\begin{array}{c}
\text{O}\ \ \text{O} \\
|\ \ \ | \\
\text{R—S—S—R}
\end{array}
\\[1em]
(1) & (2)
\end{array}
$$

Structure (1) is expected to result in two peaks of equal intensity corresponding to the two inequivalent sulphurs whereas structure (2) will result in a single line. The electron spectrum, Figure 1.28, clearly indicates structure (1). A second example illustrating the manner in which ESCA elucidates structure is provided by the C-1s spectrum of ethyltrifluoracetate (Figure 1.29). All four carbons are distinguished in the spectrum. In going from right to left in the molecule the positive charge on the carbon increases leading to a higher C-1s binding energy. In general, isomers with structure elements which differ from the valence point of view can never give rise to identical spectra. Even small differences in peak shapes may be significant in solving structure problems and valuable information can be obtained even when core lines are not resolved from one another.

Because the chemical shift is charge-dependent core level spectroscopy can be used to obtain information on the charge distributions in molecules. The question of charge transfer in transition metal carbides has been a matter of controversy. The chemical shifts observed in the ESCA spectra indicate that the charge transfer is from metal to carbon[75]. Several platinum complexes have also been investigated and the results indicate that there is a significant charge transfer from metal to ligand[76]. Buchanan et al.[77] have recently studied the ESCA spectrum of K_xFeF_3 in which the iron exists both as Fe^{2+} and Fe^{3+}, and have observed a doublet structure in the 3p and 2p lines as a result of the two valence states of the iron. Mateescu and Riemenschneider[61] have recently investigated the charge distribution in carbenium cations by means of ESCA. For the t-butyl cation, two distinct C-1s peaks were observed separated by 3.7 eV and in the ratio 3:1 indicating intensive localisation on the central carbon. On the other hand, extensive charge delocalisation as in the trityl and tropylium ions leads to a single line since all carbon atoms share the formal positive charge approximately equally. By comparing the C-1s spectra obtained from the 1-adamantyl cation and the norbonyl cation Mateescu and Riemenschneider have demonstrated that the formal charge on the norbonyl cation is delocalised which strongly suggests that the structure is the 'non-classical' one.

Another area of interest is the investigation of substituent effects using ESCA. From our discussion of chemical shifts it is clear that nearest-neighbour effects are most important. However, there are second-order effects caused by atoms or groups which are more than one atom removed. Thus, for example, in the series CCl_3X, with X = Me, CCl_3, and CF_3, the trichloromethyl carbon 1s shifts are 0.0, 1.1 and 1.6 eV respectively[78]. Aromatic substituent effects have been studied by Lindberg and Hamrin[79] and by Clark, Kilcast and Musgrave[78]. The former authors have investigated the influence on the binding energies of the atoms of the substituents themselves while the latter workers have studied the effect of the substituents on the electron binding energies of the aromatic atoms.

From the foregoing it is apparent that a large number of problems can be studied by means of core-level electron spectroscopy. For solids, the method is generally non-destructive and extremely small amounts of material are sufficient to give usable spectra. For gases this is usually not the case since the sample is normally pumped away. However, provision could be made for recovering the gas which would greatly reduce the amount of material required. The technique is applicable to all elements of the Periodic Table, except hydrogen, and, therefore, it is a versatile tool in the study of molecular

40

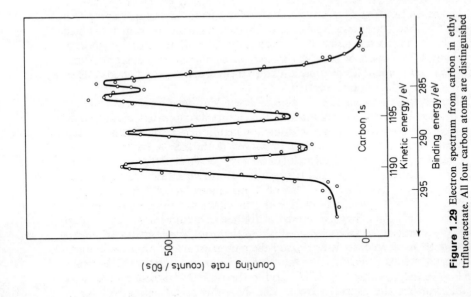

Figure 1.29 Electron spectrum from carbon in ethyl trifluoracetate. All four carbon atoms are distinguished in the spectrum.

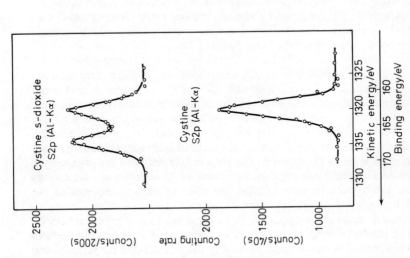

Figure 1.28 Electron spectrum from the S-2p shell of sulphur in cystine (lower part) and cystine s-dioxide showing that both oxygen atoms are bound to the same sulphur atom in the dioxide.

and solid state structure. By using monochromatic x-rays, better resolution can be obtained and this in turn increases the potential of the method. Further, as the number of workers in the field increases, the range and diversity of applications can be expected to increase.

1.5 CONCLUSION

In this article we have discussed the use of electron spectroscopy in the study of chemical problems. We have concentrated on the use of soft x-rays and have indicated some of the areas which have been explored in some detail and are now fairly well developed; we have also tried to indicate some of the areas in which profitable research might be done. Although the use of ultraviolet radiation for excitation has only been discussed briefly it is an extremely important field of research. For more information in this field and other areas the reader may consult a number of monographs and review articles[2, 3, 62, 74, 80—84].

Electron spectroscopy is now in a period of rapid development. In the next few years considerable progress in existing areas of application can be anticipated. Undoubtedly new areas of application will appear.

References

1. Siegbahn, K. (1965). *Alpha-, Beta- and Gamma-Ray Spectroscopy*, (K. Siegbahn, editor), 79. (Amsterdam: North-Holland Publishing Co.)
2. Siegbahn, K., Nordling, C., Fahlman, A., Nordberg, R., Hamrin, K., Hedman, J., Johansson, G., Bergmark, T., Karlsson, S.-E., Lindgren, I. and Lindberg, B. (1965). *ESCA Atomic, Molecular and Solid State Structure Studied by Means of Electron Spectroscopy*. (Uppsala: Almqvist & Wiksells Boktryckeri AB)
3. Siegbahn, K., Nordling, C., Johansson, C., Hedman, J., Hedén, P. F., Hamrin, K., Gelius, U., Bergmark, T., Werme, L. O., Manne, R. and Baer, Y. (1969). *ESCA Applied to Free Molecules*. (Amsterdam: North-Holland Publishing Co.)
4. Reference 3, p. 76
5. Hnatowich, D. J., Hudis, J., Perlman, M. L. and Ragaini, R. C. (1971). *J. Appl. Phys.*, **42**, 4883
6. Baer, Y., Hedén, P. F., Hedman, J., Klasson, M. and Nordling, C. (1970). *Solid State Commun.*, **8**, 1479
7. Siegbahn, K., Hammond, D., Fellner, H. and Barnett, E. F. (1971). *Science*, to be published
8. Siegbahn, K. (1971). *Asilomar Internat. Conf. on Electron Spectroscopy*. (Amsterdam: North-Holland Publishing C.), to be published
9. Clementi, E. and Davies, D. R. (1967). *J. Computational Phys.*, **2**, 441
10. Siegbahn, P. (1971). *Chem. Phys. Letters*, **8**, 245
11. Pople, J. A., Santry, D. P. and Segal, G. A. (1965). *J. Chem. Phys.*, **43**, 5129
12. McWeeny, R. and Sutcliffe, B. T. (1969). *Methods of Molecular Quantum Mechanics*. (London and New York: Academic Press)
13. Pople, J. A. and Beveridge, D. L. (1970). *Approximate Molecular Orbital Theory*. (New York: McGraw-Hill)
14. Neumann, D. B. and Moskowitz, J. W. (1969). *J. Chem. Phys.*, **50**, 2216
15. Schonland, D. S. (1965). *Molecular Symmetry. An Introduction to Group Theory and its Uses in Chemistry*. (London: D. van Nostrand Co. Ltd.)
16. Roos, B. and Siegbahn, P. (1970). *Theoret. Chim. Acta*, **17**, 199
17. Rabalais, J. W., Bergmark, T., Werme, L. O. and Siegbahn, K. (1971). *Physica Scripta*, **3**, 13

18. Gelius, U., Allan, C. J., Johansson, G., Siegbahn, H., Allison, D. and Siegbahn, K. (1971). *Physica Scripta,* **3,** 237
19. Gelius, U., Allan, C. J., Allison, D., Siegbahn, H. and Siegbahn, K. (1971). *Chem. Phys. Letters,* **11,** 224
20. Gelius, U. (1971). *Asilomar Internat. Conf. on Electron Spectroscopy.* (Amsterdam: North-Holland Publishing Co.) to be published
21. Wuilleumier, F. and Krause, M. O. (1970). *Asilomar Internat. Conf. on Electron Spectroscopy.* (Amsterdam: North-Holland Publishing Co.), to be published
22. Krause, M. O. (1971). *Chem Phys. Letters.* **10,** 65
23. Hartmann, H. and Clementi, E. (1964). *Phys. Rev.,* **133, 5A,** 1295
24. Reference 2, p. 66
25. Bagus, P. S. (1965). *Phys. Rev.,* **139,** A619
26. Schwartz, M. E. (1970). *Chem. Phys. Letters.,* **5,** 50
27. Gianturco, F. A. and Guidotti, C. (1971). *Chem. Phys. Letters,* **9,** 539
28. Turner, D. W. and May, D. P. (1966). *J. Chem. Phys.,* **45,** 471
29. Cade, P. E., Sales, K. D. and Wahl, A. C. (1966). *J. Chem. Phys.,* **44,** 1973
30. Carlson, T. A., Krause, M. O. and Moddeman, W. E. (1971). *J. de Physique,* **32,** C4–76
31. Carlson, T. A. and Krause, M. O. (1966). *Phys. Rev.,* **140,** A1057
32. Krause, M. O., Carlson, T. A. and Dismukes, R. D. (1968). *Phys. Rev.,* **170,** 37
33. Reference 3, p. 30
34. Carlson, T. A., Nostor, C. A., Jr., Tucker, T. C. and Malik, F. B. (1968). *Phys. Rev.,* **169,** 27
35. Åberg, T. (1969). *Ann. Acad. Sci. Fenn. AVI,* **308,** 1
36. Meldner, H. and Perez, J. D. (1971). *Phys. Rev. A.,* **4,** 1388
37. Manne, R. and Åberg, T. (1970). *Chem. Phys. Letters,* **7,** 282
38. Reference 3, p. 63
39. Byron, F. W., Jr. and Joachain, D. J. (1967). *Phys. Rev.,* **164,** 1
40. Novakov, T. (1971). *Phys. Rev. B.,* **3,** 2693
41. Rosencwaig, A., Wertheim, G. K. and Guggenheim, H. J. (1971). *Phys. Rev. Letters,* **27,** 479
42. Novakov, T. and Prins, R. (1971). *Asilomar Internat. Conf. on Electron Spectroscopy.* (Amsterdam: North-Holland Publishing Co.), to be published
43. Barber, M., Connor, J. A. and Hillier, I. H. (1971). *Chem. Phys. Letters,* **9,** 570
44. Wertheim, G. K. and Rosencwaig, A. (1971). *Phys. Rev. Letters,* **25,** 1179
45. Gelius, U., Hedén, P. F., Hedman, J., Lindberg, B. J., Manne, R., Nordberg, R., Nordling, C. and Siegbahn, K. (1970). *Physica Scripta,* **2,** 70
46. Thomas, T. D. (1970). *J. Chem. Phys.,* **52,** 1373
47. Reference 3, p. 113
48. Lindberg, B. J., Hamrin, K., Johansson, G., Gelius, U., Fahlman, A., Nordling, C. and Siegbahn, K. (1970). *Physica Scripta,* **1,** 286
49. Nordberg, R., Albridge, R. G., Bergmark, T., Ericson, U., Hedman, J., Nordling, C. and Siegbahn, K. (1967). *Ark. Kemi,* **28,** 257
50. Pelavin, M., Hendrickson, D. N., Hollander, J. M. and Jolly, W. L. (1970). *J. Phys. Chem.,* **74,** 1116
51. Hedman, J., Klasson, M., Nordling, C. and Lindberg, B. J. (1971). *Institute of Physics, Uppsala University* Report **UUIP-744**
52. Hollander, J. M., Hendrickson, D. N. and Jolly, W. L. (1968). *J. Chem. Phys.,* **49,** 3315
53. Hendrickson, D. N., Hollander, J. M. and Jolly, W. L. (1969). *Inorg. Chem.,* **8,** 2642
54. Hendrickson, D. N., Hollander, J. M. and Jolly, W. L. (1970). *Inorg. Chem.,* **9,** 613
55. Nordberg, R., Brecht, H., Albridge, R., Fahlman, A. and Van Wazer, J. R. (1970). *Inorg. Chem.,* **9,** 2649
56. Malmsten, G., Thorén, I., Högberg, S., Bergmark, J.-E. and Karlsson, S.-E. (1970). *Institute of Physics, Uppsala University* Report **UUIP-682**
57. Hulett, L. D. and Carlson, T. A. (1971). *Appl. Spectrosc.* **25,** 33
58. Jolly, W. L. and Hendrickson, D. N. (1970). *J. Amer. Chem. Soc.,* **92,** 1865
59. Finn, P., Pearson, R. K., Hollander, J. M. and Jolly, W. L. (1970). *Inorg. Chem.,* **10,** 379
60. Barber, M., Swift, P., Cunningham, D. and Frazer, M. J. (1970). *Chem. Commun.,* 1338
61. Mateescu, G. D. and Riemenschneider, J. L. (1971). *Asilomar Internat. Conf. on Electron Spectroscopy.* (Amsterdam: North-Holland Publishing Co.), to be published

62. Hollander, J. M. and Shirley, D. A. (1970). *Ann. Rev. Nucl. Sci.*, **20,** 435
63. References 3, p. 56
64. Fadley, C. S., Shirley, D. A., Freeman, A. J., Bagus, P. S. and Mallow, J. V. (1969). *Phys. Rev. Letters,* **23,** 1397
65. Fadley, C. S. and Shirley, D. A. (1970). *Phys. Rev. A,* **2,** 1109
66. Clark, D. T. and Adams, D. B. (1971). *Chem. Phys. Letters,* **10,** 121
67a. Hedén, P. F., Löfgren, H. and Hagström, S. B. M. (1971). *Phys. Rev. Letters,* **26,** 432
67b. Wertheim, G. K. and Cohen, R. L. (1971). *Asilomar Internat. Conf. on Electron Spectroscopy.* (Amsterdam: North-Holland Publishing Co.), to be published
68. Novakov, T. and Hollander, J. M. (1968). *Phys. Rev. Letters,* **21,** 1133
69. Reference 3, p. 14
70. Reference 2, p. 148
71. Reference 2, p. 144
72. Krause, M. O. (1969). *Phys. Rev.,* **177,** 151
73. Carlson, T. A. (1971). *Asilomar Internat. Conf. on Electron Spectroscopy.* (Amsterdam: North-Holland Publishing Co.), to be published
74. Lindberg, B. J. (1971). *23rd International IUPAC Congress,* Boston
75. Ramqvist, L., Hamrin, K., Johansson, G., Fahlman, A. and Nordling, C. (1969). *J. Phys. Chem. Solids,* **30,** 1835
76. Cook, C. D., Wan, K. Y., Gelius, U., Hamrin, K., Johansson, G., Olson, E., Siegbahn, H., Nordling, C. and Siegbahn, K. (1971). *J. Amer. Chem. Soc.,* **93,** 1904
77. Buchanan, D. N. E., Robbins, M., Guggenheim, H. J., Wertheim, G. K. and Lambrecht, V. G., Jr. (1971). *Solid State Commun.,* **9,** 583
78. Clark, D. T., Kilcast, D. and Musgrave, W. K. R. (1971). *Chem. Commun.,* 516
79. Lindberg, B. J. and Hamrin, K. (1970). *Acta Chem. Scand.,* **24,** 3661
80. Turner, D. W., Baker, C., Baker, A. D. and Brundle, C. R. (1970). *Molecular Photo-Electron Spectroscopy.* (London: Wiley-Interscience)
81. Baker, A. D. (1970). *Accounts Chem. Res.,* **3,** 17
82. Berry, R. S. (1969). *Ann. Rev. Phys. Chem.,* **20,** 357
83. Turner, D. W. (1968). *Physical Methods in Advanced Inorganic Chemistry,* (H. A. O. Hill and P. Day, editors), 74. (London: Wiley-Interscience)
84. Siegbahn, K. (1970). *Phil. Trans. Roy. Soc. London,* **258A,** 33

2
Atomic Absorption Spectroscopy

A. SEMB
Norwegian Institute for Air Research, Kjeller

2.1 INTRODUCTION: BOOKS AND REVIEWS

This review covers the two-year period 1969–1971. Since this is the first review in the series, it must inevitably be somewhat diffuse. Some selection of the papers reviewed has been necessary. Typical application papers have only been included where the subject was felt to be of particular interest. Also, recent papers have been given priority over papers from the beginning of the period.

New textbooks on atomic absorption spectroscopy (AAS) are appearing every year. This is obviously a direct result of the intense research and development work which has been carried out in various parts of the world during the last few years. It is characteristic of the present state of rapid evolution of the technique that books with nearly identical titles may not overlap significantly in their contents, so that even if we now have about ten AAS textbooks, each new book seems to fill a genuine need for information about particular aspects of the subject.

L'vov[1] treats the subject on a largely theoretical level. This book should be referred to for detailed explanations of physical principles and particularly for an authoritative account of the extensive research work on graphite cell atomisation carried out by L'vov's group.

The original Russian edition was thoroughly revised and enlarged for the English translation, so that it is also admirably up-to-date. The book is clearly intended for an academic public, particularly those who are interested in fundamental research. The practical analyst will not, however, find any detailed analysis procedures. He will probably benefit more from the book by Reynolds and Aldous[2]. This somewhat more unorthodox book is very readable, introducing atomic absorption principles and research results linked to practical analytical procedures for the various elements. It includes a useful chapter on theoretical principles by K. C. Thompson.

A recent English translation[3] of Rubeska and Moldan's book appears unfortunately to have lost some of its novelty and interest because of the time that has elapsed since the publication of the original Czech edition.

A textbook in two volumes by Pinta[4] is available in the French language.

Christian and Feldman[5] have written a commendable book on the application of AAS to problems in agriculture, biology and medicine. Beyond doubt, the most ambitious contribution to AAS literature has been the publication of *Flame Emission and Atomic Absorption Spectroscopy* edited by Dean and Rains[6]. As the title indicates, this is an attempt to link AAS with the twin technique of atomic flame emission spectroscopy. The first volume, *Theory*, consists of chapters in which selected authors discuss various aspects of flame spectroscopy, mechanisms of atom formation, interferences and so on. The authors look at the various aspects from their own points of view, and there is some overlap in the coverage of some of the subjects which can, however, only be considered advantageous. From the atomic absorption spectroscopist's point of view, it is somewhat surprising that flameless absorption cells have not been mentioned in this volume. In volume II, *Instrumentation*[7], there is a chapter on flameless AAS by Massmann. This volume contains contributions on various aspects of instrumentation, ranging from electronics to burners and nebulisers. Some of the chapters are written in the form of reviews with many references which illustrate the present state of rapid evolution in the development of AAS techniques. A third volume covering applications will appear in 1972. *Analytical Flame Spectroscopy* edited by Mavrodineanu[8] also contains much information of interest to atomic absorption spectroscopists, although flame emission methods have received somewhat more attention in this book.

The two international conferences on atomic absorption spectroscopy sponsored by the International Union of Pure and Applied Chemistry in cooperation with national societies during this period are important sources of information on active research in this field in various parts of the world. The plenary lectures at the 1969 Sheffield Conference covered such themes as atomic fluorescence spectroscopy, the relationship between atomic flame emission, atomic fluorescence and atomic absorption spectroscopy, simultaneous multi-element analysis and non-flame atom reservoirs. The papers presented in the Fundamental section were very fruitful, and several important papers have since been published in the literature. The plenary lectures have been published in book form[9]. The 1971 Paris Conference reflected a change in interest to practical applications[10]. Willis gave a lecture on biological materials with particular emphasis on non-flame methods for analysis. Also significant was the lecture on line profiles by De Galan. Only one paper on this subject was presented at the Sheffield meeting, but a large number of publications in this field has been published in the period so that this is obviously of enormous interest and will probably be a growing research field in the near future. Koirtyohann particularly stressed the problem of chemical interferences.

An international terminology of atomic absorption and flame emission methods was proposed at the Sheffield Conference. This has been revised by Mariée and Pinta[11], who have given definitions of the various terms used in French, English and German. The IUPAC V4 Commission on Spectroscopy is also deliberating on nomenclature in this area. A number of reviews stressing different aspects of AAS methodology have also appeared in the period[12-16] including a very recent review of flameless atom reservoirs[17].

Bibliographies of AAS are maintained by several instrument manufacturers. S. Slavin's yearly listing[18] of publications is probably the most widely accepted. Most instrument manufacturers also keep loose-leaf manuals of selected analytical procedures for their customers.

A special series of abstracts of atomic absorption and flame emission publications is also available[19], but it seems to suffer from difficulties in keeping abreast of the ever-increasing number of publications in this field. The Society for Analytical Chemistry (London) is producing an annual review of atomic spectroscopy, the first of which will appear in mid-1972.

2.2 SPECTROSCOPIC AND FUNDAMENTAL STUDIES

2.2.1 Line profiles

As is well known, the monochromators used in atomic absorption spectroscopy are usually capable of spectral resolutions down to a few tenths of an Ångstrom at best. The absorption and emission profiles of atomic lines have half-intensity band-widths of the order of 0.001 nm. Because of the vacuum and the low temperature, the light source emission lines are usually much narrower than the absorption profiles in the flame. However, isotope shifts and hyperfine structure caused by nuclear spin effects may still cause the light source to be far from monochromatic when seen by the combination of the atomic vapour and an optical monochromator.

Scanning Fabry–Perot interferometry has been used extensively by several research groups in the past two years to evaluate the spectral properties of various light sources for atomic absorption spectroscopy. Human and Butler[20] found half-intensity line-widths for hollow cathode lamps corresponding to Doppler temperatures below 1000 K for the emitting atoms, but the 'finesse' of their instrument was probably too low to warrant accurate interpretation of the results.

Sargent and Kirkbright[21] have described the assembly of a piezoelectric scanning interferometer for line-profile studies. The instrument was used to determine the degree of self-absorption of different atomic lines emitted from electrodeless discharge lamps and to observe changes in the intensities of various hyperfine components.

A very thorough investigation of the Ca 422.67 nm line emitted by a hollow cathode lamp was made by Bruce and Hannaford[22]. They found the line profile to be determined almost entirely by Doppler broadening. The Doppler temperature was found to be very close to the temperature on the surface of the hollow cathode which was determined with a thermocouple. It ranged between 347 and 429 K for lamp currents of 5–15 mA dc. Using a pressure-scanning Fabry–Perot interferometer, Prugger et al.[23] found a line intensity half-width for a pulsed Ca hollow cathode lamp in excellent agreement with Bruce and Hannaford's results. De Galan and Wagenaar[24] have made line profile measurements of resonance lines emitted from hollow cathode lamps and from atoms in the nitrous oxide–acetylene flame. The red shift of the lines at elevated temperatures predicted by Lindholm's theory could also be detected.

Calcium and the other alkaline earth elements are very convenient for line-profile studies because of the relative absence of isotope and nuclear spin effects. This makes it possible to study their absorption profiles in flames by the so-called Zeeman scanning[25] technique where the emission line is split by placing the light source in a magnetic field, and the components separated by polarisation filters. By varying the magnetic field the absorption profile can be scanned. A modification of this technique may also be used to evaluate spectral widths of emission lines[26].

2.2.2   Bending of analytical calibration curves

Bruce and Hannaford[22] also considered the effect of the measured source line width on the absorption by neutral atoms in an air–acetylene flame and concluded that this was less than 10 % for the Ca 422.67 nm line studied. The same would be expected to apply to other resonance lines with little or no hyperfine splitting. Extensive theoretical calculations by van Gelder[27] indicate that line broadening, hyperfine structure or self-reversal make very little, or negligible, contribution to apparent deviations from linearity. Stray light, continuous radiation and other unwanted emission from the light source reaching the photodetector were considered much more important, in agreement with the results reported by De Galan and Samaey[28].

2.2.3   Determination of atom concentrations and atomisation efficiencies from spectroscopic absorption measurements

Provided that the atomisation efficiency of the burner–nebuliser assembly and the oscillator strength or f-value of the resonance line are known, it should, at least in principle, be possible to determine the concentration of an element in a sample solution directly from atomic absorbance measurements. The practical difficulties which make this solution impractical have been reviewed in the discussion between De Galan[29] and Rann[30] following Rann's original paper, and in a later publication by de Galan and Samaey[31]. There is, however, much more interest in the use of this principle for the determination of atomisation efficiencies, i.e. free atom fractions, in various atom reservoirs used in atomic absorption analysis, particularly flames. Either a

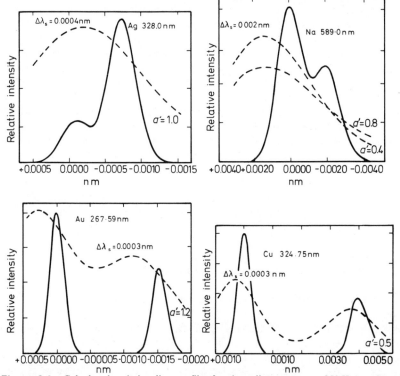

Figure 2.1   Calculated emission line profiles for sharp-line sources at 350 K (———), and absorption profiles for neutral atoms in an air–acetylene flame at 2320 K (------)
(From Willis, J. B. (1971). *Spectrochim. Acta*, **26B**, 177, by courtesy of Pergamon Press)

continuum source or a sharp line source may be used for the absorbance measurements. These two methods are usually referred to as the 'integrated-absorption' and the 'peak-absorbance' methods respectively. The first method is somewhat difficult experimentally, requiring knowledge of the spectral bandpass and spectral response of the monochromator–detector apparatus. The other method is simple and convenient experimentally, but the calculations require knowledge of the Doppler and Lorentzian broadening effects in the absorption cell for a first approximation. In more exact calculations, isotope effects, hyperfine structure, and Doppler and self-reversal broadening in the hollow cathode must also be taken into account.

Traditionally, therefore, the two methods have tended to give somewhat different results, typically by a factor of 2 or 3. Willis[32] used both methods to determine beta-values for 12 elements in both air–acetylene and nitrous oxide–acetylene flames and reproduced the discrepancies. In a later paper, the same author[33] measured atomisation efficiencies for sodium, gold, copper and silver. Using data[22] for the broadening of the light source, calculated Doppler and Lorentz broadening and red shift of the absorption line in the flame, *and* taking account of hyperfine splitting from nuclear spin effects, very satisfactory results were obtained and the discrepancies between the peak absorbance and the integrated absorption methods were practically eliminated. The calculated emission line and absorption profiles from Willis' work are reproduced in Figure 2.1. It should be stressed that these elements represent very favourable cases. Oscillator strengths are not known with satisfactory accuracy for many of the remaining elements. This limits the accuracy of determined beta-values, typically by a factor of about two[31, 33].

De Galan and Samaey[34] measured atomisation efficiencies for 27 elements in air–acetylene, nitrous oxide–acetylene and nitrous oxide–hydrogen flames by the integrated-absorption technique. Smyly, Townsend, Zeegers and Winefordner[35] measured relative free atom fractions of the elements copper, magnesium, manganese and strontium in hydrogen–oxygen–argon flames. Silver was used as reference, assuming this element to be completely atomised. Spectroscopic measurements of atomisation efficiencies and free atom fractions have been criticised by Fassel *et al.*[36], who point out that the regions of maximum absorbance are highly localised in the flame particularly for elements which form stable monoxides. L'vov[37] has used a flame as an atom reservoir for determining relative oscillator strength values.

2.2.4 Studies of atomisation efficiencies by other methods

Koirtyohann and Pickett[38] have proposed a method for estimating degrees of atomisation by simultaneously observing atomic absorption and monoxide emission intensity whilst varying the fuel to oxidant ratio. The flame temperature was kept constant by diluting with an inert gas. The quantitative displacement of the equilibrium reaction between the metal monoxide, the flame gases and free metal atoms was demonstrated for a number of elements forming refractory oxides. It is assumed that chemical equilibrium is reached in the zone of the flame which is being studied.

Fassel *et al.*[36] have studied concentration patterns of elements forming refractory oxides in the nitrous oxide–acetylene flame. The variation in concentration of free atoms in the region of maximum absorbance was studied as a function of fuel to oxidant ratio and successfully correlated with the expected degree of monoxide reduction by the flame gases from thermodynamic data and flame temperature measurements. Chester, Dagnall and Taylor[39] have made detailed calculations of concentrations of flame-gas species in the nitrous oxide–acetylene flame at different temperatures and flame compositions. The results are combined with data for monoxide dissociation energies to find theoretically obtainable atomisation efficiencies for some elements forming refractory oxides. The same approach has been used to evaluate the atomisation properties of some less common flames in atomic absorption spectroscopy[40]. These calculations showed quite clearly that only minor improvements could be expected compared with the conventional air–acetylene and nitrous oxide–acetylene flames.

Neither of these approaches takes into account the possibility of nonequilibrium atomisation processes. From the chemical interferences observed for many elements forming refractory oxides it may be concluded that solvent evaporation and the break-up of solid particles may often be important limiting factors in the formation of atoms in the flame. Chakrabarti et al.[41] have made some studies of concentration profiles in various flames with elements introduced as different chemical species in various solvents. The results are too few, however, to be interpreted in general terms. The effect of air entrainment and diffusion of oxygen from the outer parts of the flame has not been considered in any of these approaches, although the study of separated flames indeed shows that this effect is very important for flame structure. Kirkbright and Vetter[42] have studied the effect of inert gas shielding on flame temperatures. Ando, Fuwa and Vallee[43] have studied the distribution of free atoms in the long-tube burner as a function of the temperature. The complicated reaction mechanisms which occur in low temperature hydrogen flames have been studied for chromium compounds by Bulewicz and Padley[44].

2.3 INSTRUMENTATION AND METHODOLOGY

2.3.1 Light sources

Marked improvements in the design of ordinary commercial hollow cathode lamps have probably had the result that these light sources will continue to be used into the foreseeable future for most of the elements commonly determined by atomic absorption spectroscopy. Further developments of hollow cathode lamps include the pulsed HCls which have been investigated by Katskov', Lebedev and L'vov[45] and by Prugger et al.[23] Because the formation of atomic vapour by the sputtering effect is slower than the excitation mechanism, significantly higher instantaneous emission intensities can be achieved without the appearance of self-reversal effects than with ordinary a.c. modulated hollow cathode lamps. Similar effects are utilised in selectively modulated hollow cathode lamps, the use of which has been reviewed by Sebestyen[46]. The excitation mechanism in Sullivan–Walsh-type high-intensity hollow cathode lamps has been investigated by Bueger and Fink[47]. Another type of high-intensity hollow cathode lamp has been developed by Human and Butler[48]. A microwave cavity was used to transfer additional excitation energy to the atomic cloud at the face of the cathode.

Prugger[49] has investigated the radiation intensity of a number of light sources used in AAS, including high-intensity hollow cathode lamps and electrodeless discharge lamps. In terms of radiation temperature (Kirchhoff's law), he concluded that ordinary hollow cathode lamps were 'hot' enough for most elements in the air–acetylene flame, but only just sufficient for the nitrous oxide–acetylene flame. For this flame then, high-intensity lamps and other intense light sources might represent an advantage. Induction plasmas and other high-temperature atom reservoirs were not considered useful for AAS because of the lack of sufficiently intense light sources.

In agreement with this, Hildon and Sully[50] found that the use of lens masks (to isolate a defined section of the flame), which results in less light transmitted to the photodetector did not result in increased electronic noise. Interest in high-intensity light sources, such as electrodeless discharge lamps (EDLs) has, therefore, mainly been for their use in atomic fluorescence spectroscopy. However, these light sources also have properties which make them attractive for certain uses in atomic absorption spectroscopy. They are particularly advantageous in connection with the determination of elements which are volatile in the elemental state. Goleb[51] used electrodeless discharge tubes for the determination of argon and neon in helium. The discharge was modulated electronically by an external electromagnetic field. Electronic modulation of EDLs was described by Wildy and Thompson[52]. Woodward[53] prepared useful lamps for the alkali metals by introducing the elements in the form of silicates, borates or phosphates.

Cooke, Dagnall and West studied the performance in atomic absorption spectroscopy of modulated mercury, thallium and selenium EDLs in various microwave cavities under different operating conditions[54]. EDLs are also being used in connection with atomic absorption spectroscopy in the USSR[55]. Thompson[56] used the Ge 422.657 nm atomic line emission from a microwave powered EDL to determine calcium with reduced sensitivity by virtue of the spectral overlap of the germanium line with the calcium resonance line at 422.67 nm.

2.3.2   Burners and flames

Mansell[57] studied the possible advantages of the use of the nitrous oxide–MAPP (stabilised methylacetylene propadiene) flame for AAS. The burning velocity of the mixture is significantly lower than for nitrous oxide–acetylene so that the demands on burner design are somewhat less exacting. The flame is, however, less effective than the nitrous oxide–acetylene flame for atomising refractory oxide elements. Rankin and Bailey have investigated the analytical[58] and physical[59] properties of various liquid-fuel flames. Sargent, Kirkbright and West[60] described an atomic absorption burner for nitrous oxide–acetylene with inert gas shielding and studied the performance of this arrangement for a number of elements. Lowering of detection limits by a factor of about two was obtained for many elements. The use of a separated nitrous oxide–acetylene flame for the determination of arsenic and selenium was studied by Kirkbright and Ranson[61]. A mechanically separated flame has been used for AAS by Ure and Berrow[62]. Aldous et al.[63] described the construction of burners for air–acetylene and nitrous oxide–acetylene made from stainless steel capillary tubing. The burners could be made with

rectangular or circular cross-sections and with provision for inert gas shielding. The construction is exceptionally safe against flashback of the nitrous
oxide–acetylene mixture. Another interesting property is the nearly flat
reaction zone which may be obtained. This may make this construction of
value in flame-atomisation studies.

Various modifications of the long tube burner technique for easily atomis-
able elements have been used by workers in Japan[64, 65], Czechoslovakia[66] and
the USSR[67]. The review by Fuwa[12] gives a very thorough account of the
development of this technique. A very interesting paper by Goguel[68] describes the function of a thermostatted burner head for nitrous oxide–acetylene
flames. The author calls attention to the effect of the burner temperature on
the flow resistance and gas velocity distribution in a slot burner.

2.3.3 Sample introduction

In the conventional indirect nebuliser–burner system only about 5% of the
sample solution aspirated actually reaches the flame. Uny et al.[69] have
investigated the use of a heated spray chamber and a condenser to increase
sensitivity and found a 20-fold increase for some elements. No comparative
interference study was carried out. Uny and Spitz[70] have reported the use of

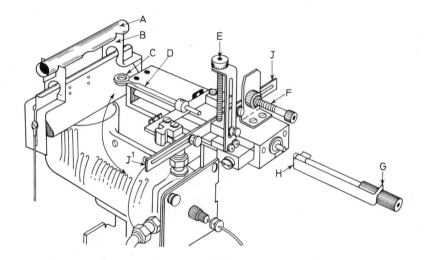

Figure 2.2 Delves' sample introduction arrangement:
A = Nickel absorption tube 100 mm long × 12.5 mm o.d., made from 0.006 inch Ni foil
B = Nickel supports for A, made from 0.020 inch Ni foil
C = Nickel crucible, 10 mm o.d. × 5 mm tall, made from 0.006 inch Ni foil
D = Platinum-wire holder, 0.5 mm meter, for C, sealed in a glass tube
E = Vertical adjustment screw
F = Horizontal angular adjustment screw
G = Adjustment screw for slide stop
J and J^1 = Rotating limb
(From Delves[76], by courtesy of the Society for Analytical Chemistry)

a radioactive tracer method to determine nebulisation efficiencies based on the collection of aerosol droplets on a filter. This procedure is only valid if the nebuliser is not affected by the change in the spray chamber pressure when the flame is burning. Duckworth and Coleman[71] have shown that useful determinations may be made from the spikes obtained when aspirating limited (100 µl) sample volumes.

Modulation of the analytical signal by attenuation of the sample aspiration with piezoelectrically operated valves has been described by Mosotti, Abercrombie and Eakin[72]. There has for a long time been a strong interest in the direct atomisation of solid samples for AAS. Kashiki and Oshima[73] described a method for the aspiration of fine powders into a premixed nitrous oxide–acetylene flame and applied this successfully to the determination of cobalt and molybdenum in alumina catalysts. Winge, Fassel and Kniseley[74] applied the principle of Jones and Dahlquist in preparing a sample aerosol from polished metal samples by a spark discharge. The design of a nebuliser and burner assembly for viscous samples and samples with high concentrations of dissolved solids has been discussed by Hell and Ramirez-Munoz[75].

Delves[76] has developed a very efficient method for direct introduction of evaporated samples, which is a synthesis of Kahn's tantalum boat technique and Robinson's T-piece adapter. The principle of the method is shown in Figure 2.2. For certain easily atomised elements, including lead, the sensitivity of this method rivals flameless atomic absorption with the graphite cell and the method is very suitable for clinical analysis, although a high degree of operational skill is required for obtaining reproducible results. Among other methods of sample introduction resulting in increased sensitivity, the determination of selenium and arsenic by reduction with zinc and flushing the volatile hydrides into an argon (entrained air)–hydrogen flame may be mentioned[77].

2.3.4 Measurement and computer techniques

The precision and accuracy of atomic absorption analysis has been discussed by Roos[78, 79]. For some elements, the results agreed well with the Twyman–Lothian error function, but this was not the case for all elements studied. The use of sensitivity diagrams (Ringbom–Ayres plots) has been proposed[80] by Ramirez-Munoz, who has also determined the design of a slide rule to assist in atomic absorption calculations[81]. A modified standard addition procedure was described by Leirtie and Mattson[82]. Feldman[83] discussed various methods of internal standardisation. Booher et al.[84] described a rapid procedure for determination of detection limits using a specially designed exponential dilution flask. The terms 'sensitivity' and 'detection limit' in flame spectroscopic methods have been discussed by de Galan[85]. A number of papers[86–89] describe various computer techniques in connection with routine analysis of large numbers of samples. Integrated circuits are used in many commercial instruments to process the analytical signal before the numerical or analogue readout. A simple log converter which can be used to give linear concentration readout has been described[90].

2.4 INTERFERENCES IN FLAME ATOMIC ABSORPTION SPECTROSCOPY

2.4.1 Spectral interferences

No more spectral interferences from atomic species in the flame have been reported, which may be taken as definite evidence that such interferences are extremely rare.

Broad-band molecular absorption and scattering effects from samples with a high concentration of dissolved solids are of much more genuine concern. These effects are now customarily corrected for by the use of hydrogen or deuterium lamps. Smith and Robinson[91] have investigated the absorption spectra from flame species formed when aspirating organic solvents in an oxyhydrogen flame. Analysis of samples with high sodium contents have been studied by Ramirez-Munoz[92] and Skurnik-Sarig et al.[93].

2.4.2 Ionisation interferences

Ionisation problems are very pronounced in the nitrous oxide–acetylene flame. Alkali salts are customarily added to effect ionisation buffering and avoid interference from other easily ionisable elements. Complete suppression of ionisation is however not always achieved by this procedure. Paus[94] studied the effect of alkali salts on the atomic absorption of 50 μg cm^{-3} of aluminium in the nitrous oxide–acetylene flame. The results are shown in

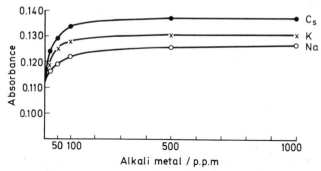

Figure 2.3 Effect of added alkali metal salts on the absorbance of aluminium (50 μg cm^{-3}) in a nitrous oxide–acetylene flame (From Paus[94], by courtesy of Elsevier)

Figure 2.3. It is seen that at least 1 μg cm^{-3} of caesium must be added to suppress the ionisation completely. Woodward[95] has tabulated expected ionisation for 48 elements in the nitrous oxide–acetylene, air–acetylene and air–propane flames.

2.4.3 Chemical interferences

The rediscovery of the interference effect of phosphate and aluminium on calcium and magnesium has apparently at last ceased! Recent reports of chemical interferences in the air–acetylene flame have mainly been in connection with the determination of chromium and the iron group of elements[96–104]. Interelemental depressive interferences become more pronounced under fuel-rich flame conditions. Citric acid and probably also

Table 2.1 Effect of various interferents on the absorbance of ytterbium (60 µg/cm^{-3}) in a nitrous oxide–acetylene flame

(From van Loon, Galbraith and Aarden[120], by courtesy of The Society for Analytical Chemistry)

Interference	Concentration	Absorbance change, %
Fe	500 p.p.m.	+20
K	500 p.p.m.	+78
Mg	5 p.p.m.	0
	50 p.p.m.	+ 4
	500 p.p.m.	+20
Na	5 p.p.m.	+ 7
	50 p.p.m.	+22
	500 p.p.m.	+90
Mn	500 p.p.m.	+20
Ca	500 p.p.m.	+50
Ti	500 p.p.m.	−35
Other rare earths	200 p.p.m.	+13 to +21
Al	10 p.p.m.	− 7
	100 p.p.m.	−22
	500 p.p.m.	−60
Zr	1000 p.p.m.	−60
HNO$_3$	0.1 M	0
	1.0 M	0
	3.0 M	−15
HCl	0.1 M	0
	1.0 M	− 2
	3.0 M	−10

other chelating organic acids also exert depressive effects[101] and chromium is less subject to interference when present as chromates[98] in the sample solution. According to Roos[103], the depressive effects cannot be explained entirely on the basis of formation of mixed metal oxides. He considers the physical properties of the reduced metal droplets to be a more likely explanation. Coker and Ottaway[105] have also suggested the possibility of chemical interference due to over-excitation phenomena. Lanthanum salts[106] and sodium sulphate[107] have some effects as interference releasing agents, as well as the alkaline earths and 8-hydroxyquinoline[104]. Even more serious interference effects are encountered in the analysis of the platinum group metals[108–111]. Again, the low volatility of the metals seems to be the main cause of the difficulties. Apart from the fact that many substances interfere, either by depression or enhancement, the degree of the interference is very dependent on flame operating conditions. The interferences can be suppressed by addition of suitable releasing agents, such as sodium sulphate, copper, cadmium, lanthanum, vanadium or uranium salts[110, 111]. The interferences are much less pronounced in the hotter and more reducing nitrous oxide–acetylene flame. Juliano and Harrison[112] studied the effect of various interferents on the determination of tin in the air–hydrogen flame. The interferences were severe and the authors suggested that air–acetylene was a better alternative even if the sensitivity in this flame is considerably lower. The nitrogen (entrained air)–hydrogen flame was shown to be even more subject to interferences. The effect of potassium salts in the determination of tin was studied by Levine et al.[113]. The use of low temperature flames for AAS work is generally not advisable.

Chemical interferences are often considered to be insignificant in the nitrous oxide–acetylene flame. In spite of this a number of interferences have now been reported, particularly for the elements which form refractory oxides. Thomas and Pickering[114] studied the effects of various acids and cations on the determination of niobium, tantalum, titanium and tungsten and found very marked effects from fluorides, no doubt because of the complex-forming ability of this ion. Bond[115] investigated the use of ammonium fluoride as a releasing agent for zirconium and some other elements. Interferences in the determination of titanium, zirconium and hafnium have been investigated by Panday[116]. Marks and Welcher[117] studied the effect of various additives in the determination of aluminium, titanium and chromium and nickel in the nitrous oxide–acetylene flame. The effect was most pronounced in the flame region immediately above the primary reaction zone. Interferences in the determination of aluminium have also been studied by Urbain and Varlot[118] and by Feris et al.[119].

Chemical interferences occur also in the determination of rare earths and yttrium[120]. Table 2.1 shows the effect of some cations and acids on the relative absorbance of ytterbium. The depressive effects of aluminium and zirconium are perhaps not quite unexpected, but the enhancements are somewhat more puzzling. Ytterbium has a relatively high ionisation potential, so that the enhancements are not primarily due to ionisation effects.

It is seen that the effects of sodium and of potassium salts are equal. Addition of excess lanthanum (1 mg cm^{-3}) was shown to remove most of the interferences. The effects of alkali salts and acids in the determination of yttrium have been investigated by Cattrall and Slatter[121]. The effects on beryllium, which have been investigated by Fleet, Liberty and West[122], are very similar to the rare earth elements. Addition of 8-hydroxyquinoline enhanced the absorbance of 8 µg cm^{-3} of beryllium by about 8% and was shown to be effective in removing the interference from aluminium at concentrations up to 4 mg cm^{-3}. Again, enhancement effects occurred which could not be explained satisfactorily from our present knowledge of flame-reaction mechanisms.

Chemical interferences have usually been considered more serious when organic solvents are nebulised because of the increased nebulisation efficiency. The results obtained by Panday and Ganguly[123] cast some doubt on this however. Pitts and Beamish also observed no increased interference effects in the determination of platinum in 80% acetone or ethyl acetate. Using the nitrous oxide–acetylene flame[124] Culp et al.[125] found that organic solvents had a favourable effect on the determination of copper in the air–

hydrogen flame, but this may very well be due to the change in flame chemistry. Chakrabarti et al.[126, 127] have continued their investigations of the effects of various chelating agents in combination with organic solvents. Addition of lanthanaum naphthenate as a releasing agent in organic solvents has been proposed by Mostyn et al.[128].

The determination of elements present as volatile organometallic compounds in organic solvents is not easy. The addition of halogens may be used to break down the compounds. Kashiki et al.[129, 130] have shown, for example, that the addition of iodine in methyl isobutyl ketone to gasoline with tetramethyl or tetraethyl lead is effective in removing the differences in nebulisation efficiencies of these two formulations. The halogen addition method is also recommended for other organometallic compounds. A few publications deal with the atomisation of samples in the form of liquid suspensions[131—133].

2.5   FLAMELESS ATOMIC ABSORPTION

2.5.1   Graphite cells and filament atom reservoirs

The development of these atomisation techniques has without doubt been the most important extension of the analytical potential of atomic absorption in the period of this review. By avoiding both the rejection of a major part of the sample and the dilution of the atomised sample with flame gas, very low absolute detection limits have been obtained, typically down to 10^{-10}–10^{-12} g. In addition to this, elemental carbon at elevated temperatures is an even more powerful reducing agent than the flame radicals found in the fuel-rich nitrous oxide–acetylene flame. Work with graphite cell atom reservoirs was pioneered by L'vov and his group in the USSR, and a thorough account of the technique and theory of the method is given in L'vov's recent book[1].

An interesting possibility of enclosed atom reservoirs is the detection of elements with resonance lines in the vacuum ultra-violet. L'vov and Khartsyzov[134] were able to determine nanogram amounts of sulphur, iodine, phosphorus and mercury by the atomic absorption of their resonance lines between 160 and 180 nm in a graphite cell flushed with argon at atmospheric pressure. L'vov and Khartsyzov[135] have also studied the radial loss of atomic vapour in the graphite cell. The graphite cell used by Massmann is now available commercially[136] and has already been used in several important applications[137—139]. Baudin, Chaput and Feve[140] have studied sample matrix interferences in a Massmann-type graphite furnace under different operating conditions.

It appears that different drying, ashing and 'booster' atomisation temperatures should be chosen with care, depending on the type of the matrix and the anions present. Contrary to experience with flame atomisation methods, 'chemical interferences' are to be expected when volatile molecular compounds of the element can be formed. Proper choice of operating conditions is also important in order to avoid obscuring the atomic absorption signal by molecular absorption from species formed during decomposition of the sample, even if deuterium or hydrogen arcs are used for correction of spectral interferences.

One of the disadvantages of the L'vov and Massmann type of furnace is its large power consumption which requires special installations. The carbon filament atom reservoir developed by West and co-workers is considerably less power-consuming and significant progress has been made in developing this type of atom reservoir during the period. The original surrounding glass envelope has been replaced by a sheath of inert gas[141]. The concentration profiles of neutral atoms have been studied for typical elements[142] and the matrix effects of thousandfold excesses from 42 metal cations and a number of anions have been studied for eight elements[143]. The results indicate that most of the interferences occur in the gaseous phase. Chemical interference from formation of refractory oxides is not experienced. The influence of organic solvents and extraction agents was studied by Aggett and West[144]. Only the 8-hydroxyquinoline compound with copper gave rise to a slightly anomalous behaviour when compared with aqueous standards.

Dipiero and Tessaro[145] used an original West and Williams type of filament reservoir to determine nanogram quantities of nickel and found that anions, particularly chloride, had marked effects on the analytical signals. Brodie and Matousek[146] determined metals in petroleum products, including lubricating oils. Aqueous solutions could be used as standards.

Several publications[147-149] deal with the determination of various elements in blood and serum samples with the carbon filament atomiser and a modification of this in which the sample is deposited in a transverse hole in the carbon rod. Surrounding the graphite rod with hydrogen instead of argon may also be advantageous, particularly when the amplifier is that of a conventional flame spectrometer. Glenn et al.[149] used argon shielding combined with a surrounding argon–hydrogen diffusion flame in connection with the determination of copper in serum with a carbon filament atomiser.

A filament atom reservoir has also been described by Donega and Burgess[150]. In this modification the filament sample boats are cut from graphite sheet or tantalum or tungsten foil. The filaments are heated electrically to c. 2200 °C in less than 0.1 s. The atomiser is enclosed in a quartz tube and operated with inert gas filling at reduced pressures. Headridge and Smith[151] used an induction-heated graphite oven to determine traces of cadmium in solutions and zinc base alloys. Morrison and Talmi[152] developed an induction-heated furnace where atomisation was effected in small graphite crucibles containing solid or evaporated liquid samples.

2.5.2 Other methods

Massmann[153] has described a demountable graphite hollow cathode lamp for atomisation of liquid and solid samples. The hollow cathode was operated in pulses of 50 Hz half-waves, and the absorption recorded in the 'off' periods in order to avoid the intense hollow cathode emission. Integration of the absorbance signal was necessary in order to obtain linear calibration graphs for extended concentration ranges. The method was successfully applied to the determination of elements in alloy samples, but the reproducibility was not high. Hollow cathodes have also been investigated by Korovin and Kuchumov[154] as atomising units, and they came to similar conclusions.

Cathodic sputtering was also used by Gandrud and Skogerboe[155] who evaporated liquid samples on disc-shaped aluminium and graphite cathodes (platrodes). Taylor, Gibson and Skogerboe[156] were able to determine sulphur by atomising samples in an r.f.-induced plasma. Sulphur could be determined by emission or absorption of a previously unknown resonance line at 216.9 nm. Robinson et al.[157–158] have used a graphite furnace to atomise lead, cadmium and mercury aerosols in air. Ambient air is drawn through a heated graphite furnace and the elements are determined by atomic absorption in the exhaust gases which are passed through a T-shaped absorption cell. A number of publications[161–169] deal with the determination of mercury in various sample materials by flameless atomic absorption, mainly by variations of Hatch and Ott's procedure. Marinkovic and Vickers[159] used a stabilised d.c. arc to atomise samples. The method is probably of academic interest only. A laser was used for the atomisation of solid samples by Krivchikova and Demin[160]. The laser served also as a continuum light source and the elements were determined by their Fraunhofer lines on a spectrographic plate.

2.6 APPLICATIONS

The applications of flameless atomic absorption methods were covered in the preceding section. Both the graphite cell and the filament atom reservoirs are still so new that more evaluation work is needed before the potential and restrictions of these atomisation methods are fully realised, even if their usefulness in some applications, particularly in the analysis of biological samples, is now well documented. General considerations indicate, however, that these techniques will serve as an extension of AAS in flame media rather than as a replacement atom reservoir in established applications. As procedures for the determination of most common elements by flame-based atomic absorption spectroscopy have been available for some time now, most effort in the past two years has been directed towards the development of more rational analyses and sample preparation methods. Some years ago, atomic absorption spectroscopy in many laboratories was a new and suspect black box and was used only for a limited number of elements which could not readily be determined with established 'classical methods'. There is now a growing tendency to use atomic absorption spectroscopy for an increasing number of elements and for comprehensive analysis schemes, avoiding unnecessary and tedious separations and sample matching procedures.

A number of applications involving flame atomisation methods were cited in the preceding sections. As mentioned in the introduction, no attempt will be made to give a comprehensive survey of application publications. The examples given below have been chosen mainly to illustrate the general trends referred to above.

A number of publications deal with the determination of precious and noble metals[170-180]. Because of the serious interferences encountered with the determination of gold and platinum group metals, particularly in the air–acetylene flame, rather selective separation methods such as coprecipitation and solvent extraction have to be used, as well as the more conventional fire-assay method. These procedures also serve to concentrate the elements in low grade ores.

Another group of 'rare' elements which have received some attention are the rare earths[181], including yttrium. There is also a considerable interest in the use of atomic absorption for the determination of non-metals, such as arsenic[182], selenium[183], antimony[184] and tellurium[185].

Comprehensive analysis schemes for the analysis of silicates by AAS after fusion of the sample with lithium borates have been given by van Loon and Parissis[186] and by Boar and Ingram[187]. Langmyhr and Paus[188] have applied their hydrogen fluoride decomposition technique to the analysis of sulphide minerals and ores. Lucas and Ruprecht[189] analysed chromium ores and chrome magnesite refractories by atomic absorption after dissolution in phosphoric acid.

Thomerson and Price[190] gave a detailed procedure for the determination of ten elements in a wide range of steels based on a simple dissolution of the steel sample in perchloric acid. An alternative dissolution method was needed for the determination of tungsten and molybdenum in the presence of tungsten. The same authors discussed critically the determination of chromium and molybdenum in steel samples[191]. Vienney[192] determined 21 trace elements in plutonium reactor fuel material by atomic absorption without prior separation.

Solvent extraction remains the most popular technique for concentrating and separating trace constituents for subsequent determination by AAS. Recent applications include the determination of lead and bismuth in steels[193-195], cobalt and zinc[196], and zinc, iron and lead[197] in high purity metals, vanadium in titanium tetrachloride[198], and brines[199]. Ammonium hexamethylene dithiocarbamate remains the most popular of the extraction agents used for the concentration of sulphide group cations[200] for AAS. Sachdev and West[201] have proposed a mixture of three extraction agents for the simultaneous concentration of ten metal ions.

Solvent extraction also finds some use in indirect AAS methods. Lau and Lott[202] determined selenium by extraction of a ternary complex with 2,3-diaminonaphthalene and $PdCl_2$ and determined the amount of coextracted palladium. LeBihan and Courtot-Coupez[203] determined cationic and anionic detergents by their synergistic effect on the extraction of copper and iron o-phenanthrolines. Pinta[204] has given a useful review of indirect atomic absorption methods. Fritz and Latwesen[205] have published a separation scheme for multi-element samples in which AAS is used to determine minor constituents. Ion exchange was used by Galle to determine trace constituents in brines[206] and Burke[207] used coprecipitation with manganese dioxide to determine microgram quantities of antimony, bismuth, lead and tin in nickel metal. Among indirect methods the atomic absorption inhibition titration methods for sulphate[208] and silicate[209] proposed by Looyenga and Huber deserve mention though this is of course not the first indirect method that has been proposed for sulphate. In a way, these methods illustrate the general usefulness of AAS, but direct methods are generally to be preferred.

References

1. L'vov, B. V. (1970). *Atomic Absorption Spectral Analysis.* (London: Adam Hilger)
2. Reynolds, R. J. and Aldous, K. M. (1970). *Atomic Absorption Spectroscopy, A Practical Guide.* (London: Griffin)
3. Rubeska, I. and Moldan, B. (1969). *Atomic Absorption Spectroscopy.* (London: Iliffe)
4. Pinta, M. (1971). *Spectrometrie d'Absorption Atomique.* (Paris: Masson)
5. Christian, G. D. and Feldman, F. J. (1970). *Atomic Absorption Spectroscopy Applications in Agriculture, Biology and Medicine.* (New York: Wiley)
6. Dean, J. A. and Rains, T. C. (editors) (1960). *Flame Emission and Atomic Absorption Spectrometry, Volume I: Theory.* New York: Dekker)
7. Dean, J. A. (editor) (1971). *Flame Emission and Atomic Absorption Spectrometry Volume II: Instrumentation.* (New York: Dekker)
8. Mavrodineanu, R. (1970). *Analytical Flame Spectroscopy.* (London: Macmillan)
9. Dagnall, R. M. and Kirkbright, G. F. (editors) (1970). *Atomic Absorption Spectroscopy— Plenary Lectures presented at the International Conference, Sheffield 1969.* (London: Butterworths)
10. Proceedings. *3 ème Conférence Internationale d'Absorption et de Fluorescence Atomique.* Paris, 1971. Plenary Lectures. Meth. Phys. Anal. (GAMS) special issue
11. Mariée, M. and Pinta, M. (1970). *Meth. Phys. Anal.,* **6,** 361, 368
12. Fuwa, K. in Reilley, C. N. and McLafferty, F. W. (editors). *Advances in Analytical Chemistry and Instrumentation* (1971), **9,** Winefordner, J. D. *"Spectrochemical Methods of Analysis",* 189. (New York: Wiley Interscience)
13. West, T. S. (1970). *Chem. and Ind. (London),* 387
14. De Galan, L. (1971). *Chem. Weekbl.,* **67,** A24
15. L'vov, B. V. (1971). *Zh. Anal. Khim.,* **26,** 590
16. Winefordner, J. D. (1970). *Anal. Chem.,* **42,** 206R
17. Kirkbright, G. F. (1971). *Analyst,* **96,** 609
18. Slavin, S. (1971). *Atom. Absorpt. Newslett.,* **10,** 17
19. Masek, P. R. and Sutherland, I. (editors) (1969) *Atomic Absorption and Flame Emission Abstracts.* (London: Science and Technology Agency)
20. Human, H. G. C. and Butler, L. R. P. (1970). *Spectrochim. Acta,* **25B,** 647
21. Sargent, M. and Kirkbright, G. F. (1970). *Spectrochim. Acta,* **25B,** 577
22. Bruce, M. F. and Hannaford, P. (1971). *Spectrochim. Acta,* **26B,** 207
23. Prugger, H., Grosskopf, R. and Torge, R. (1971). *Spectrochim. Acta,* **26B,** 191
24. De Galan, L. and Wagenaar, H. (1971). *3rd Atom. Abs. and Fluorescence Spectroscopy Conference*
25. Hollander, T., Jansen, B. J., Plaat, J. J. and Alkemade, C.Th.J. (1970). *J. Quantit, Spectrosc. Rad. Transfer.,* **10,** 1301
26. Heek, H. F. van (1970). *Spectrochim. Acta,* **25B,** 107
27. Gelder, Z. van (1970). *Spectrochim. Acta,* **25B,** 669
28. De Galan, L. and Samaey, G. F. (1969). *Spectrochim. Acta,* **24B,** 679
29. De Galan, L. (1969). *Spectrochim. Acta,* **24B,** 629
30. Rann, C. S. (1969). *Spectrochim. Acta,* **24B,** 685
31. De Galan, L. and Samaey, G. F. (1970). *Anal. Chim. Acta,* **50,** 39
32. Willis, J. B. (1970). *Spectrochim. Acta,* **25B,** 487
33. *Idem.* (1971). *Spectrochim. Acta,* **26B,** 117
34. De Galan, L. and Samaey, G. F. (1970). *Spectrochim. Acta,* **25B,** 245
35. Smyly, D. S., Townsend, W. P., Zeegers, P.Th.J. and Winefordner, J. D. (1971). *Spectrochim. Acta,* **26B,** 531
36. Fassel, V. A., Richardson, J. O., Knisely, R. N. and Cowley, Th.B. (1970). ibid., **25B,** 559
37. L'vov, B. V. (1970). *Opt. Spektrosk.,* **28,** 18
38. Koirtyohann, S. R. and Pickett, E. E. (1971). *Spectrochim. Acta,* **26B,** 349
39. Chester, J. E., Dagnall, R. M. and Taylor, M. R. G. (1970). *Analyst,* **95,** 702
40. Chester, J. E., Dagnall, R. M. and Taylor, M. R. G. (1971). *Anal. Chim. Acta,* **55,** 47
41. Chakrabarti, C. L., Katyal, M. and Willis, D. E. (1970). *Spectrochim. Acta.,* **25B,** 629

42. Kirkbright, G. F. and Vetter, S. (1971). *Spectrochim. Acta*, **26B**, 505
43. Ando, A., Fuwa, K. and Vallee, B. L. (1970). *Anal. Chem.*, **42**, 818
44. Bulewicz, E. M. and Padley, P. J. (1971). *Proc. Roy. Soc. (London) Ser. A.*, **323**, 377
45. Katskov, D. A., Lebedev, G. C. and L'vov, B. V. (1969). *Zh. Prikl. Spektrosk.*, **10**, 215
46. Sebestyen, N. A. (1970). *Spectrochim. Acta*, **25B**, 261
47. Bueger, P. A. and Fink, W. (1971). *Spectrochim. Acta*, **26B**, 1359
48. Human, H. G. C. and Butler, L. R. P. (1970). *Spectrochim. Acta*, **25B**, 647
49. Prugger, H. J. (1969). *Spectrochim Acta*, **24B**, 197
50. Hildon, M. A. and Sully, G. R. (1971). *Anal. Chim. Acta*, **53**, 192
51. Goleb, J. A. (1970). *Anal. Chim. Acta*, **51**, 343
52. Wildy, P. C. and Thompson, K. C. (1970). *Analyst*, **95**, 562
53. Woodward, C. (1970). *Anal. Chim. Acta*, **51**, 548
54. Cooke, D. O., Dagnall, R. M. and West, T. S. (1971). *Anal. Chim. Acta*, **56**, 17
55. Baranova, S. V., Ivanov, N. P., Pofralidi, L. G., Knyazev, V. V., Talaev, B. M. and Vasil'ev, E. N. (1971). *Zh. Prikl. Spektrosk.*, **14**, 592
56. Thompson, K. C. (1970). *Analyst*, **95**, 1043
57. Mansell, R. E. (1970). *Spectrochim. Acta*, **25B**, 219
58. Bailey, B. W. and Rankin, J. M. (1971). *Anal. Chem.*, **43**, 216
59. *Idem.* (1971). ibid., **43**, 219
60. Kirkbright, G. F., Sargent, M. and West, T. S. (1969). *Talanta*, **16**, 1467
61. Kirkbright, G. F. and Ranson, L. (1971). *Anal. Chem.*, **43**, 1238
62. Ure, A. M. and Berrow, M. L. (1970). *Anal. Chim. Acta*, **52**, 247
63. Aldous, K. M., Browner, R. F., Dagnall, R. M. and West, T. S. (1970). *Anal. Chem.*, **42**, 939
64. Iida, C. and Yamasaki, K. (1970). *Anal. Lett.*, **3**, 251
65. Yamada, J., Iida, C. and Yamasaki, K. (1970). *Bunseki Kagaku*, **19**, 1259
66. Moldan, B., Rubeska, J. and Miksovsky, M. (1970). *"Anal. Chim. Acta*, **50**, 342
67. Zelynkova, Y. V., Nikonova, M. P. and Poluektov, N. S. (1969). in Egorov, Y. P. (editor). *Spektrosk. At. Mol.* (Moscow)
68. Goguel, R. (1971). *Spectrochim. Acta*, **26B**, 313
69. Uny, G., Loftin, J., N'Guea, Tardif, J. P. and Spitz, J. (1971). ibid., **26B**, 151
70. Uny, G. and Spitz, J. (1970). ibid., **25B**, 391
71. Duckworth, H. W. and Coleman, J. E. (1970). *Anal. Biochem.*, **34**, 382
72. Mosotti, V. G., Abercrombie, F. N. and Eakin, J. A. (1971). *Appl. Spectrosc.*, **25**, 331
73. Kashiki, M. and Oshima, S. (1970). *Anal. Chim. Acta*, **51**, 387
74. Winge, R. K., Fassel, V. A. and Kniseley, R. N. (1971). *Appl. Spectrosc.*, **25**, 636
75. Hell, A. and Ramirez-Munoz, J. (1970). *Anal. Chim. Acta*, **51**, 141
76. Delves, H. T. (1970). *Analyst*, **95**, 431
77. Fernandez, F. and Manning, D. C. (1971). *At Absorpt. Newslett.*, **10**, 86
78. Roos, J. T. H. (1969). *Spectrochim. Acta*, **24B**, 255
79. *Idem.* (1970) ibid., **25B**, 539
80. Ramirez-Munoz, J. (1969). *Talanta*, **16**, 1037
81. *Idem.* (1970). ibid., **17**, 279
82. Leiritie, M. and Mattson, B. (1970). *Anal. Lett.*, **3**, 315
83. Feldman, F. J. (1970). *Anal. Chem.*, **42**, 719
84. Booher, T. R., Elser, R. C. and Winefordner, J. D. (1970). *Anal. Chem.*, **42**, 1677
85. De Galan, L. (1970). *Spectrosc. Lett.*, **3**, 123
86. Pearton, Donald, C. G. (1970). *Natt. Inst. Met.*, Republ. S. Afr. Rep **No. 1163**, 7pp
87. Blouin, L. T., Dostie, C. V., Bloom, W. L. and Low, F. J. (1970). *Anal. Chem.*, **42**, 1298
88. Boyle, W. G. and Sunderland, W. (1970). *Anal. Chem.*, **42**, 1403
89. Butler, L. R. P., Jackson, P. F. S. and Kroeger, K. (1971). *Spectrosc. Lett.*, **4**, 195
90. Morris, M. D. and Orenberg, J. B. (1969). *Talanta*, **16**, 539
91. Smith, V. J. and Robinson, J. W. (1970). *Anal. Chim. Acta*, **49**, 161
92. Ramirez-Munoz, J. (1970). *Anal. Chem.*, **42**, 517
93. Skurnik-Sarig, S., Glasner, A., Zidon, M. and Weiss, D. (1969). *Talanta*, **16**, 1488
94. Paus, P. E. (1971). *Anal. Chim. Acta*, **54**, 164
95. Woodward, C. (1971). *Spectrosc. Lett.*, **4**, 191
96. Curtis, K. E. (1969). *Analyst*, **94**, 1068
97. Terashima, S. (1969). *Bunseki Kagaku*, **18**, 1259

98. Yanagisawa, M., Suzuki, M. and Takeuchi, T. (1970). *Anal. Chim. Acta,* **52,** 386
99. Toshihisa Maruta, Suzuki, M. and Takeuchi, T. (1970). *ibid.,* **51,** 381
100. Yanagisawa, M., Kihara, H., Suzuki, M. and Takeuchi, T. (1970). *Talanta,* **17,** 888
101. Roos, J. T. H. and Price, W. J. (1971). *Spectrochim. Acta,* **26B,** 279
102. Roos, J. T. H. (1971). ibid., **26B,** 285
103. Roos, J. T. H. and Price, W. J. (1971). *Spectrochim. Acta,* **26B,** 441
104. Ottaway, J. M., Coker, D. T., Rowston, W. B. and Bhattarai, D. R. (1970). *Analyst,* **95,** 567
105. Coker, D. T. and Ottaway, J. M. (1970). *Nature,* **227,** 831
106. Martin, Margaret, J. (1971). *Chem. Ind.,* (London), **19,** 514
107. Hurlbut, J. A. and Chriswell, C. D. (1971). *Anal. Chem.* **43,** 465
108. Kallmann, S. and Hobart, E. W. (1970). *Anal. Chim. Acta,* **51,** 120
109. Montford, B. and Cribbs, S. C. (1971). ibid., **53,** 101
110. Janssen, A. and Umland, F. (1970). *Zh. Anal. Khim.,* **251,** 101
111. Mallett, R. C., Pearton, D. C. G., Ring, E. J. and Steele, T. W. (1970). *Natt. Inst. Met.,* Republ. S. Africa. Rep. **No. 1108,** 20pp
112. Juliano, P. O. and Harrison, W. W. (1970). *Anal. Chem.,* **42,** 84
113. Levine, J. R., Moore, S. G. and Levine, S. L. (1970). *Anal. Chem.,* **42,** 412
114. Thomas, P. E. and Pickering, W. F. (1971). *Talanta,* **18,** 127
115. Bond, G. M. (1970). *Anal. Chem.,* **42,** 932
116. Panday, V. K. (1971). *Anal. Chim. Acta,* **57,** 31
117. Marks, J. Y. and Welcher, G. G. (1970). *Anal. Chem,* **42,** 1033
118. Urbain, H. and Varlot, M. (1970). *Meth. Phys. Anal.,* **6,** 373
119. Ferris, A. P., Jepson, W. B. and Shapland, R. C. (1970). *Analyst,* **95,** 574
120. Van Loon, J. C., Galbraith, J. H. and Aarden, H. M. (1971). *ibid.,* **96,** 47
121. Cattrall, R. W. and Slater, S. J. E. (1971). *Anal. Chim. Acta,* **56,** 143
122. Fleet, B., Liberty, K. V. and West, T. S. (1970). *Talanta,* **17,** 203
123. Panday, J. K. and Ganguly, A. K. (1970). *Anal. Chim. Acta,* **52,** 417
124. Pitts, G. E. and Beamish, F. E. (1970). ibid., **52,** 405
125. Culp, J. H., Windham, R. L. and Whealy, R. D. (1971). *Anal. Chem.,* **43,** 1321
126. Sastri, V. S., Chakrabarti, C. L. and Willis, D. E. (1969). *Talanta,* **16,** 1093
127. Chakrabarti, C. L. and Singal, S. P. (1969). *Spectrochim. Acta,* **24B,** 663
128. Mostyn, R. A., Newland, B. T. N. and Hearn, W. E. (1970). *Anal. Chim. Acta,* **51,** 520
129. Kashiki, M., Yamazoe, S. and Oshima, S. (1971). ibid., **53,** 95
130. *Idem.* (1971). ibid., **54,** 533
131. Harrison, W. W. and Juliano, P. O. (1971). *Anal. Chem.,* **43,** 248
132. Lee, R. F. and Pickering, W. F. (1971). *Talanta,* **18,** 1083
133. Lacour, A., Teinturier, C. and Isabelle, D. B. (1971). *Method Phys. Anal.,* **7,** 49
134. L'vov, B. V. and Khartsyzov, A. D. (1969). *Zh. Prikl. Spektrosk.,* **10,** 413
135. *Idem.* (1970). *Zh. Anal. Khim.,* **25,** 1824
136. Manning, D. C. and Fernandez, F. (1970). *At. Absorpt. Newslett.,* **9,** 65
137. Welz, B. and Wiedeking, E. (1970). *Fresenius' Z. Anal. Chem.,* **252,** 111
138. Omang, S. H. (1970). *Anal. Chim. Acta,* **55,** 439
139. *Idem.* (1970). ibid., **56,** 470
140. Baudin, G., Chaput, M. and Feve, L. (1971). *Spectrochim. Acta,* **26B,** 425
141. Alder, J. F. and West, T. S. (1970). *Anal. Chim. Acta,* **51,** 365
142. Anderson, R. G., Johnson, H. N. and West, T. S. (1971). ibid., **57,** 281
143. Alger, D., Anderson, R. G., Maines, J. S. and West, T. S. (1971). ibid., **57,** 271
144. Aggett, J. and West, T. S. (1971). ibid., **57,** 15
145. Dipiero, S. and Tessari, G. (1971). *Talanta,* **18,** 707
146. Brodie, K. G. and Matousek, J. P. (1971). *Anal. Chem.,* **43,** 1557
147. Amos, M. D., Bennett, P. A., Brodie, K. G., Lung, P. W. Y. and Matousek, J. P. (1971). *Anal. Chem.,* **43,** 211
148. Matousek, J. P. and Stevens, B. J. (1971). *Clin. Chem.,* **17,** 363
149. Glenn, M., Savory, J., Hart, L., Glenn, T. and Winefordner, J. (1971). *Anal. Chim. Acta,* **57,** 263
150. Donega, H. M. and Burgess, T. E. (1970). *Anal. Chem.,* **42,** 1521
151. Headridge, J. B. and Smith, D. R. (1971). *Talanta,* **18,** 247
152. Morrison, G. F. and Talmi, Y. (1970). *Anal. Chem.,* **42,** 809
153. Massmann, H. (1970). *Spectrochim. Acta,* **25B,** 393

154. Korovin, Yu I. and Kuchumov, V. A. (1971). *Zh. Prikl. Spektrosk.*, **14**, 778
155. Gandrud, B. and Skogerboe, R. K. (1971). *Appl. Spectrosc.*, **25**, 243
156. Taylor, H. E., Gibson, J. H. and Skogerboe, R. K. (1970). *Anal. Chem.*, **42**, 1569
157. Loftin, H. P., Christian, C. M. and Robinson, J. W. (1970). *Spectrosc. Lett.*, **3**, 161
158. Christian, C. M. and Robinson, J. W. (1971). *Anal. Chim. Acta*, **56**, 466
159. Marinkovic, M. and Vickers, T. J. (1971). *Appl. Spectrosc.*, **25**, 345
160. Krivchikova, E. P. and Demin, V. S. (1971). *Zh. Prikl. Spektrosk.*, **14**, 592
161. Manning, D. C. (1970). *At. Absorpt. Newslett.*, **9**, 97
162. *Idem.* (1970). ibid., **9**, 109
163. Lindstedt, G. (1970). *Analyst*, **95**, 264
164. Lindstedt, G. and Skare, I. (1971). ibid., **96**, 223
165. Anderson, D. H., Evans, J. H., Murphy, J. J. and White, W. W. (1971). *Anal. Chem.*, **43**, 1511
166. Bailey, B. W. and Lo, F. C. (1971). *Anal. Chem.*, **43**, 1525
167. Omang, S. H. (1971). *Anal. Chim. Acta*, **53**, 415
168. Omang, S. H. and Paus, P. E. (1971). ibid., **56**, 393
169. Stainton, M. P. (1971). *Anal. Chem.*, **43**, 625
170. Michailova, T. P. and Rezepina, V. A. (1970). *Analyst*, **95**, 769
171. Kallmann, S. and Hobart, E. W. (1970). *Talanta*, **17**, 845
172. Hildon, M. A. and Sully, G. R. (1971). *Anal. Chim. Acta*, **54**, 245
173. Groenewald, T. and Jones, B. M. (1971). *Anal. Chem.*, **43**, 1689
174. Fishkova, N. L. (1970). *Zavod. Lab.*, **36**, 1461
175. Jarsic, M. L. (1970). *Nucl. Sci. Abstr.*, **24**, 18873
176. Luecke, W. and Zielke, H. J. (1971). *Fresenius' Z. Anal. Chem.*, **253**, 20
177. Rowston, W. B. and Ottaway, J. B. (1970). *Anal. Lett.*, **3**, 411
178. Pitts, A. E., Van Loon, J. C. and Beamish, F. E. (1971). *Anal. Chim. Acta*, **50**, 181
179. *Idem.* (1970). ibid., **50**, 195
180. Montford, B. and Cribbs, S. C. (1971). *Anal. Chim. Acta*, **53**, 101
181. Melamed, Sh. G., Saltykova, A. M. and Chelidze, L. F. (1971). *Zavod. Lab.*, **37**, 166
182. Ando, A., Suzuki, M., Fuwa, K. and Vallee, B. L. (1969). *Anal. Chem.*, **41**, 1974
183. Nakahara, T., Munamori, M. and Mushia, S. (1970). *Anal. Chim. Acta*, **50**, 51
184. Nicolas, D. J. (1971). ibid., **55**, 59
185. Uny, G., Tardif, J. P. and Spitz, J. (1971). ibid., **54**, 91
186. Van Loon, J. C. and Parissis, C. M. (1969). *Analyst*, **94**, 1059
187. Boar, P. L. and Ingram, L. K. (1970). *Analyst*, **95**, 124
188. Langmyhr, F. J. and Paus, P. E. (1970). *Anal. Chim. Acta*, **50**, 515
189. Lucas, R. P. and Ruprecht, B. C. (1971). *Anal. Chem.*, **43**, 1013
190. Thomerson, D. R. and Price, W. J. (1971). *Analyst*, **96**, 825
191. *Idem.* (1971). ibid., **96**, 321
192. Vienney, J. (1971). *Anal. Chim. Acta*, **55**, 37
193. Kisfaludi, G. and Lenhof, M. (1971). *Anal. Chim. Acta*, **54**, 83
194. Hofton, M. E. and Hubbard, D. P. (1970). *Anal. Chim. Acta*, **52**, 425
195. Headridge, J. B. and Richardson, J. (1970). *Analyst*, **95**, 968
196. Donaldson, E. M., Charette, D. J. and Rolko, V. H. E. (1969). *Talanta*, **16**, 1305
197. Uny, G., Mathieu, C., Tardif, J. P. and Tran Van Danh (1971). *Anal. Chim. Acta*, **53**, 109
198. Rudnevskii, N. K., Demarin, V. T. and Gruzdeva, T. M. (1971). *Zh. Prikl. Spektrosk.*, **14**, 567
199. Crump-Wiesner, H. J., Feltz, H. R. and Purdy, W. C. (1971). *Anal. Chim. Acta*, **55**, 29
200. Busev, A. I., Byrko, V. M., Tereshchenko, A. P. and Novikova, N. N. (1970). *Zh. Anal. Khim.*, **25**, 665
201. Sachdev, S. L. and West, P. W. (1970). *Environ. Sci. Technol.*, **4**, 749
202. Lau, H. K. Y. and Lott, P. F. (1971). *Talanta*, **18**, 303
203. LeBihan, A. and Courtot-Coupez, J. (1970). *Bull. Soc. Chim. Fr.*, 406
204. Pinta, M. (1970). *Meth. Phys. Anal.*, **6**, 268
205. Fritz, J. S. and Latwesen, G. L. (1970). *Talanta*, **17**, 81
206. Galle, O. K. (1971). *Appl. Spectrosc.*, **25**, 664
207. Burke, K. E. (1970). *Anal. Chem.*, **42**, 1536
208. Looyenga, R. W. and Huber, C. O. (1971). *Anal. Chim. Acta*, **55**, 179
209. *Idem.* (1971). *Anal. Chem.*, **43**, 498

3
Atomic Fluorescence Spectroscopy

R. M. DAGNALL* and B. L. SHARP
Imperial College of Science and Technology, London

* Present address: Biochemical Division, Huntingdon Research Centre, Huntingdon, U.K.

3.1   INTRODUCTION

Atomic fluorescence spectrometry is a relatively new trace element analytical technique and because of its resemblance to atomic absorption spectrometry it has received more than a passing interest since the initial analytical studies of Winefordner *et al.* in 1964. However, the almost unprecedented adoption of atomic absorption in the 'inorganic' analytical laboratory has meant that atomic fluorescence has been treated with some degree of caution and there has been little need on the part of the instrument manufacturers to carry out the necessary new development. Hence much of the work to date on atomic fluorescence has been carried out in academic institutions and has been usually of a purely research nature. In consequence, the more important advantages in the analytical sense and the possible applications of the technique have not always been conveyed satisfactorily to the potential user. The purpose of this chapter is not to review the technique extensively, this has been dealt with on many occasions previously and the reader is referred to these[1-4] for a more comprehensive survey of the literature. At this time an attempt is made to indicate the advantages and disadvantages in relation to atomic absorption spectrometry and to point out possible applications and the future trend of the atomic fluorescence technique. In this context it is relevent to describe the theoretical considerations in some detail, although even here an attempt has been made to up-date the material involved in keeping with present trends and practices in atomic spectrometry. The remainder of the chapter is devoted to a discussion of the essential experimental parameters and applications with respect to both dispersive and non-dispersive arrangements.

Atomic fluorescence spectrometry may be defined as the measurement of radiation from discrete atoms excited by the absorption of radiation from a source which is not seen by the detector; this is true of most experimental arrangements and the technique may be considered to be the atomic analogue of molecular spectrofluorimetry. The phenomenon of atomic fluorescence was first reported by Wood[5] in 1905, almost 100 years after the first observation of atomic absorption. This was achieved by vaporising sodium metal in an evacuated test tube, illuminating the tube with radiation from a coal gas flame containing sodium chloride and visually observing the yellow sodium D lines. Wood called this fluorescence resonance radiation because it was predicted by the classical theory of a light wave vibrating with the same frequency as the dipole oscillations of the medium. Following this work, the atomic fluorescence of lithium[6], caesium[7], mercury[8], cadmium[9], zinc[10], thallium, lead, bismuth, antimony[9] and manganese[11] were all observed. In each instance the element concerned was sealed into an evacuated quartz vessel and its vapour pressure was controlled by the temperature of a containing oven. The sources of excitation were generally vacuum arcs of either the type described by Ellett[12], in which the metal is held in a side-arm and its vapour pressure maintained by an oven, or the Schüler tube[13], which is a close precursor of the present-generation hollow cathode lamp. However, vapour discharge lamps[5, 9, 10] and flame sources[5, 6] were also used. Although the scope of this early work was limited to those elements with vapour pressures about 10^{-4} Torr at the maximum furnace temperatures available ($<1200\,°C$), it allowed a detailed examination of the atomic fluorescence process to be made. An excellent summary of much of this work has been made by Mitchell and Zemansky[14].

The first reference to atomic fluorescence of metal vapours in flames was reported in 1923 by Nichols and Howes[15] and then by Badger[16] and Mankopff[17]. These workers obtained weak signals from high concentrations of elements such as sodium, lithium, calcium, strontium, barium, cadmium, magnesium, mercury and silver. More recently, Robinson[18] observed weak atomic fluorescence of the 285.2 nm magnesium line in an oxy-hydrogen flame using a magnesium hollow cathode lamp. In 1962, Alkemade[19] used the atomic fluorescence of sodium to obtain mechanisms of excitation and deactivation of atoms in flames and to measure the quantum efficiency of the sodium 589.0 nm line. He was also the first to point out the possible analytical applications of this technique. The first analytical method as such was developed by Winefordner and his co-workers in a series of papers[20-23] published in 1964–1965. A year later Dagnall, West et al.[24] utilised commercially available atomic absorption equipment for measuring atomic fluorescence. Since then the technique has been more completely characterised and developed, mainly by these two research groups at the

University of Florida (U.S.A.) and Imperial College of Science and Technology (U.K.) and it is now possible to observe atomic fluorescence for nearly every element capable of being determined by atomic absorption spectrometry. At the present time, many instrument manufacturers offer atomic fluorescence accessories to their range of atomic absorption equipment although as far as the authors know only one manufacturer has attempted to design an instrument specifically for the measurement of atomic fluorescence. It is hoped that this chapter will explain these two methods of approach and, in the light of present knowledge, the realisation of the preferred atomic fluorescence instrument.

3.2 THE THEORY OF ATOMIC FLUORESCENCE ANALYSIS

3.2.1 The origin of fluorescence

The occurrence of atomic spectra is associated with the transitions of atoms between stable energy states. Where k is an upper energy state and i is a lower energy state, the energy of the transition is given by the equation:

$$E_k - E_i = hv$$

where E_k is the energy of kth level (J), E_i is the energy of the ith level (J), h is Planck's constant (J s) and v is the frequency of the emitted radiation (s^{-1}). Einstein's theory of radiation[14] classifies transitions as being of three types each having a probability in unit time F_{ik} of occurring thus:

(a) Absorption (B_{ik}) of a quantum enabling the atom to move from a lower to a higher energy state.

(b) Emission (A_{ki}) spontaneously of a quantum having random phase and enabling the atom to move from an upper to a lower energy state.

(c) Emission (B_{ki}) stimulated by external radiation of a quantum whose phase is the same as that of the stimulating quantum.

Atomic fluorescence at present is concerned only with processes (a) and (b). Phenomenon (c) has already been exploited in the laser, but as yet this form of fluorescence has not been considered in an analytical context.

3.2.2 The concept of spectral line width

Spectral lines are not monochromatic, but consist of a spread of frequencies about a mean value (v_0). This broadening can be attributed to the following processes.

3.2.2.1 Natural broadening

The Dirac quantum mechanical description of radiative processes replaces the discrete energy levels of atoms by probability maxima in a continuous energy spectrum. The energy transition between two states can therefore have an infinite number of values within the range $v_0 \pm \frac{1}{2}(\Delta E_i + \Delta E_k)/h$ where

ΔE_i and ΔE_k are the widths of the atomic energy levels i and k defined by the appropriate probability density function. The natural half-width of a spectral line is directly related to the half-life of the excited state τ by the equation

$$\Delta v_N = 1/2\pi\tau \tag{3.1}$$

Typical natural half-widths are Hg 253.7 nm, $\Delta v_N = 0.53 \times 10^{-4}$ cm^{-1}; Cd 228.8 nm, $\Delta v_N = 2.7 \times 10^{-3}$ cm^{-1}. The line profile exhibited when only natural broadening is present is represented by the Lorentz function[14]

$$L(v) = \frac{\text{Const}}{1 + \left[\dfrac{2(v - v_0)}{\Delta v_N}\right]^2} \tag{3.2}$$

3.2.2.2 Doppler broadening

This arises due to the motion of the emitting or absorbing atoms relative to the observer. When a monochromatic light source of frequency v_0 is propagated having a component of velocity along the observer's line of sight of V_x, then there is an apparent shift in the frequency equal to $v_0 V_x/c$ (c is the velocity of light). Provided that the atoms exhibit random motion a broadening about v_0 occurs. If thermodynamic equilibrium exists, then the intensity distribution follows the Maxwell distribution law, that is,

$$dN(V_x) \propto \exp\left[-(MV_x^2/2RT)\right] dV_x \tag{3.3}$$

where M is the molecular weight, R is the universal gas constant and T is the absolute temperature. Now

$$v = v_0 - v_0 V_x/c$$

and therefore

$$G(v) = I_0 \exp\left\{-(Mc^2/2RT)\left[(v_0 - v)^2/v_0^2\right]\right\} \tag{3.4}$$

because

$$I_G(v)dv \propto N(V_x)\, dV_x$$

Equation (3.4) shows that a pure Doppler-broadened line-profile will be Gaussian in shape and that the half-width is given by the equation:

$$\Delta v_D = \frac{2v_0}{c}\sqrt{\left(\frac{2\pi RT \ln 2}{M}\right)} \tag{3.5}$$

Typical Doppler half-widths are Na 589.0 nm (293 K), $\Delta v_D = 0.034$ cm^{-1}, B 249.8 (200 K), $\Delta v_D = 0.12$ cm^{-1}.

3.2.2.3 Lorentz broadening

This form of line broadening is often referred to as pressure broadening because it results from the interation of the emitting or absorbing species with molecules of a foreign gas. When the pressure of the gas is increased, a threefold effect is observed involving shift of the line maximum, broadening of the profile and asymmetry of the profile. Two approaches have been made to explain Lorentz broadening; one is based on the Collision theory (study of the instantaneous changes that take place during a collision) and the other on Statistical theory (by considering the effect of a number of molecules on an atom in a quasi-steady state). It has been shown that the centre of the line profile is well explained by Collision theory and that the line wings are best explained statistically. Here we shall present the results obtained by Collision theory because they relate to the part of the line having most analytical use. The early Lorentz collision theory yielded an expression for the half-width:

$$\Delta v_L = 2N_A \sigma^2 P \bigg/ \sqrt{\left(\frac{2}{\pi R T}\left[\frac{1}{A}+\frac{1}{M}\right]\right)}$$

where N_A is the Avogadro number, P is the pressure of gas, $\pi\sigma^2$ is the collision cross-section between emitting atom and gas molecule, A is the atomic weight of the atoms and M is the molecular weight of gas molecules.

The equation successfully predicts the linear relation between broadening and pressure, but does not explain the shift that occurs. Later, Lindholm and Weisskopf[25] succeeded in explaining both half-width and shift from a collisional approach; the relevant equations are:

$$\Delta v_L = 1.30\, C_6^{2/5} V^{3/5} N$$

$$\Delta v_{SHIFT} = 0.47\, C_6^{2/5} V^{3/5} N$$

where C_6 is the van der Waal's constant for the interaction between particles, V is the relative velocity of particles and N is the number of particles per unit volume. We may note that $\Delta v_L/\Delta v_{SHIFT} = 2.77$ and experimental values for various systems of gas and emitting atom fall in the range 1.8–6.0. The relation between Δv_L and temperature is different for the Lorentz and Lindholm theories:

$$\left.\begin{array}{ll} \text{for Lorentz} & \Delta v_L \propto T^{1/2} \\ \text{for Lindholm} & \Delta v_L \propto T^{3/10} \end{array}\right\} \text{ at } 0\,°\text{C, 1 atm.}$$

Work by L'Vov and Plyushch[26] tends to suggest that the Lindholm relation is nearer the truth. Typical values of the Lorentz half-width and shift are Ar $\Delta v_S = -2.77\,\text{cm}^{-1}$, $\Delta v_L = 2.21\,\text{cm}^{-1}$, He $\Delta v_S = +0.93\,\text{cm}^{-1}$, $\Delta v_L = 2.77\,\text{cm}^{-1}$ (all at 0 °C and 1 atm) where $+$ indicates a violet shift and $-$ a red shift. The line profile for Lorentz broadening is the same as that for natural broadening and is therefore described by equation (3.2).

3.2.2.4   Other sources of broadening

Here one can mention Stark broadening which results from interactions between the emitting atoms and charged particles (electrons, ions etc), and is therefore of particular interest in plasma spectroscopy. There is also Resonance (Holtsmark) broadening which is a special case of pressure broadening arising from the interaction between atoms of the same kind in their ground and excited states.

3.2.2.5 The observed spectral profile

The observed spectral profile represents a graphical addition of each broadening process and therefore will depend on the nature of the emitting species, the pressure, the temperature and the nature of the foreign species present (Figure 3.1). Now for the conventional atom reservoirs used in atomic spectrometry, that is flames (N_2O/C_2H_2, air/C_2H_2, $Ar/O_2/H_2$ etc.) and non-flame devices such as the L'Vov furnace[35], Massmann furnace[120] and West filament atom reservoir[49] where the pressure is usually atmospheric, the temperature and degree of ionisation are not excessive ($T \leqslant 3\,200$ K) and to a first approximation thermodynamic equilibrium can be said to exist in the analytical volume. Therefore the shape of the spectral line is mainly determined by Doppler and Lorentz processes. Other contributions will be present, but their effect will be small in comparison. Mathematically then the observed spectral profile for analytical systems can be represented by the folding integral for the Lorentz and Gaussian shape functions:

$$P'(v) = \int_{-\infty}^{+\infty} L(v)G(v)\,dv$$

This is known as the Voigt[27] profile and is usually written in the form appropriate to an absorption line:

$$P(y) = \frac{K_v}{K_0} = \frac{a}{\pi} \int_{-\infty}^{+\infty} \frac{\exp(-t^2)\,dt}{a^2 + (y-t)^2} \tag{3.6}$$

where K_v is the absorption coefficient at frequency v (from Beer's law), K_0 is the absorption coefficient at the line centre v_0 for a purely Doppler-broadened line given by:

$$K_0 = \frac{2\sqrt{(\ln 2)}\lambda_0^2 A N_0}{\Delta v_D \sqrt{\pi 8\pi}} \frac{g_k}{g_i} \tag{3.7}$$

where

$$a = \frac{\Delta v_L}{\Delta v_D}(\ln 2)^{1/2} \qquad \text{(the damping parameter)},$$

$$y = \frac{2(v-v_0)}{\Delta v_D}(\ln 2)^{1/2}$$

$$t = \frac{2\delta(\ln 2)^{1/2}}{\Delta v_D}$$

and δ is the frequency displacement from $(v-v_0)$. The integral expressed in equation (3.6) does not yield to analytical solution, but it has been tabulated by several authors, e.g. Davies and Vaughan[28].

3.2.2.6 Self-absorption

Self-absorption can be defined here as the fluorescent radiation re-absorbed by ground-state atoms of the emitting species and it is this factor which is mainly responsible for the limiting curvature of analytical curves in atomic fluorescence and atomic emission spectrometry. It is appropriate to describe it in this section because quantitatively it is related to the absorption coefficient K_v and, in the limit, causes a broadening of the spectral profile.

Orthmann and Pringsheim[29] in 1927 showed that when fluorescence radiation is observed at right angles to the exciting beam its frequency distribution is independent of that of the exciting beam and depends only on the form of the absorption coefficient. This is also the case even at reduced pressure provided that sufficient collisions occur between the excited atoms and the flame-gas molecules during the lifetime of the excited state. Thus, if the total absorption factor A_T, defined as the ratio of the absorbed energy to the incident energy given by:

$$A_T = \int_0^\infty [1 - \exp(-K_v L)] \, dv \qquad (3.8)$$

is plotted as a function of v for different values of $N_0 L$ (as in Figure 3.1) the effect of self-absorption on the intensity and frequency distribution of the fluorescence radiation becomes evident. It should be realised that in real situations the flame is only partially illuminated and therefore considerable self-absorption occurs as the fluorescent beam passes out through non-irradiated sections of the bulk of the flame. Figure 3.1 (curves (A)) shows that when $N_0 L$ is not too large the line-width remains constant while the area under the curve A_T increases in direct proportion to $N_0 L$. It has been shown that[30]:

$$A_T = \int_0^\infty [1 - \exp(-K_v L)] \, dv \, (s^{-1})$$

$$\dot{=} \int_0^\infty K_v L \, dv$$

$$= \frac{K_0 L \Delta v_D \sqrt{\pi}}{2\sqrt{(\ln 2)}}$$

i.e. $A_T \propto N_0 L$.

For large values of $N_0 L$ the radiation at the line centre is completely absorbed, see Figure 3.1 curves (B) (i.e. K_{max} reaches the classical limit given by the Planck black-body radiator law) and further absorption can only occur at the line wings (where K_v reaches measurable proportions because of the large number of atoms present) causing an apparent increase in the line-width. Under these conditions it has been shown[31] that:

$$A_T = \sqrt{\left[\frac{K_0 L \Delta v_D^2 a \sqrt{\pi}}{\ln 2}\right]}$$

i.e. $$A_T = \sqrt{(N_0 L)}$$

This discussion has invoked the independence of the frequency distributions of the exciting and emitted radiation; however, this does not imply that the shape of the analytical curve will be independent of the exciting-beam spectral distribution because the efficiency of the initial absorption process depends on the relative spectral overlap of the exciting and absorbing profiles.

3.2.3 Derivation of the intensity of the radiation emitted by atomic fluorescence

The following treatment is based on that due to H. P. Hooymayers[32]. The expression derived relates the integrated resonance fluorescence intensity emitted through a small solid angle $d\omega$ at right angles to both the incident beam and the flame axis to the ground-state atom concentration in the flame. The model is based on a Voigt profile and is therefore valid for pure Doppler profiles as well as those including Lorentz contributions. For simplicity, effects due to hyperfine structure and re-emission of self-absorbed radiation are ignored.

Consider an ideal flame-shielded flame of square cross-section with the following properties: (a) uniform temperature distribution throughout the analytical volume and (b) uniform ground-state atom distribution throughout the analytical volume; This is traversed by a parallel homogeneous beam of rectangular cross-section $\sigma = 2l \times 2l'$ (Figure 3.2). The energy absorbed per unit frequency interval in the volume $2l'\,du\,dx$ per second is: (Intensity at plane x) × (fraction of energy absorbed at frequency v when the beam travels a distance dx) × (the cross-sectional area of beam) $= [I_v \exp(-K_v x)] + [K_v\,dx] \times [2l'\,du]$. The integrated intensity for all v is therefore:

$$2l'\,du\,dx \int_0^\infty I_v K_v \exp(-K_v x)\,dv \qquad (3.9)$$

The intensity of fluorescence radiation emitted by this small volume over the solid angle $d\omega$ at right angles to the exciting beam is:

$$\frac{d\omega}{4\pi}\left[2l'\,du\,dx \int_0^\infty I_v K_v \exp(-K_v x)\,dv\right]\left[\int_0^\infty p\alpha_{v'} \exp\{-K_{v'}(L-l+u)\}\,dv'\right]$$
$$(3.10)$$

The second term allows for self-absorption within the flame because p is the probability that an excited atom will emit a photon before suffering a quenching collision. The term $\alpha_{v'}\,dv'$ is the probability that the photon emitted has a frequency within the range $v'+dv'$, and $\exp\{-K_{v'}(L-l+u)\}$ is the fluorescence beam intensity at the flame edge if both the probabilities could take values of unity.

In fact, $\alpha_{v'}$ represents the normalised shape of the spectral line when not distorted by self-absorption because

$$\int_0^\infty \alpha_{v'}\,dv' = 1$$

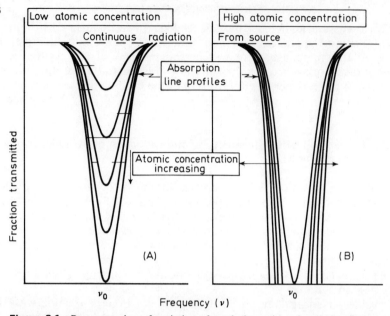

Figure 3.1 Representation of variation of total absorption of radiation from a continuous source with atomic concentration
(From Winefordner, J. D., Parzons, M. L., Mansfield, J. M. and McCarthy, W. J. (1967). *Spectrochim. Acta*, 23**B**, 37, by courtesy of Microfilms International Marketing Corporation)

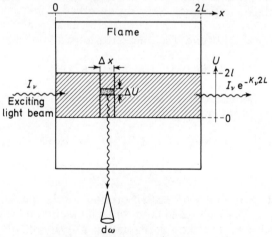

Observed fluorescence radiation

Figure 3.2 Part of square flame cross-section, as seen from above. The co-ordinate x, varying between 0 and $2L$, measures the distance (in cm) from the flame border ($x = 0$) to the considered element of volume $2l'$ du dx (indicated with a black spot in the figure) along the direction of the exciting beam. $2l'$ is the beam dimension perpendicular to the plane of drawing. The variable u, varying between the integration limits 0 and $2l$, fixes the position of the element of volume in the direction of the observed fluorescence radiation.
(From Hooymayers[32], by courtesy of Microfilms International Marketing Corporation)

Classical theory[14, 29] shows that

$$\alpha_{v'} = 8\pi \frac{1}{\lambda_0^2} \frac{g_i}{g_k} \frac{1}{A_{k \to i}} \frac{K_{v'}}{N_0} \tag{3.11}$$

where λ_0 is the central wavelength of emitted fluorescence and $A_{k \to i}$ is the spontaneous transition probability s^{-1}. Substituting for $\alpha_{v'}$ and integrating with respect to x over the limits $0-2L$ and with respect to u over the limits $0-2l$, we obtain the total radiant flux produced by fluorescence:

$$E_F = \frac{d\omega}{4\pi} \frac{2l'p8\pi g_i}{N_0 A \lambda_0^2 g_k} \int_0^\infty I_v (1 - \exp\{-K_v 2L\}) \, dv \int_0^\infty \exp[-K_{v'}(L-l) - \exp$$
$$\{-K_{v'}(L+l)\}] \, dv' \tag{3.12}$$

Substituting for K_0 in equation (3.14) one obtains:

$$E_F = \frac{d\omega}{4\pi} \frac{2l'p2\sqrt{\ln(2)}}{\Delta v_D \sqrt{\pi K_0}} \int_0^\infty I_v (1 - \exp\{-K_v 2L\}) \, dv \int_0^\infty \exp[-K_{v'}(L-l) -$$
$$\exp\{-K_{v'}(L+l)] \, dv' \tag{3.13}$$

where

$$\Delta v_D = \frac{1}{\lambda_D} \left(\frac{8\pi kT \ln(2)}{m} \right)^{\frac{1}{2}}$$

(following directly from equation (3.5)) and k is the Boltzmann constant; T is the flame temperature and m is the mass of the emitting atoms in the flame. It is now convenient to introduce the dimensionless variable

$$\chi = K_0 L$$
$$\propto N_0 L$$

and by substituting for the cross-sectional area σ of the exciting beam one obtains:

$$E_F = \frac{d\omega}{4\pi} \frac{\sigma p (\ln 2)^{\frac{1}{2}}}{\Delta v_D \sqrt{\pi}} \frac{L}{l} \frac{1}{x} \int_0^\infty I_v (1 - \exp\{-2\chi(K_v/K_0)\}) \, dv$$

$$\times \int_0^\infty \left[\exp\left\{ -\frac{K_{v'}}{K_0} x \left(1 - \frac{l}{L}\right) \right\} - \exp\left\{ -\frac{K_{v'} x}{K_0} \left(1 + \frac{l}{L}\right) \right\} \right] dv' \tag{3.14}$$

However, from equation (3.6):

$$P(y) = \frac{K_v}{K_0} = \frac{a}{\pi} \int_{-\infty}^\infty \frac{\exp(-t^2) dt}{a^2 + (y-t)^2}$$

Substituting into (3.14) yields:

$$E_F = \frac{d\omega}{4\pi} \frac{\sigma p(\ln 2)^{\frac{1}{2}}}{\Delta v_D \sqrt{\pi}} \frac{L}{l} \frac{4}{\chi} \int_0^\infty I(y)\,(1-\exp\{2\chi P(y)\})$$

$$\times \int_0^\infty \left[1-\exp\left\{-P(y')\chi\left(1+\frac{l}{L}\right)\right\} - \left(1-\exp\left\{-P(y')\chi\left(1-\frac{l}{L}\right)\right\}\right)\right] dy'$$

$$(3.15)$$

$P(y)$ is expressed for the limits $-\infty$ to $+\infty$ and therefore in keeping the limits between 0 and $+\infty$ for the product of the two integrations (dv, dv') it is necessary to multiply the right-hand side of equation (3.14) by 4.

Equation (3.15) then represents the total radiant flux emitted by fluorescence in terms of a Voigt profile for the absorption line. When the exciting line profile and the emitted line profile are similar, equation (3.15) is difficult to evaluate because it is necessary to know the spectral distribution of the exciting line in order to obtain an explicit expression for $I(y)$. However, in the limiting cases of a continuum source and a narrow line source $I(y)$ can be approximated.

3.2.3.1 Use of continuum source ($\Delta v_{\text{Source}} \gg \Delta v_{\text{Absorption}}$)

In this instance, $I(v)$ can be approximated by I_0 because it is essentially constant over the frequency range of the absorption profile.

Equation (3.15) can thus be written:

$$E_F \bigg/ \left[\frac{d\omega}{4\pi} \frac{I_0 \sigma p(\pi \ln 2)^{\frac{1}{2}}}{\Delta v_D}\right] = \frac{1}{\pi} \frac{L}{l} \frac{4}{\chi} \int_0^\infty \{1-\exp[-2\chi P(y)]\}\,dy$$

$$\times \int_0^\infty \left[1-\exp\left\{-P(y')\chi\left(1-\frac{l}{L}\right)\right\} - \left(1-\exp\left\{-P(y')\chi\left(1-\frac{l}{L}\right)\right\}\right)\right] dy'$$

$$(3.16)$$

Now the plot of

$$\log\left(E_F \bigg/ \left[\frac{d\omega}{4\pi} \frac{I_0 \sigma p(\pi \ln 2)^{\frac{1}{2}}}{\Delta v_D}\right]\right) v.\ \log f(\chi)$$

represents the theoretical analytical curve because $\chi = \text{const } N_0 L$.

The theoretical and experimental curves can be brought into coincidence by a shift of the theoretical curve parallel to the plotting axes. It must be noted that ionisation, incomplete dissociation, self-reversal and other causes of non-linearity are not considered in this treatment. Hooymayers, in plotting the ordinate, used the growth curve calculations of Van Trigt et al.[33] who computed values of the integral

$$\int_0^\infty \{1-\exp[-K_0 LP(y)]\}\,dy$$

over the range

$$\frac{N_0 f L(\ln 2)^{\frac{1}{2}}}{\pi \Delta v_D} = 2\text{–}9600 \text{ s cm}^{-2}$$

for a series of the damping parameter (a) values of 0.0–5.0. In order to relate Van Trigt's abscissa values to the variable χ, the following equality was used:

$$\frac{N_0 f L(\ln 2)^{\frac{1}{2}}}{\pi \Delta v_D} = 10.63\chi$$

where f is the oscillator strength[14]. The theoretical analytical curves obtained with a continuum excitation source are shown in Figure 3.3. The curves exhibit two main characteristics, they are linear with atomic concentration

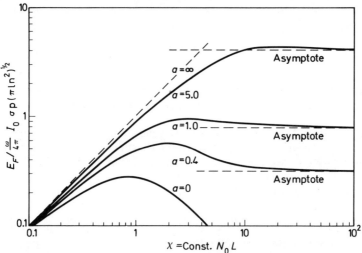

Figure 3.3 Some theoretical analytical curves for a single spectral line in the case of a continuum background source. The value of l/L is equal to 0.25 and the a-parameter values are 0, 0.4, 1.0 and 5.0 respectively (as indicated in the figure). The final asymptotes are calculated from equation (11) of the reference. The ordinate values in this figure are proportional to the meter deflections corresponding to the fluorescence radiation power while the abscissa values are proportional to the metal concentration N_0 in the flame
(From Hooymayers[32], by courtesy of Microfilms International Marketing Corporation)

when the atomic density is low, and they reach a horizontal asymptote when the atomic density is high. Hooymayers describes these limiting cases in terms of approximations to equation (3.15) which also show the importance of the damping ratio in determining the final ordinate value of the asymptote. Here an empirical approach will be adopted to emphasise the physical processes involved and no account will be taken of the damping ratio since this is effectively a constant of the atom reservoir which is usually chosen on the criteria of atomisation efficiency and spectral characteristics. The value of N_0 appears three times in equation (3.15), once inside each integral and once in the variable χ. The factor $1/\chi$ was introduced by substituting initially for $\alpha_{v'}$ and will be common to all approximations. The limiting shape of the analytical curve can, therefore, be determined by considering the primary absorption and the self-absorption processes as represented by the two integrals. Physically the two processes are identical, the only difference being the origin of the exciting radiation, and thus each can be described in terms of analogies to Section 3.2.2.6. For the low density case then, as N_0 tends to 0 both the primary absorption and the self-absorption will be proportional to N_0 yielding:

$$E_F \propto (N_0 L)(N_0 l)/N_0$$

i.e. $E_F \propto N_0 Ll$.

When the atomic density is high, i.e. $N_0 \to \infty$, analogy with Section 3.2.2.6 shows that both the primary absorption and self-absorption will be proportional to $\sqrt{N_0}$ yielding:

$$E_F \propto (\sqrt{(N_0 L)} \times (\sqrt{(N_0 P)})/N$$

i.e. $E_F \propto (Ll)^{\frac{1}{2}}$.

Thus at high atomic density the emitted fluorescence intensity becomes independent of the concentration and the curve yields no further analytical information.

These results have been confirmed experimentally by Veillon[82] et al. who plotted analytical curves for Zn and Cd using a 150 W xenon arc as the excitation source. Measurements were taken in an air–H_2 diffusion flame; limits of detection were Cd 0.08 and Zn 0.6 p.p.m. with curvature starting at about 10 p.p.m. and reaching an asymptote at c. 100 p.p.m.

3.2.3.2   Use of a narrow line source ($\Delta v_{Source} \gg \Delta v_{Absorption}$)

It must be noted that in applying the limiting conditions, the assumption is made that the absorption and emission profiles are the same. This assumption has been shown to be valid for normal flames where a large number (15–100) of adiabatic perturbing collisions occurs during the lifetime of the excited state.

For a narrow line, the integral

$$\int_0^\infty I(v)(1 - \exp[-K_v 2L])\,dv$$

can be approximated by

$$I(1 - \exp[-2P_0\chi]) \tag{3.17}$$

where

$$I = \int_0^\infty Iv\,dv$$

which equals the radiant flux over all frequencies of the exciting beam in units of energy per unit time per unit area.

Also

$$P_0 = \frac{K(v=0)}{K_0} = \frac{a}{\pi} \int_{-\infty}^{+\infty} \frac{\exp(-y^2)\,dy}{a^2+y^2} = \frac{1}{\pi^{\frac{1}{2}}} \int_0^\infty \exp(az-z/4)\,dZ$$

$$= \exp(a^2)(1-\operatorname{erf} a)$$

where

$$\operatorname{erf} a = \frac{2}{\pi^{\frac{1}{2}}} \int_0^a \exp(-t^2)\,dt$$

Now substituting (3.17) into (3.14) yields

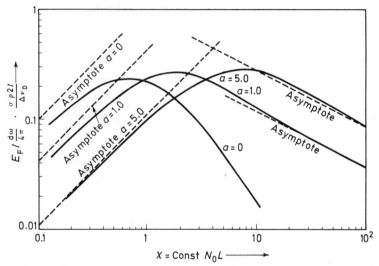

Figure 3.4 Some theoretical analytical curves for a single spectral line in the case of a narrow line source. The value of l/L is equal to 0.25 and the a-parameter values are 0, 1.0 and 5.0 (as indicated in the figure). The initial and final asymptotes are computed from equation (17) and (19), respectively, of the reference. The ordinate values are proportional to the meter deflections corresponding to the fluorescence radiation power while the abscissa values are proportioned to the metal concentration N_0 in the flame
(From Hooymayers[32], by courtesy of Microfilms International Marketing Corporation)

$$E_F \Big/ \frac{d\omega}{4\pi} \cdot \frac{\sigma p 2I}{\Delta v_D} = \frac{1}{2\sqrt{\pi}} \cdot \frac{4}{\chi} \cdot \frac{L}{l} (1 - \exp[P_0 2\chi])$$

$$\times \int_0^\infty [(1-\exp\{-P(y)\chi(1+l/L)\}) - (1-\exp\{-P(y)\chi(1-l/L)\})]\,dy$$

Again, using the tabulated values of Van Trigt *et al.*, curves of

$$E_F \Big/ \frac{d\omega}{4\pi} \cdot \frac{p\sigma 2I}{\Delta v_D} \ v. \ \chi$$

were plotted (see Figure 3.4). The curves obtained exhibit the same essential features as those with a continuum source except that the final asymptote has a negative gradient. At low atomic density we have once again both the primary absorption and the self-absorption processes proportional to N_0 thus:

$$E_F \propto N_0 Ll$$

However, at high atomic density all the exciting radiation is absorbed so that the first term becomes constant, but the self-absorption term continues to increase as $\sqrt{N_0}l$ yielding:

$$E_F \propto \sqrt{(N_0 l)}/N_0$$

i.e. $$E_F \propto N_0^{-\frac{1}{2}} l^{\frac{1}{2}}.$$

The graphs show that in both the continuum and line-source cases the linear range and position of the maxima are affected by the 'a' values, and that generally the most useful analytical curves are obtained when 'a' is high. The fraction of the flame illuminated by the source and observed by the detector (i.e. the l/L ratio) affects the curve shape in a similar way to the variation in 'a' value, but in this instance the ratio can be controlled experimentally (the limits being set by the source and detector dimensions). For the continuum source case, increasing the l/L ratio (i.e. the fraction illuminated/observed) tends to extend the linear range and make the maximum that occurs less pronounced. With a line source the most marked effect of increasing the l/L ratio is to decrease the slope of the final asymptote since the result is to reduce the number of atoms not in the exciting beam (i.e. capable of self-absorption). It should be noted that all references to the flame are to that part of the real flame which yields optimum analytical signals. For a further discussion and treatment of the shapes of analytical curves obtained in atomic fluorescence spectrometry the reader is referred to the paper given by Alkemade[19].

3.2.4   Other possible atomic fluorescence mechanisms

The discussion of atomic fluorescence so far has been limited to resonance fluorescence where the frequencies of the exciting and emitted lines are essentially the same. Several other mechanisms are possible and are represented schematically in Figure 3.5. In general, resonance fluorescence (a) will be of most importance analytically because in normal flames ground state atom concentrations are high compared with those of excited states and the transition probabilities $B_{0\rightarrow1}$ and $A_{1\rightarrow0}$ are often high compared with those for other transitions. When $B_{0\rightarrow1}$ and $A_{1\rightarrow0}$ are not significantly high and possibly metastable states exist, then processes such as (b) and (d) may become more favourable and produce useful atomic fluorescence signals. Thermally assisted mechanisms (e) generally do not produce intense fluorescence signals; however, this may not be a handicap when the aim is to remove an interference and not to achieve high sensitivity. The radiational deactivation in normal direct-line fluorescence does not terminate in the ground state and, therefore can be particularly useful (when suitable low lying states are available) when high values of the self-absorption term are expected. An example of normal direct-line fluorescence is the emission of the Tl 535.5 nm line after excitation with the Tl 377.6 nm line, the atoms undergoing deactivation to the $6^2P_{\frac{1}{2}}$ metastable state. Another practical advantage of non-resonance fluorescence is in the removal of interference due to scatter.

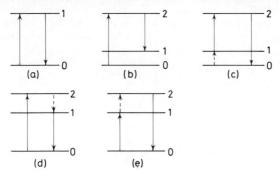

Figure 3.5 Types of atomic fluorescence: (a) resonance fluorescence; (b) normal direct-line fluorescence; (c) thermally-assisted direct-line fluorescence; (d) normal step-wise-line fluorescence; (e) thermally assisted direct-line fluorescence. The solid lines represent radiational processes and the dashed lines represent non-radiational processes.

There is no significance to the relative positioning of energy levels from one diagram to the next

(From Winefordner, J. D. and Elser, R. C. (1971). *Anal. Chem.*, **43**, 24A, by courtesy of The American Chemical Society)

For example, use could be made of the step-wise fluorescence observed when atomic sodium is irradiated with the 330.3 nm line; emission of the 589.0 doublet occurs following collisional deactivation to the $3P_{\frac{1}{2}}$ and $3P_{\frac{3}{2}}$ levels.

3.2.5 Factors influencing the atomic fluorescence intensity

The relationship between intensity and atomic concentration has already been fully discussed in Section 3.3 with reference to the importance of the 'a' parameter and the l/L ratio. In this section we shall discuss briefly parameters such as the fluorescence yield factor and others implicit in the general intensity expression such as temperature and line oscillator strength.

3.2.5.1 The fluorescence yield factor (p)

From the definition of p given earlier it can be seen that at a given pressure and temperature the value of p will depend on the effective quenching cross-section for the excited atoms and the surrounding gas molecules. Experiments have shown that processes involving deactivation and conversion of the energy into translational modes have low cross-sections (i.e. low occurrence probability), whereas those involving conversion of energy into molecular vibrational and rotational modes have high cross-sections. This result is to be expected and we may extend the argument and say that the greater the number of closely spaced levels the molecule possesses, the more effective it will be as a quencher because the probability of it having an internal transition closely approximating in energy to the deactivation step is then higher. Any energy deficit, which will often be a fraction of an electron volt, can then be dissipated easily in translational modes. Species possessing high quenching cross-sections are CO_2, CO, N_2 and O_2 whereas H_2, H_2O and Ar possess low quenching cross-sections. Experimental values of the quenching cross-sections for these common flame species with respect to the quenching of sodium and potassium atomic fluorescence have been tabulated by Hooymayers[32, 33]. A knowledge of the quenching ability of various gases is essential for the use of most non-flame atom reservoirs, because analytical signals are greatly enhanced by the choice of suitable shielding gases such as argon. Both temperature and pressure will affect collisional quenching, but molecular structure is still the dominant feature.

3.2.5.2 The temperature

Temperature is the most difficult parameter to discuss with respect to atomic fluorescence intensity because it is an implicit factor in several of the intensity function terms, i.e. K_0, Δv_0, p, N_0 etc. To assess its effect on each term and then to combine the effects to yield an overall picture is beyond the scope of this study and therefore instead we shall make some general observations about the likely effect of temperature on analytical signals.

First let us examine the equation:

$$E = [A_{k-i}] [N_0(g_i/g_a) \exp \{-\varepsilon/kT\}] \tag{3.18}$$

where ε is the energy difference.

This equation states that under equilibrium conditions the intensity of a spectral line depends only on the product of the spontaneous transition probability and the population of the excited state. The exponential term shows that the emission intensity will be strongly dependent on temperature. Calculation of the population of states giving rise to lines in the ultraviolet region of the spectrum (i.e. below 350 nm) for normal analytical flames with temperatures not exceeding 3000 K shows the number of excited atoms to be extremely small. Thus, when intense atomic fluorescence of cadmium at 228.8 nm is observed in an air–Ar–H$_2$ diffusion flame we must conclude that the population of the 5^1P_1 excited state is controlled almost entirely by the intensity and spectral distribution of the exciting source. When the resonance state approaches the ground state more closely (c. 3.5 eV), then the thermal population can become significant and a need arises to distinguish between the thermal and fluorescence contributions to the spectral radiance. Similarly, we may expect that direct line fluorescence will be insensitive to temperature except where the first excited state is close enough to the ground state to have appreciable thermal population. Temperature is certainly a more critical factor in step-wise fluorescence because translational activation or deactivation is a prerequisite to the radiative step.

Finally, the effect of temperature on the atomising efficiency of the flame must be considered. Droplets entering a flame have to proceed through desolvation, volatilisation and molecular dissociation before free atoms are produced and it is well known that for elements forming refractory compounds, e.g. oxides of Al, Si etc., that flames such as N_2O—C_2H_2 ($T = 3000$ K) are required to yield useful free atom concentrations. It should of course be realised that provided the temperature is sufficient to produce the analyte rapidly in a gaseous form then the chemical environment will be the major factor determining the production of free atoms. In particular, reducing conditions[34] should be maintained in the flame to avoid the formation of stable oxide species.

3.2.5.3 The oscillator strength

The oscillator strength (f) is defined by the equation:

$$f = \frac{mc}{8\pi e^2} \frac{g_k}{g_i} \lambda_0^2 A_{k \to i} = 1.51.10^{-16} \frac{g_k}{g_i} \lambda_0^2 A_{k \to i} \tag{3.18a}$$

where e is the charge on the electron.

Inspection of equation (3.18a) shows that the most intense spectral lines will in general have high f values. Now considering equation (3.15) in the limit $N_0(K_0) \to 0$, that is at the detection limit, we can make the approximation that

$$\int (1 - \exp\{-x\})\, dx = \int x\, dx$$

and noting that

$$\int_0^\infty K_v\, dv = \frac{\lambda_0^2}{8\pi} \frac{g_k}{g_0} A_{k \to i} N_0 = \frac{1}{2}\sqrt{(\pi/\ln 2)} K_0 \Delta v_0$$

so that

$$\int_0^\infty P(y)\, dy = \sqrt{\pi/2}$$

Using this approximation for both primary and self-absorption terms equation (3.15) then becomes:

$$E_F \approx \frac{d\omega}{4\pi} \frac{I_0 \sigma}{\Delta v_D} P \sqrt{\left(\frac{\pi}{\ln 2}\right)} \chi$$

i.e. $E_F \propto \chi$

i.e. $E_F \propto \lambda_0^2 (g_k/g_i) A$

and thus $E_F \propto f$

Therefore, experimentally, we may expect that lines having large f values will yield the most intense atomic fluorescence signals and indeed the initial choice of the analytical line can be based on this criterion.

3.3   THE ANALYTICAL SIGNAL IN ATOMIC FLUORESCENCE SPECTROMETRY

The analytical signal may be defined as the response of the measuring device to the presence of the analyte. This definition is obvious, but it includes information often ignored in the course of analytical experiments, i.e. the recorder trace or the meter deflection is as much a function of the measuring device as the physical process causing it. We may identify time as the important factor and proceed to discuss the work of L'Vov[35] for the analysis of analytical signals. The signals obtained in analytical spectrometry may for convenience be divided into two types, steady state signals and transient signals. The former are the most commonly encountered and they are produced with continuous sampling devices such as pneumatic nebulisers which are standard on nearly all commercial atomic absorption and emission instruments. Transient signals produced by one shot sampling devices are now becoming more familiar with the commercial availability of devices such as the Massmann furnace[120] and West[49] filament atom reservoir.

3.3.1   Steady-state signals

If N_0 is the number of analyte atoms in the sample, N the number of analyte atoms in the analytical cell (exciting light path for fluorescence) at time t, τ_1 the duration of transfer of analyte atoms into the cell and τ_2 the average time spent by analyte atoms in the cell, then the balance between atoms entering and leaving the cell can be written in the form:

$$\frac{dN}{dt} = n_1(t) - n_2(t)$$

where $n_1(t)$ and $n_2(t)$ are functions of the atomisation system used. For a flame cell with a pneumatic nebuliser and steady flow rate we can write:

$$n_1(t) = N_0/\tau_1$$

$$n_2(t) = N/\tau_2$$

$$\therefore \quad \frac{dN}{dt} = \frac{N_0}{\tau_1} - \frac{N}{\tau_2} \tag{3.19}$$

By separating the variables and integrating, two equations are obtained:

i.e. $$N(t) = \frac{N_0\tau_2}{\tau_1}(1 - \exp\{-t/\tau_2\}) \qquad\qquad t \leqslant \tau_1 \tag{3.20}$$

and

$$N(t) = \frac{N_0\tau_2}{\tau_1}(1 - \exp\{-\tau_1/\tau_2\} \exp\{-(t-\tau_1/\tau_2)\} \qquad t \geqslant \tau_2 \tag{3.21}$$

These completely describe the shape of the radiation pulse received by the detection system. Figure 3.6 (after L'Vov[35]) shows the pulse shapes from a flame cell for various values of the ratio τ_1/τ_2. In practical situations the pulse shape will approach that shown for $\tau_1/\tau_2 = 5$ where an equilibrium value is reached for a finite period. At equilibrium it can be seen that:

$$dN/dt = 0 \qquad (3.22)$$

Thus from equations (3.19) and (3.22)

$$N_{eq} = N_0\tau_2/\tau_1$$

Measurement of N_{eq} is the usual technique employed to estimate N_0; calibration is used to eliminate the necessity of knowing (τ_2/τ_1). This method is widely employed because interpretation of results is easy and almost any

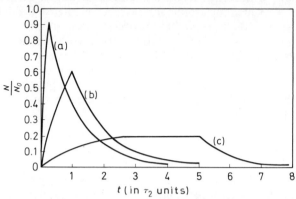

Figure 3.6 Variation in the number of atoms in an analysis volume for different ratios τ_1/τ_2:(a) $\tau_1/\tau_2 = 0.2$, (b) $\tau_1/\tau_2 = 1$, (c) $\tau_1/\tau_2 = 5$
(From L'Vov[35], by courtesy of Adam Hilger Ltd.)

d.c. or a.c. (in the case of chopped radiation beams) current measuring device (assuming the prime detector to be a photomultiplier) will effectively monitor N_{eq}. The use of a modulated radiation source and a tuned a.c. amplifier enables automatic subtraction of the flame background from the signal read-out. Pulses (a) and (b) (Figure 3.6) cannot be characterised by a value N_{eq}, but instead a value N_{peak} could be used. From Figure 3.6 it follows that N_{peak} is reached at $t = \tau_1$, i.e. when all the sample has been transferred into the cell so that substituting into equation (3.20) yields: $N_{peak} = (N_0\tau_2/\tau_1)$ $(1 - \exp\{-\tau_1/\tau_2\})$
For

$$\tau_1/\tau_2 \ll 1$$

$$1 - \exp(-\tau_1/\tau_2) = \tau_1/\tau_2$$

$$\therefore N_{peak} \doteqdot N_0$$

Thus, N_{peak} may also be used as a measure of N_0 and indeed provided that the condition $\tau_1/\tau_2 \ll 1$ is met, it is not sensitive to the actual values taken by τ_1, τ_2, e.g. $\tau_1/\tau_2 = 0.1$, $N_{peak} = 0.95\ N_0$; $\tau_1/\tau_2 = 0.2$, $N_{peak} = 0.90\ N_0$. It should be noted that the special case of a pneumatic spray device producing a peaked profile as presented here is seldom met in practice.
Finally, each of the three curves could be characterised by the area it

encloses (Q_1) which is the basis of the third measuring technique, namely, the integral method. Thus performing the integration:

$$Q_1 = \int_0^{\tau_1} \frac{N_0\tau_2}{\tau_1} \left(1 - \exp\left\{-t/\tau_2\right\}\right) dt$$

$$+ \int_{\tau_1}^{\infty} \frac{N_0\tau_2}{\tau_1} \left(1 - \exp\left\{\tau_1/\tau_2\right\}\right) \exp\left\{-(t-\tau_1)\tau_2\right\} dt = N_0\tau_2 \qquad (3.23)$$

shows that Q_1 can also be used to measure N_0, the subscript I being introduced to indicate that the integration should be carried out in units proportional to N_0 which for atomic fluorescence is intensity. Equation (3.23) shows that as with the peak method processes of atomisation which retain the atoms in the analytical cell for a long period will yield the highest sensitivity, but also with the integral method the sensitivity is independent of the rate of sample introduction into the cell. Comparison of the sensitivities of the three methods reveals that the peak method is $c.\ \tau_1/\tau_2$ times as sensitive as the equilibrium method and the integral method is τ_2 times more sensitive than the peak method.

3.3.2 Transient signals

For most of the devices which produce a transient signal at the photo-multiplier, e.g. L'Vov furnace[35], platinum loop[112], West filament atom reservoir[49] and Massmann furnace[120], the rate of introduction of atoms into the cell is a linear function of the evaporation temperature which can be approximated to a linear function of time such that:

$$n_1(t) = At$$

However, because the normalisation condition

$$\int_0^{\tau_1} n_1(t)\, dt = N_0$$

must apply it follows that:

$$n_1(t) = 2N_0t\tau_1^2$$

Therefore

$$\frac{dN}{dt} = \frac{2N_0t}{\tau_1^2} - \frac{N}{\tau_2}$$

This equation can be solved by applying the formula:

$$N(t) = \exp(-h)\left[\int \exp(h)r\, dt + C\right]$$

where

$$h = \int f(t)\, dt \qquad \text{(Ref. 36)}$$

which then yields the two equations:

$$N'(t) = \frac{N_0 2\tau_2^2}{\tau_1^2}\left(\frac{t}{\tau_2} - 1 + \exp\{-t/\tau_2\}\right) t \leqslant \tau_1 \qquad (3.24)$$

$$N'(t) = \frac{N_0 2\tau_2^2}{\tau_1^2}\left(\frac{\tau_1}{\tau_2} - 1 + \exp\{-\tau_1/\tau_2\}\right)\exp\{-(t-\tau_1)/\tau_2\} \, t \geqslant \tau_1 \qquad (3.25)$$

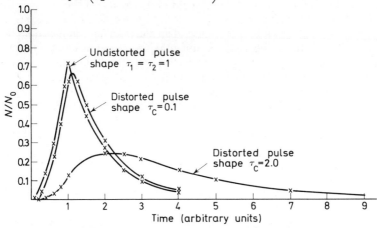

Figure 3.7 Variation in pulse shape with different time constants

Equations (3.24) and (3.25) represent (as do equations (3.20) and (3.21)) the signal arriving at the detector, but we must now consider the effect of the detector on the pulse shape. From the theory of electrical networks[37] the response of a circuit can be calculated from the equation:

$$i(t) = \int_0^\infty A(t-\tau)N(\tau)\,d\tau \qquad (3.26)$$

where $A(t)$ is the impulse response of circuit, $N(t)$ is the signal pulse shape and τ is the integration variable. Thus, for (Resistance/Capacitance) circuits (all photomultiplier outputs correspond to a parallel RC network):

$$A(t) = \frac{1}{C}\exp(-t/RC)$$
$$= \frac{1}{C}\exp(-t/\tau_c)$$

where τ_c is the time constant for the circuit. Thus, calculating the distortion caused by the network based on equations (3.23)–(3.25), we find that:

$$i(t) = \frac{N_0 2\tau_2}{\tau_1^2}\frac{1}{(\tau_2-\tau_c)}\left[\tau_2^2\left(\frac{t}{\tau_2} - 1 + \exp\{-t/\tau_2\}\right)\right.$$
$$\left. - \tau_c^2\left(\frac{t}{\tau_c} - 1 + \exp\{-t/\tau_c\}\right)\right] t \leqslant \tau_1 \qquad (3.27)$$

$$i(t) = \frac{N_0 2\tau_2}{\tau_1^2}\frac{1}{(\tau_2-\tau_c)}\left[\tau_2^2\left(\frac{\tau_1}{\tau_2} - 1 + \exp\{-\tau_1/\tau_2\}\right)\exp\{-(t-\tau)/\tau_2\}\right.$$
$$\left. - \tau_c^2\left(\frac{\tau_1}{\tau_2} - 1 + \exp\{-\tau_1/\tau_c\}\right)\exp\{-(t-\tau_1)/\tau_2\}\right] t \geqslant \tau_1 \qquad (3.28)$$

Figure 3.7 shows the pulse shape plotted for $\tau_2 = \tau_1 = 1$ and the resulting distortion when the pulse is fed through networks having time constants of 0.1 and 2.0. The plots illustrate that when it is desired to preserve the shape of the pulse (perhaps because it contains useful physical information) circuit time constants or response times should be short compared with the pulse-width (in general the response time of the system should be c. 5 times less than the half width of the event). Measurement of N_0 based on the N_{peak} estimate requires different criteria because it is desired to maximise the signal/noise ratio and it is well known that the noise of an electrical circuit is proportional to $1/\tau_c$. L'Vov[35] have shown that for maximum signal/noise (under the assumption that $\tau_1/\tau_2 = 0$) the criterion $\tau_2 = \tau_c$ should be met. Practically, provided that $\tau_1/\tau_2 \leqslant 1$ the criterion will hold, but above values of unity τ_c needs to be somewhat greater than τ_2.

Much of the recent work in our own laboratory has been in connection with the filament atom reservoir[49] and the platinum loop atomiser[112] which produce transient type signals. Here we may present some of our findings and illustrate how they relate to the cuvette theory. The simplicity of the peak recording method makes it an obvious first choice, but as the theory has shown there are criteria to be met in the circuit design to avoid over distortion and inferior noise characteristics. Our work[38] has shown also that pulse widths are often concentration dependent, the variation originating in the period τ_1, thus making it impossible to optimise the measurement circuit completely. The practical outcome of this is poor linearity of the working curve and low precision of the analytical results. The same problems are encountered when a.c. systems are used having a period within an order of magnitude of the pulse width. The solution to the problem is to use the integral method of recording which is independent of τ_1 (since $\int_0^{\tau_1} N'(t) + \int_{\tau_1}^{\infty} N'(t) = N_0\tau_2$) and does, as our results confirm, yield linear working curves and good precision.

3.3.3 Integration methods for transient and steady-state signals

Two approaches to the problem of integration have been examined in our laboratory. The first method[39] involves the circuit shown in Figure 3.8 in which the signal is current amplified by one operational amplifier and then passed to a second having a capacitor in its feedback loop. Current trimming allows the background signal to be set to zero so that only the analytical signal is integrated by the capacitor, the resulting voltage (step V_c) is measured by a potentiometric recorder. The device operated well in conjunction with a platinum loop nebuliser producing signals of c. 250 ms duration, but was limited in sensitivity by base line drift. The longer the integration period required, the more serious is the drift problem so that this type of circuit is not suited to long period integration at low signal levels.

Photon counting offers an alternative to d.c. integration and incorporates a number of advantages[38, 40–42]. Figure 3.9 is a schematic diagram of a simple photon-counting system capable of detecting steady signals down to c. 50 photons s^{-1} and having an upper linear counting rate of c. 310 000 s^{-1}. The upper rate is determined by the onset of pulse pile-up, that is the

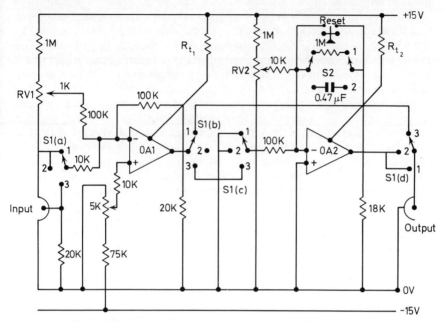

Figure 3.8 Amplifier/integrator circuit:
OA1 = operational amplifier No. 1
OA2 = operational amplifier No. 2 }Philbrick PF85AU
S1 = switch No. 1–3 way 4 pole
S2 = switch No. 2–2 way 4 pole
RV1 = current off-set
RV2 = integrator off-set
Resistors, 0.5 W Hystab (Radiospares Ltd.)
Power supply, -15, 0, $+15$ V, Coutant OA10 stabilised supply unit
(From Aldous, Dagnall, Sharp and West[39], by courtesy of Elsevier)

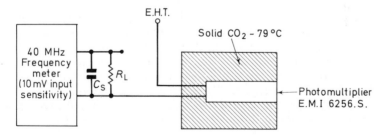

Figure 3.9 Schematic diagram of a simple photon-counting system.

point at which the finite response time of the system prevents resolution of individual photo-electron pulses. Cooling of the photomultiplier is employed which reduces the dark count rate to c. three counts per second and obviates the need for a pulse height discriminator. In practice, some discrimination exists due to the finite pulse counting sensitivity of the frequency meter which in this case was 10 mV. Photon-counting systems are not subject to drift and are therefore suitable for both short and long period integration. The simple logic step of counting each pulse (regardless of size) as 1 unit accounts for a reduction in a number of noise sources encountered in d.c. and a.c. systems, for example, the component of shot noise arising from the statistical conversion of photo-electrons into anode charge pulses, band-width noise and Johnston noise. Unfortunately, background pulses are also counted as one and integrated and it is this fact combined with the limited linear counting range that makes photon counting most suitable for low background systems, e.g. non-flame atomic fluorescence. It should be stressed that the signal/noise ratio will always increase with increasing signal strength and therefore no advantage will result from attenuating intense signals to bring them into the linear range of a photon counter.

Under ideal conditions then (ignoring source and flame flicker), the only noise in the fluorescence signal should be that due to the random time distribution of the emitted photons, i.e. Poisson noise. It follows from Poisson statistics that if the steady signal count rate is N then the variance is also N and the standard deviation or noise will be $N^{\frac{1}{2}}$. When a number of independent variables are contributing to the noise, the total noise can be found by adding the variances and taking the root of the sum. Thus, in an atomic fluorescence experiment where N_0 is the background count rate, N_s the signal count rate (i.e. total count − background count) and τ the count period, the total noise is

$$\text{Noise} = \sqrt{[(N_s\tau + N_0\tau) + N_0\tau]} \qquad (3.29)$$

The extra term $N_0\tau$ appears because the background has to be measured independently of the total count (in a single beam system).

Therefore the signal:noise ratio is

$$\frac{S}{N} = \frac{N_s\tau^{\frac{1}{2}}}{(N_s + 2N_0)^{\frac{1}{2}}} \qquad (3.30)$$

This equation illustrates the method of calculating the signal/noise ratio, but can hardly be expected to apply in real experiments where other sources of noise, particularly flame and source flicker, may be of greater importance. Because it is difficult, if not impossible, to estimate or measure the variance contributions of each noise source independently, the only valid measure of noise is to determine experimentally the standard deviation of the analytical signal. The detection limit follows, therefore, as the concentration yielding a signal/noise ratio of two, the noise being measured at, or as near as possible, to the detection limit.

Consider first Table 3.1 showing the noise figures for different sources calculated from equation (3.30) and measured experimentally. As expected, the dark pulses obey Poisson statistics, but the deviation from ideality increases with the instability of the experimental system. The limiting form of the Poisson distribution is the normal distribution which applies when the count rate is high, i.e. > 100. Normal populations have the property that the standard deviation of the sample mean decreases as the root of the number of samples taken. In counting experiments the number of samples taken is equivalent to the count period and therefore it might be expected that the signal/noise ratio would increase with the root of the integration period as in equation (3.30). The following signal/noise ratios were obtained for different integration periods from 5 p.p.m. Hg sprayed

Table 3.1 Comparison of standard deviations measured experimentally and obtained from counting statistics

(From Alger, Dagnall, Sharp and West[40], by courtesy of Elsevier)

Photon-counting system	Average count/s^{-1}	Standard deviation[a]	
		Repetitive readings	Counting statistics
Typical dark count	468.7	20.33	21.63
Typical count from W lamp at 550.0 nm	26 654	218.9	166.1
Typical count from microwave plasma at 228.8 nm	81 019	661.4	290.9

[a]Obtained from 30 repetitive counts.

into an argon–hydrogen–air diffusion flame illuminated by a mercury electrodeless discharge lamp:1 s count period $S/N = 1784$, 10 s count period $S/N = 2373$, 100 s count period $S/N = 2397$. The fact that the predicted increase in signal/noise ratio does not occur can be attributed to three causes:

(i) The presence of flicker noise.

(ii) During long period integration the probability of interference from electrical, vibrational, cosmic and other sources increases. Such noise is difficult to assess statistically because even if it were Poisson in time (which is unlikely since causes such as switching of equipment tend to occur periodically) the time scale would be so different to that of photon arrival that a particular interference would be a unique event in terms of the integration period. Furthermore, these interferences result in bursts of counts which are related to the type of source.

(iii) Cumulative interferences such as monochromator drift and source intensity drift alter the sampling population and thus adversely affect the precision theoretically obtainable by integration. In certain circumstances, this type of error can be minimised by corrective feedback mechanisms or double beam instrumentation.

The discussion of photon counting so far has been related to the ability of the technique to integrate steady signals over long periods; however, when transient signals are obtained the criteria of operation are slightly different. Consider a typical transient signal of duration τ which is an unknown function of time, i.e. $N(t)$, then photon counting enables the measurement of the total photon output N_s:

as
$$N_s = \int_0^\tau N(t)\, dt \qquad (3.31)$$

where τ may depend on the concentration. Now, since the background radiation adds to the on-signal noise it is evident that the best precision will be obtained when the count period is set approximately to τ. (It may be that for certain functions $N(t)$ integration periods of less than τ might prove optimal). Practically, when a range of concentrations has to be covered, it suffices to set the count period equal to the longest anticipated signal duration. Figure 3.10 shows four analytical curves obtained from transient

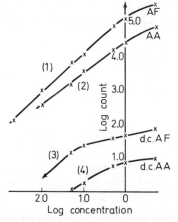

Figure 3.10 Analytical working curves for atomic fluorescence (AF) and atomic absorption (AA) of cadmium using photon counting and d.c. measurement
(From Dagnall, Sharp and West[38], by courtesy of Macmillan Ltd.)

signals having durations of 0.1–5.0 s. Curves (1) and (2) were from photon counting measurements, curves (3) and (4) from a d.c. system comprising a potentiometric recorder having a response time for full scale deflection of $c.$ 1 s. The superiority of the integration technique is obvious from the linearity of the curves. There is a limit to the minimum signal duration that photon counting can handle because of the limited linear counting range. This is best illustrated by the hypothetical example of a signal comprising 100 photons arriving in one microsecond which would require linear counting of 10^8 per second for measurement without loss due to pulse pile-up. Before photon counting can become as widely applicable as d.c. or a.c. methods the upper range must be extended and there are a number of possible ways of achieving this namely

(i) the use of high-gain fast photomultipliers capable of pulse pair resolution down to nanosecond or sub-nanosecond times,

(ii) employing upper discriminator levels which distinguish pile-up pulses from single photon pulses and then place two in the count register instead of one, and

(iii) an effective increase in the linear counting range which can be achieved by employing mathematical correction techniques.

3.4 THE ANALYTICAL TECHNIQUE OF ATOMIC FLUORESCENCE SPECTROMETRY

The theoretical discussions given above describe atomic fluorescence in detail and the factors influencing the intensity of the emitted fluorescence radiation. A study of the atomic fluorescence mechanism shows that it is a combination of atomic absorption and atomic emission phenomena and therefore it is to be expected that the three analytical techniques deriving from these mechanisms will have certain similarities. For the analyst, the advantages of the atomic fluorescence technique result from the differences between atomic fluorescence, absorption and emission and it remains to point out these practical differences, to indicate their analytical significance and to describe the necessary instrumentation.

3.4.1 A comparison of atomic fluorescence spectrometry with atomic absorption and atomic emission spectrometry

3.4.1.1 The advantages of atomic fluorescence in relation to atomic absorption spectrometry

(i) Atomic fluorescence is essentially an emission technique and the analytical sensitivity may be increased by either increasing the excitation source intensity or the gain of the amplifier/detector system. In atomic absorption the absorbance value is a ratio ($\log I_0/I_t$ where I_0 is the intensity of the incident radiation and I_t is the intensity of the transmitted radiation) and any increase in the value I_0 is accompanied by a corresponding increase in I_t; hence the above ratio is largely unchanged. Occasionally there are advantages in sensitivity to be gained in utilising a source possessing high resonance line intensity in atomic absorption, although in the main this serves to overcome the inadequacies of the instrumentation used.

(ii) Because the detector system does not 'see' the source of excitation in atomic fluorescence (in most studies the experimental arrangement involves a source of excitation mounted so that it subtends a right angle with the entrance slit of the detector system at the atom reservoir, i.e. an atomic absorption spectrometer bent at a 90 degree angle around the absorption cell axis), it is possible to use a continuum source of excitation. Such sources can be used in atomic absorption, but a very high resolving power monochromator is necessary together with a high amplification system if a satisfactory analytical sensitivity is to be obtained.

(iii) Atomic fluorescence spectrometry does not require a sophisticated dispersive system because the detector does not view the source of excitation directly. This is in fact the most significant difference from the instrumental requirements of atomic absorption spectrometry. The consequence of this is that it is possible to construct an atomic fluorescence instrument which need not contain a conventional dispersion system; the selectivity and effective dispersion is achieved within the absorption cell as a result of the atomic fluorescence process. In such instances it is, however, usually necessary to utilise line sources of excitation rather than continuous sources, assuming that more than one analyte element atom is present.

(iv) For many analyte element atoms it is possible to observe more fluorescence lines than absorption lines and this gives a greater choice in the wavelength of measurement. For example, the major resonance lines for arsenic lie at 189.0, 193.7 and 197.2 nm, the region in which there is considerable absorption due to atmospheric oxygen, flame gases and the instrument optics. In addition, it is not uncommon for photomultipliers to have a poor response in this region. However, although these wavelengths must be used in atomic absorption measurements, the atomic fluorescence resulting from the absorption at these lines can be measured at 228.8 or 235.0 nm[25] where the above effects are minimised.

(v) The present trend towards the use of non-flame cells is advantageous in both atomic absorption and fluorescence as a result of the efficient production of atoms, usually in a confined space. The major cause of noise in atomic absorption is source flicker and source shot noise produced at the photomultiplier and this is not altered significantly by the use of a non-flame cell. In fact the situation may be worse due to non-flame background radiation. In atomic fluorescence, however, the major noise contribution results from the atom reservoir and the use of a non-flame cell can give significant improvements in the signal/noise ratio provided it is effectively screened from the detector. This fact of course imposes certain restrictions on the use of flames as atom cells in atomic fluorescence studies. Hence the future of atomic fluorescence spectrometry may rest with the development of alternative methods of atomisation to the flame, especially if the basis of comparison concerns itself with limits of detection (see below).

(vi) Simultaneous analysis for many elements is readily accomplished by atomic emission spectrometry, but the situation is complicated for both atomic absorption and fluorescence spectrometry. The techniques described in the past for atomic absorption[43–45] have not been widely accepted for a variety of reasons, mainly because these systems appear to be a combination of several individual atomic absorption instruments each with its own source and detector. In addition, the use of resonance detectors or selective modulation methods[43] in some of these instruments was not widely applicable. In contrast, the proposed multi-channel atomic fluorescence spectrometers[43,46,47] are more elegant in that they utilise a single detector. Usually these instruments have been constructed about non-dispersive systems such as filters and a conventional photomultiplier or a single solar blind photomultiplier. Although both atomic absorption and fluorescence multi-channel spectrometers have certain basic limitations the implications are that the atomic fluorescence technique is preferable. This approach is dealt with in more detail later in the chapter.

3.4.1.2 The disadvantages of atomic fluorescence in relation to atomic absorption spectrometry

efficiency: atomic fluorescence measurements are directly proportional

(i) Atomic absorption measurements are independent of the quantum efficiency; atomic fluorescence measurements are directly proportional to p and any change in this value will be reflected by a change in the value of the observed atomic fluorescence intensity. However as long as the flame temperature and composition are kept constant and the partial pressures of the nebulised species are low (which is generally the case), then the inter-element effects on the value of p should be negligible. Alkemade[19] found that the quantum efficiency for the resonance fluorescence of sodium was unchanged when the temperature of an argon–oxygen–propane flame was raised by 100 °C; however a sixfold decrease in p was noted when the argon was changed to carbon dioxide. In the application of non-flame cells in atomic fluorescence such effects may be used to advantage, i.e. the system is usually purged with argon.

(ii) The determination of elements that are prone to form refractory oxides in flames is more difficult by atomic fluorescence than by atomic absorption spectrometry. Atomic fluorescence is essentially an emission technique and the high flame radiative background from the use of the nitrous oxide–acetylene flame can cause excessive noise at the photomultiplier when large monochromator slit widths are used. In atomic absorption this radiation may be minimised by the use of narrow slit widths without a great loss of analytical sensitivity.

(iii) The determination of the alkali elements with resonance lines in the visible region is difficult by flame atomic fluorescence because of the high thermal excitation produced, even in relatively low temperature flames. As will be shown later, atomic fluorescence is not generally considered to be a useful technique for elements with major resonance lines at wavelengths greater than c. 350 nm.

(iv) Scattering of source radiation by solid particles present in the flame gases is a serious problem in both atomic absorption and atomic fluorescence spectrometry. The authors consider that scatter in atomic fluorescence is often more apparent than in atomic absorption, especially when using a continuam source of excitation. The main difficulty is in determining the extent of scatter, although the majority of pneumatic-type nebulisers and analytically useful flames do not give rise to a high proportion of undissociated solid particulate matter. In general, the problem arises when using solutions containing a high solid content rather than when the analyte element is present together with material which is prone to the formation of undissociated particulate matter. The use of total consumption (or direct injection) burners in conjunction with low temperatures flames is not to be recommended because even with pure solutions of the analyte element it is possible to observe scatter.

It should be pointed out again, however, that one is not necessarily limited to resonance atomic fluorescence measurements. Direct-line, step-wise or thermally assisted direct-line atomic fluorescence measurements may be made at different wavelengths from the exciting radiation and, hence, scatter will not be observed in these instances. In this respect, therefore, atomic fluorescence is even less subject to scatter than atomic absorption measurements.

3.4.1.3 The advantages of atomic fluorescence in relation to flame emission spectrometry

(i) Atomic fluorescence depends on the absorption of radiation by ground state atoms and thus, unlike flame emission processes, will not be exponentially dependent on the flame temperature.

(ii) Atomic fluorescence is analytically very much more sensitive for elements with resonance lines in the far ultraviolet region (e.g. arsenic, selenium, tellurium, cadmium, zinc, bismuth, antimony, etc.).

(iii) Relatively low temperature, low burning velocity flames which give negligible thermal emission for most elements can be used for atomic fluorescence studies if matrix effects permit.

3.4.1.4 The disadvantages of atomic fluorescence in relation to flame emission spectrometry

(i) Atomic fluorescence like atomic absorption measurements requires the use of a source of excitation and, usually, a different source for each analyte element.

(ii) Atomic fluorescence is less sensitive than flame emission spectrometry for readily excited elements such as the alkali and alkaline earth elements with resonance lines in the visible region of the electromagnetic spectrum.

3.4.2 Factors influencing the limit of detection in atomic fluorescence spectrometry

Although the above facts may be interpreted in various ways, the most publicised advantage of atomic fluorescence spectrometry made by the early workers was the ability to achieve very low limits of detection. Indeed, attempts to demonstrate this experimentally were made by Winefordner et al.[20-23] for a few elements, but unfortunately their use of the direct-injection burner also suggested the possibility of a scatter contribution. Following this, much of the work was directed towards comparing the limits of detection in atomic absorption and atomic fluorescence spectrometry. Because of the different types of instrumentation used by various researchers it was not surprising that this line of approach did not meet with much general agreement. However, this aspect is of some importance when considering the potential of atomic fluorescence and especially because it is possible to make certain predictions based on entirely theoretical considerations. Winefordner et al.[48] have dealt extensively with the comparison of atomic emission, absorption and fluorescence spectrometry using flames as a source of atom production. Various assumptions were made such as:

(i) The same metal resonance line is used for each model.

(ii) The spectral line is affected only by Doppler and Lorentz broadening to allow application of conventional intensity expressions.

(iii) The flame has the same temperature and atomic concentration at all points in the analytical volume and this volume is in thermodynamic equilibrium.

(iv) The flame region of interest is uniformly illuminated by the source in atomic absorption and atomic fluorescence and the resulting atomic fluorescence over the entire flame segment and in the direction of the monochromator is imaged on the monochromator entrance slit.

(v) The instrumental system, including the nebuliser, burner, flame, photodetector, monochromator and measurement electronics is identical in all three methods; in addition, the solid angle of radiation collected from the source in atomic absorption and atomic fluorescence is the same as the solid angle of radiation collected from the flame in atomic emission, absorption and fluorescence. Also the luminous area of the source of excitation is identical in atomic absorption and atomic fluorescence.

(vi) The monochromator is of sufficient resolution that spectral interferences are negligible in all three methods.

(vii) The atomic concentration of the measured species is low (linear region of growth curves, i.e. slope of unity on a log–log plot).

Obviously such assumptions are essential in a strict theoretical comparison, but equally obviously few of these assumptions are likely to hold in practice. In consequence the value of a comparison on these lines is again not likely to be very convincing, but certain facts do become apparent. For example, on the basis of a comparison of the signals received at the photodetector in each of the three techniques, it is clear that the signal in atomic absorption is about 10^3 greater than the fluorescence signal. This arises because the ratio of the measured signals for dilute atomic vapours in atomic absorption with respect to atomic fluorescence is:

$$\frac{AA \text{ signal}}{AF \text{ signal}} \propto \frac{1}{p\Lambda/4\pi}$$

where p is the fluorescence quantum efficiency and $\Lambda/4\pi$ is the fractional solid angle of radiation from the excitation source incident upon the flame in atomic absorption or fluorescence. The value of p for resonance lines of many metals is $c.$ 0.1 and for a typical flame spectrometer incorporating a prism or diffraction grating $\Lambda/4\pi$ is about 0.01. When a large aperture filter system is used to select the wavelength of fluorescence, then the value of $p\Lambda/4\pi$ may of course be increased considerably.

In addition, the signal magnitude in atomic absorption (usually a continuum or line source) is larger than the atomic thermal emission signal for all resonance lines within the optical spectral range of $c.$ 160.0–900.0 nm for all analytical flames having temperatures below $c.$ 3500 K. In comparison with atomic thermal emission, the atomic fluorescence signal is likely to be larger for all resonance lines with wavelengths below $c.$ 340.0 and 260.0 nm for flames of temperatures 2600 and 3200 K respectively.

However, in considering sensitivity of determination it is not the signal magnitude that is of prime importance in atomic spectrometry, but the signal/noise ratio. When the noise level is considered, the above calculations are subject to modification in terms of the obtainable limit of detection. For example, the noise ratio AA_N/AE_N is always much greater than unity because the noise in atomic emission is primarily due to phototube-shot noise and flame-flicker noise, whereas the noise in atomic absorption is primarily due to source-flicker noise, assuming the source is relatively intense (with a smaller contribution due to flame-flicker and phototube-shot noise). The noise in atomic fluorescence at the limit of detection is of course essentially the same as the noise in atomic emission, although because atomic fluorescence can be carried out in the non-luminous portion of the flame (or even above the flame) then the noise in atomic fluorescence can be smaller than that in atomic emission. The only assumption here being that source scatter noise (from unevaporated solvent and solute particles) is negligible. The source flicker contribution to noise in atomic fluorescence at the limit of detection is negligible, in other words, assuming a typical flame spectrometric system it can be shown that atomic fluorescence (using a continuum or line source) can give lower limits of detection than atomic absorption

Table 3.2 Comparison of experimental limits of detection (p.p.m.) in flame atomic spectrometry

Element	AA*	AF†	AE†	Wavelength/nm	Technique giving lowest limit of detection‡
Ag	0.005	0.000 1	0.02	− 328.1 −	AF
Al	0.1	0.1	0.005	− 396.2 −	AE
As	0.1	0.1	50	193.7, 193.7, 235.0	AA, AF
Au	0.02	0.005	4	242.8, 267.6, 267.6	AF
Be	0.002	0.008	0.1	− 234.9 −	AA
Bi	0.05	0.005	2	− 223.1 −	AF
Ca	0.002	0.02	0.000 1	− 422.7 −	AE
Cd	0.001	0.000 001	2	222.8, 222.8, 326.1	AF
Co	0.005	0.001 5	0.05	240.7, 240.7, 345.4	AF
Cr	0.005	0.001 5	0.005	357.9, 357.9, 425.4	AF
Cu	0.005	0.001	0.01	324.7, 324.7, 327.4	AF
Fe	0.005	0.003	0.05	248.3, 248.3, 372.0	AF
Ga	0.07	0.01	0.01	287.4, 487.2, 417.2	AF, AE
Ge	1.0	0.1	0.5	− 265.2 −	AF
Hg	0.5	0.000 2	40	− 253.7 −	AF
In	0.05	0.1	0.005	303.9, 451.1, 451.1	AE
Mg	0.000 3	0.000 2	0.005	− 285.2 −	AF
Mn	0.002	0.006	0.005	279.5, 279.5, 403.1	AA
Mo	0.03	0.5	0.1	313.3, 313.3, 390.3	AA
Ni	0.005	0.002	0.6	232.0, 232.0, 341.5	AF
Pb	0.03	0.01	0.2	283.3, 405.8, 405.8	AF
Pd	0.02	0.04	0.05	274.6, 340.5, 363.5	AA
Pt	0.1	0.15	2	− 265.9 −	AA
Rh	0.03	0.3	0.3	343.5, 369.2, 369.2	AA
Sb	0.1	0.04	20	217.5, 231.1, 259.8	AF
Se	0.1	0.04	−	− 196.1 −	AF
Si	0.1	0.6	5	251.6, 204.0, 251.6	AA
Sn	0.02	0.05	0.3	224.6, 303.4, 284.0	AA
Sr	0.01	0.03	0.000 2	− 460.7 −	AE
Te	0.1	0.005	200	214.3, 214.3, 238.3	AF
Ti	0.1	4	0.2	364.3, 319.9, 399.8	AA
Tl	0.025	0.008	0.02	276.8, 377.6, 377.6	AF
V	0.02	0.07	0.01	318.4, 318.4, 437.9	AE
Zn	0.002	0.000 02	50	− 213.8 −	AF

*From Kahn, H. L. (1968). *Advances in Chemistry Series*, **73**, 183
†From Winefordner, J. D. and Elser, R. C. (1971). *Analyt Chem.*, **43**, 24A and Refs. 131 and 132
‡Only in two instances (Si and Rh) does AA give a significantly lower limit of detection than either AF or AE (i.e. better than 4 fold)

(continuum or line source) and atomic thermal emission at all wavelengths *below c.* 300.0 nm; and atomic emission can give lower limits of detection than the other techniques *above c.* 400.0 nm. All methods should give similar results in the intermediate region of 300.0–400.0 nm. When it is necessary to use a high temperature reducing flame, for example to achieve sufficient atomisation of elements prone to formation of refractory oxides, then atomic absorption will give similar results to atomic fluorescence. An examination of Table 3.2 which lists the experimental limits of detection for all three flame atomic spectrometric techniques indicates that these theoretical observations appear to hold in practice.

With non-flame atom reservoirs, atomic fluorescence should be more favoured than atomic absorption (assuming all other factors are again equal) because the noise level due to non-flame background radiation and radiation flicker can be made negligible. The major source of noise in atomic fluorescence hence becomes shot noise, which is often less than flame-flicker noise; in atomic absorption the major source of noise is still source-flicker noise and phototube-shot noise. Non-flame cells will of course also be beneficial in atomic absorption because of the elimination of flame flicker noise.

At present, there is insufficient experimental evidence available in non-

Table 3.3   Limits of detection in non-flame atomic fluorescence spectrometry

Element	Line/nm	Massmann*/g	FAR†/g	Pt loop‡/g	Pt furnace/ p.p.m. n§
Ag	328.1	1.5×10^{-12}	1.0×10^{-12}	2.0×10^{-12}	
Au	267.6		4.0×10^{-12}		
Be	234.9			4.0×10^{-9}	
Bi	306.8		1.0×10^{-11}	4.0×10^{-11}	
Ca	324.7			2.0×10^{-10}	
Cd	228.8	2.5×10^{-13}	1.0×10^{-15}	2.0×10^{-14}	2.0×10^{-4}
Co	240.7		2.0×10^{-11}		
Cu	324.7	4.5×10^{-10}	2.0×10^{-12}		2.0×10^{-2}
Fe	248.3	3.0×10^{-9}	2.0×10^{-12}		1.0×10^{-1}
Ga	417.2		1.0×10^{-9}	2.0×10^{-8}	
Hg	253.7		5.0×10^{-11}	2.0×10^{-11}	8.0×10^{-3}
Mg	285.2	3.5×10^{-12}	1.0×10^{-12}	4.0×10^{-9}	
Mn	279.5		5.0×10^{-12}		
Ni	232.0		5.0×10^{-12}		
Pb	405.8	3.5×10^{-11}	1.5×10^{-11}	4.0×10^{-9}	
Sb	231.2	2.0×10^{-10}	1.0×10^{-9}		
Tl	377.6	2.0×10^{-9}	5.0×10^{-11}	2.0×10^{-9}	
Zn	213.9	4.0×10^{-14}	2.0×10^{-14}	2.0×10^{-11}	4.0×10^{-4}

*From Massmann[20]
†From West, T. S., (1971). *Pure and Applied Chemistry*, 25, 47, and private communication
‡From Bratzel *et al.*[127]
§From Black *et al.*[125]
Line sources of excitation used in all instances

flame atomic fluorescence to evaluate the situation critically, but from the information available (Table 3.3) it appears that the theoretical expectations are likely to be fulfilled.

Based upon the above conclusions and indications and assuming the presence of intense, stable line sources and high temperature, stable flames then the *optimum flame spectrometer for trace metal measurement* would be an instrument designed for atomic fluorescence (using a line source of excitation) and atomic thermal emission. Both methods require the same instrumental features, e.g. large monochromator aperture, a photomultiplier with low dark current, etc.; in addition the instrument would be quite simple and rapid to operate and no change in optics would be required in changing from atomic fluorescence to atomic emission. Furthermore, such a system would be suitable for multi-channel analysis. Assuming that the system is used to measure low radiation levels, detection techniques such as photon counting can be employed. As shown earlier, the relatively high radiation levels common to atomic absorption impose certain restrictions upon photon counting techniques.

Perhaps the simplest flame spectrometric system for measurement of low levels of all elements would be an instrument designed for atomic fluorescence, using a continuum source of excitation, and atomic emission. However, the introduction of such an instrument is by no means imminent because of the dearth of continuum sources of radiation with appreciable outputs below 300.0 nm. In such an instrument scanning techniques could be used with advantage to overcome the effects of 'scattering-type' interferences, to select the appropriate lines for measurement and to achieve optimisation of experimental conditions.

3.4.3   Interference in atomic fluorescence spectrometry

Possible interferences include spectral, physical and chemical, temperature effects, scattering of source radiation, band absorption and quenching of excited atoms.

3.4.3.1   Spectral

In general, the measurement of radiation characteristics of the interferent rather than, or as well as, the analyte radiation is relatively more likely in atomic emission than in atomic absorption or atomic fluorescence. If a continuum source of excitation is used instead of a line source (although this is not likely in practice), then a greater degree of spectral interference must be expected. However, the problems encountered for some elements in atomic absorption due to spectral interference from the emission of non-absorbing lines of analyte element or filler gas by the source is not encountered in atomic fluorescence. Such an instance exists when the source emits a non-resonance line close to the major resonance line, both of which are within the spectral band-width of the monochromator system. In atomic absorption this situation results in low sensitivity (expressed in p.p.m. for 1 % absorption) and premature curvature of the calibration graph for the element concerned. In atomic fluorescence, however, the non-absorbing radiation passes through the flame cell and only the re-emitted atomic fluorescence radiation is detected.

3.4.3.2   Physical, chemical

In this instance the reduction or enhancement in the concentration of analyte atomic concentration by both physical and chemical interferences is the same for all the three flame spectrometric techniques. Such interferences include changes in sample transport, solvent evaporation and flame shape (which influence the efficiency of nebulisation) and changes in solute vaporisation, solute dissociation, compound formation and ionisation (which influence the efficiency of atomisation). In general, the latter type result in a decrease in the measured signal.

The use of non-flame cells does not eliminate these interferences, except for those peculiar to the use of a flame, e.g. changes in sample transport, flame shape and possibly ionisation; in fact such devices may drastically modify the conventional interference effects and introduce completely new types of interferences. For example, studies in the author's laboratories have shown that solute vaporisation and dissociation using non-flame cells (such as the Massmann furnace in the form of the Perkin–Elmer H.G.A. 70 and the West filament atom reservoir[49]) is greatly dependent on the associated anion of the analyte element. In addition, effects such as excessive 'soaking' of solvent and solute into the graphite cell material (particularly in the West filament atom reservoir) and condensation of vaporised material in the cooler regions of graphite tubes (in the Perkin–Elmer H.G.A. 70) can lead to a distorted signal or the occurrence of multiple transient signals. However, as before, the effects are similar for both atomic absorption and atomic fluorescence.

3.4.3.3   *Temperature variation*

The interference effect of temperature variation is twofold because as stated above it affects both the degree of excitation and the efficiency of atomisation. Changes in the efficiency of atomisation, such as those encountered with elements that form, for example, stable monoxides in the flame gases, will affect emission absorption and fluorescence equally. However, changes in the degree of excitation will in particular affect emission measurements.

3.4.3.4   *Scattering of radiation and band absorption*

The presence of band absorption results in a reduction in source radiance owing to non-atomic absorption; however, such effects are not of great importance in either flame atomic absorption or atomic fluorescence methods. It is anticipated however that the effects of band absorption with the use of non-flame cells is likely to be more obvious. This can arise when material is volatilised in the molecular state from the hot graphite surface. An associated interference with the use of non-flame cells is the increase in background radiation during the measurement period. This will only occur when the furnace walls are not completely screened from the detector; however, with the introduction of 'non-flame attachments' for conventional flame spectrometric equipment it will be surprising if such effects are not reported frequently in the future. Undoubtedly the use of a background corrector source will become increasingly important to overcome these interferences in atomic absorption.

The scatter of incident radiation from the source by particulate matter in the flame is as much a problem in atomic absorption as in atomic fluorescence. However, little mention has been made in the past of this effect in atomic absorption, presumably because the resulting interference effect has not been large. In consequence, the effects in atomic fluorescence must also be small in spite of expectations by some early commentators. This of course assumes that similar analytically useful pre-mixed flames and pneumatic nebulisers are used in both techniques. Scatter of incident source radiation was observed originally using direct-injection burners and it may even be observed using a relatively cool pre-mixed flame when a high concentration of a matrix element, which may not be efficiently vaporised in the flame, is present in the sample solution. However, as indicated above, scatter signals may be overcome by measuring the atomic fluorescence intensity at a longer wavelength than the exciting radiation, that is using a direct-line or step-wise fluorescence processes. Such a possibility exists for many elements and provided that any incident source radiation at the longer wavelength is filtered out before entering the atom cell then atomic fluorescence may be considered to be somewhat more free from scatter interference than atomic absorption spectrometry. The measurement of a non-resonance fluorescence signal in this way also results in a much reduced tendency to deviate from linearity in calibration graphs where this is caused by self-absorption effects.

3.4.3.5 Quenching

This is always present in any dynamic medium such as a flame and it results in a reduction in fluorescence intensity. In general, it exhibits a greater effect in fluorescence than in absorption or thermal emission, but the effect is a constant factor and it presents no interference as such. Quenching arises because of the delay between the absorption and re-emission of radiation and qualitative expressions can be derived, based on a kinetic approach, to describe such effects and to predict the optimal flame conditions for atomic fluorescence spectrometry with minimal quenching. In general, the flame gases should contain a high concentration of monatomic species (such as argon). Also the flame temperature should be as low as possible compatible with efficient atomisation and the flame fuel gas should wherever possible be hydrogen rather than a hydrocarbon or carbon monoxide. In consequence a suitable flame would be expected to be a stoichiometric oxygen–hydrogen flame diluted with argon; indeed such flames have been shown to give the highest atomic fluorescence intensities in practice although a fuel rich flame is usually used for reasons of atom formation[50, 51]. Experiments in which the nitrogen in air–hydrogen or air–hydrocarbon flames has been replaced by argon show a distinct improvement in fluorescence intensity due to the much lower quenching cross-section of argon. In general, two- to four-fold improvements in sensitivity have been noted; in no instance has inter-element quenching been reported, i.e. a decrease in sensitivity of determination caused by the presence of another element in the solution.

Quenching can easily be avoided in the case of non-flame atom cells by appropriate choice of the sheathing gas, for example it has been demonstrated that argon is the most satisfactory inert gas (with respect to the sensitivity of determination) to use with the filament atom reservoir provided that resonance radiation (involving singlet–singlet transitions) is measured[49]. When the fluorescence process involves a collisional conversion of the singlet to a triplet state before the radiational step, then shielding with a gas such as nitrogen yields greater intensities, e.g. measurement of the atomic fluorescence intensity of Cd at 326.1 nm after excitation at 228.8 nm. This is because the emission at 326.1 nm arises due to atomic phosphorescence and in this instance the formation of the triplet state is more favoured with nitrogen (a diatomic molecule) as sheathing gas.

Interferences with non-flame cells will be similar to those predicted for flame cells, but are difficult to compare in detail without specifying the exact non-flame type and, perhaps even more important, the method of sample introduction.

3.5 ATOMIC FLUORESCENCE INSTRUMENTATION

The essential instrumental components for atomic fluorescence spectrometry are as follows:

(i) An intense source of radiation, particularly in the ultraviolet region of the electromagnetic spectrum.

(ii) An efficient atom reservoir with a low level of emission from background radiation.

(iii) A dispersive or filtering arrangement with high light gathering power.

(iv) A detecting/amplifying system with high gain, but with low noise.

3.5.1 The optical system

An important aspect of atomic fluorescence instrumentation is the utilisation of lenses and mirrors to maximise the amount of exciting radiation reaching the atom cell and furthermore to increase the amount of fluorescence radiation reaching the detector. Figure 3.11 shows schematically the various arrangements which have appeared in the literature for focusing radiation from source to atom cell and from atom cell to entrance slit of the dispersion/detection system. In each instance the conventional 90 degree configuration is used. The arrangement (a) is the most popular although it is about the least efficient because the sensitivity of determination depends greatly on the f number of the lenses employed and hence upon the size of the lenses. A superior arrangement would seem to be that depicted in (b) which utilises mirrors rather than lenses. Mirrors are superior to lenses in several respects, in particular they have a greater light collecting power. However, it must be appreciated that the use of such mirrors will firstly alter the shape of atomic fluorescence calibration curves depending of whether the measurement involves resonance or direct-line fluorescence signals etc.[51]. In general, the analytical curves obtained with the source mirror alone have initially a greater ordinate (than in its absence), but with a gradient less than unity.

This is found again using the field mirror (i.e. the mirror directly opposite the entrance slit), but it is more marked for resonance than for direct-line atomic fluorescence measurements. Furthermore, the use of these mirrors will influence the limit of detection to a degree which varies depending on the level of the background radiation from the atom cell. The use of both mirrors is likely to be advantageous to the limit of detection, but unless particular care is exercised in adjusting the field mirror its usefulness will be nullified

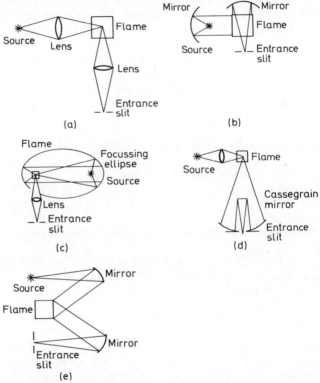

Figure 3.11 Schematic diagram of entrance optics used in atomic fluorescence spectrometry. (a) All lens system—right-angle illumination—measurement. (b) All mirror or mirror-lens combination—right-angle illumination—measurement. (c) Elliptical mirror system—right-angle illumination—measurement. (d) Systems with Cassegranian mirror—right-angle illumination—measurement. (e) All mirror systems—front-surface illumination—measurement
(From Winefordner, J. D. and Elser, R. C. (1971). *Analyt. Chem.* **43**, 24A, by courtesy of The American Chemical Society)

by the increased background radiation reaching the detector. Arrangements (c) and (e) are less common, more complex and have little advantage over the above systems. In general, (c) provides an efficient means for collecting source radiation although care must be taken to avoid scattered radiation. In arrangement (e) scattered source radiation is less likely.

Arrangement (d) is the most expensive and complex system depicted, but at the same time it has the advantage of being capable of collecting a large, solid angle of atomic fluorescence radiation and is of considerable use with a non-dispersive detection system. This system is dealt with in more detail later in the chapter.

3.5.2 The source of excitation

The intensity of emitted fluorescence radiation is proportional to the source intensity over the profile of the absorption line (see Section 3.2) and therefore sources of high-intensity yield the highest analytical sensitivity. The essential source requirements are as follows:

(i) The source must be intense and free of self-reversal over the absorption profile of the analyte atoms. The radiation at the line centre is most useful for producing atomic fluorescence. Radiation in the wings of the line contributes towards background scatter and noise. Hence continuum sources at the present state of development are of little real value in atomic fluorescence spectrometry.

(ii) The source fluctuations (long-term drift and short-term flicker) should be minimal.

(iii) The source should be adaptable to a wide variety of elements and require a minimum of preparation (if it is necessary to construct sources of rare elements or radioactive isotopes, then this could be an important consideration).

(iv) The lifetime of the source should be long.

(v) The manufacturing cost should be low because many individual line sources are necessary.

(vi) The source should be simple and safe in operation and be adaptable to a variety of methods of operation (e.g. pulsed or continuous).

Unfortunately, at the present time, no single source meets all (or even a majority) of these requirements and in consequence many source types have been used to evaluate the atomic fluorescence process.

3.5.2.1 Metal-vapour arc lamps

Metal-vapour arc lamps were among the first source type to be utilised in atomic fluorescence flame spectrometry[53], and they have been used since by several workers[54-59]. They consist of a sealed tube containing the metal and an inert gas at a low pressure (usually 1–10 Torr) with an oxide-coated filament at each end of the tube. The tube is surrounded by an evacuated jacket so that a sufficiently high vapour pressure of the metal is maintained. The lamp is initiated by sending a current of a few amperes through both the filaments until they are red hot; then the filament current is terminated and an alternating voltage is applied across the two filaments, producing an arc which keeps the filaments hot and the metal partially vaporised.

Commercial metal-vapour arc lamps have been used extensively for the atomic fluorescence of zinc, cadmium and thallium. Although these sources are available for other metals, only these three are useful analytically. All metal-vapour arc lamps, especially those of mercury and the alkali metals, suffer from self-reversal and resonance broadening[53-55] and hence their use in atomic fluorescence spectrometry is somewhat limited. In addition, some lamps, e.g. indium and gallium lamps, emit extremely weak line radiation and are therefore not very useful[55].

In operation, the lamps must be carefully thermostatted in order to preserve stability and to minimise self-reversal. Mansfield *et al.*[55] achieved this by optimising and monitoring the operating current. Dagnall *et al.*[56] effected the same by directing a controlled air stream upon the lamp.

3.5.2.2   *Hollow cathode discharge lamps (HCLs)*

The present HCLs are second generation developments of the Paschen lamp[60] and the Schüler tube[61, 62], although most of the HCLs used today are modifications of the design by Walsh and co-workers[63, 64]. Hollow cathode lamps have been constructed for virtually every element and are available commercially. They are used almost exclusively in atomic absorption spectrometry because of the narrowness of the emitted spectral lines. One disadvantage is the cost of these sources.

Dinnin[65], Dinnin and Heltz[66, 67], and Rossi and Omenetto[68] have used demountable HCLs for atomic fluorescence studies. While these lamps are somewhat elaborate in construction, they have the desirable feature that the cathodic material is interchangeable. Thus, one lamp lends itself to several elements. Using the demountable HCL, reasonably low limits of detection have been obtained for several elements.

Hollow cathode lamps are generally operated at low currents, *c.* 10 mA, and under such conditions are stable and long-lived, but they usually lack the intensities necessary for atomic fluorescence applications[69]. At higher currents, very little useful line intensity is gained, because metal is sputtered from the cathode rapidly producing self-reversal, resonance broadening and instability of the resonance lines as well as window blackening and cathodic deterioration.

Sullivan and Walsh[70] have overcome some of these problems with the so-called 'high intensity' HCL. These are similar to the conventional HCL, except that an auxiliary pair of electrodes are mounted in front of the hollow cathode. The atomic vapour sputtered from the primary cathode is further excited in the positive column of the auxiliary electrode discharge. An additional power supply is required for the added electrodes, making lamp operation less straightforward and more expensive.

The recently developed shielded HCL[71, 72] now manufactured by Westinghouse and Perkin–Elmer gives a similar intensity to the high intensity HCL[73] while not requiring a secondary power supply. Sources of this type have been used by several workers[74-81] for the study of atomic fluorescence processes.

3.5.2.3   *Continua*

The xenon arc is one of the most intense of continuum sources of ultraviolet radiation. The emission from such a source is approximately continuous because of resonance and Doppler broadening of the emitted spectral lines. However, because of the continuum nature of the emission, the intensity per unit frequency interval is relatively low. In addition, the energy available in the ultraviolet region (and especially at wavelengths less than 300 nm) drops off rapidly. Because of the theoretical and practical superiority of continuum sources in atomic fluorescence, several workers have evaluated the analytical application of the xenon arc[82-85].

In general, the limits of detection obtained are about an order of magnitude down on those obtained using an intense line source, and furthermore the interference effects (spectral) are somewhat worse[101]. Also the effects of scattered source radiation are more serious with a continuum than a line source.

3.5.2.4 Electrodeless discharge lamps (EDLs)

The most intense source presently available for atomic fluorescence spectrometry is the electrodeless discharge lamp. From the standpoint of simplicity and cost, EDLs would appear to be the ideal source. They consist of a sealed-off reservoir of glass, or more usually quartz, containing an inert gas (a few Torr) and a small amount of a metal salt or the metal whose spectrum is desired. EDLs are operated at RF or microwave (usually 2450 MHz) frequencies, the power being coupled by suitable antennae or wave-guide cavities. The absence of any internal electrodes leads to long lifetimes as no electrode burn-away occurs and there are no quartz to metal seals which are prone to gas leaks. The real potential of EDLs as primary sources for atomic absorption and fluorescence was realised by Dagnall et al.[86-91] and Winefordner and co-workers[92, 93]; they have described the preparation of EDLs for several elements.

Radiofrequency and microwave excited EDLs have been used for many years dating back to the work of Jackson[94], who used an EDL operated from a 10 MHz oscillator to obtain the arc spectrum of caesium in order to study its hyperfine structure. EDLs have been used since then in numerous spectroscopic applications, although most of the earlier applications did not require either high intensity or stability. A renewed interest in EDLs was kindled in the late 1940s when Meggers[95-97] carried out his wavelength studies of mercury-198 in search of an ultimate standard of length. By the early 1960s, EDLs had been constructed for nearly every chemical element, and microwave cavities for the operation of these lamps had undergone considerable development[98, 99].

Until recently there has been little interest in EDLs as sources for atomic absorption and atomic fluorescence spectrometry where stability and intensity are of paramount importance. Ivanov et al.[100] reported the use of EDLs as sources in atomic absorption and Goodfellow[58] carried out some interference studies for atomic fluorescence using microwave excited EDLs. Winefordner and co-workers[53, 55] used commercially available microwave excited EDLs for atomic fluorescence studies; their initial results were rather poor, probably because these lamps were designed for interferometry and wavelength calibration where high intensity and stability are not essential.

Since 1966, Winefordner and co-workers[92, 93, 102] and Dagnall, West and co-workers[52, 86–91, 103, 104] have examined the effects of various parameters on the operation and use of EDLs as sources for atomic fluorescence flame spectrometry.

The most recent studies in this area[105–108] should be consulted for a more detailed account of the method of preparation. Techniques of electronic modulation have also been dealt with by Dagnall and co-workers[108, 109, 110].

At the present time, EDLs can be made to operate satisfactorily for atomic fluorescence purposes, but they cannot be considered to be suitable for routine application. The sources require to be tuned into the microwave field to achieve optimum performance and the nature of the lamp fill material, the method of preparation, microwave power, lamp temperature, etc. are all factors which are likely to affect the intensity and stability of the resonance radiation.

Undoubtedly this type of source is worthy of a more detailed examination for use in atomic absorption and fluorescence spectrometry and especially because it is capable of giving intense resonance radiation for highly volatile elements, e.g. As, Se, Te, Sb, Zn, Cd, S, P, halogens etc. The intensity of the radiation emitted by an EDL can be judged from the use of an iodine EDL to excite ground state bismuth atoms[89]. In this instance, there is sufficient iodine line intensity at 206.63 nm and overlap with the bismuth absorption line at 206.170 nm to give an adequate detection limit by atomic fluorescence spectrometry. Precise methods of lamp preparation and operation in well thermostatted cavities would certainly give considerable improvements in lamp stability. However, EDLs should not be regarded as a general source of excitation for all elements; in our experience it is difficult to produce very reliable sources for relatively involatile elements (or involatile compounds of these elements).

3.5.2.5 Other sources of excitation

The principal aim in atomic fluorescence is to obtain a source exhibiting intense resonance line radiation. In this respect the use of pulsing techniques would seem obvious because sources which operate continuously at some arbitrary radiance can provide a greater peak radiance by pulsing them at higher current levels than is obtained by continuous operation. The only limitation is that the average pulsing current be about the same as the average continuous operating current to obtain long lamp-life. This method of source operation of course imposes some restrictions on the detector system, but it is possible to achieve low noise levels.

Although it is possible to pulse-operate most types of source, the only application in atomic fluorescence at the present time is that due to Fraser and Winefordner[111]. In this work the atomic fluorescence of several elements is produced using a tunable dye laser pumped with a N_2 laser. Unfortunately the emission from the laser was such that only elements with principal resonance lines above c. 350 nm could be investigated. However, this work shows that the use of a laser is possible for this purpose and, more important, the results indicate that atomic fluorescence is likely to become a major analytical technique with the continuing development of laser systems. Such possibilities must surely encourage further evaluation and interest in atomic fluorescence spectrometry.

3.5.3 Flame and non-flame atomisers

Until recently the most widely used method of atomisation has been that of the pre-mixed flame. Flames have several advantages:

(a) They are convenient, reliable and relatively free from memory effects.

(b) They are relatively inexpensive and can be interfaced with quite simple nebuliser systems for both liquid and powder samples.

(c) A variety of flame types is available to allow selection of optimum conditions for different analytical problems, although only the nitrous oxide–acetylene flame can be regarded as being a particularly useful analytical flame.

(d) In most instances, the flames in common use allow adequate analytical sensitivity and precision to be obtained.

For atomic fluorescence the atomiser requirements may be summarised as follows: (i) good atomiser efficiency, (ii) low background radiation and background flicker; (iii) low concentration of quenching species; (iv) long residence time of analyte atoms; (v) simple operation and inexpensive.

Requirements (ii) and (iii) mean that somewhat different flame types are preferred in atomic fluorescence compared with atomic absorption spectrometry. In early studies non-hydrocarbon flames were used to eliminate the presence of high background radiation and quenching species (e.g. CO and CO_2). The most popular flame became air-hydrogen or variations such as nitrogen– or argon–hydrogen–entrained air and argon–hydrogen–oxygen flames burning on both direct injection burners and pre-mixed burner systems.

The use of direct injection burners, although popular in initial studies of atomic fluorescence, is now only considered to be of use in certain instances. They frequently give rise to low limits of detection[112, 113] as a result of the high solution up-take rate, but they are prone to severe chemical and physical interference effects. Furthermore, source scatter problems arise because the drop size is somewhat large.

Laminar-flow pre-mixed flames have been used extensively in atomic fluorescence studies in conjunction with almost every conceivable flame composition. In general, the cooler flames, e.g. air–hydrogen, hydrogen diffusion flames, air–propane etc. result in the lowest detection limits, but at the expense of considerable chemical and physical interferences. Higher temperature flames such as the air–acetylene and nitrous oxide–acetylene flames exhibit a high background radiation, particularly above c. 250 nm, and in conventional form they have been avoided. An important contribution in this respect has been the work of Kirkbright and co-workers on the use of sheathed or separated flames.

Teclu[114] and Smithells and Ingle[115] showed that the outer mantle of a flame could be separated from the primary cone by preventing access of atmospheric oxygen at the base. In this way the chemiluminescence produced in the outer mantle of a pre-mixed hydrocarbon flame by reactions of the type shown in equation (3.32) is removed from the field of view of the

$$CO + \tfrac{1}{2}O_2 \longrightarrow CO_2^* \longrightarrow CO_2 + h\nu \qquad (3.32)$$

detector system. In general, atomic fluorescence signals originate mainly from the interconal region in the flame centre just above the primary combustion zone. Hence, removal of excess background radiation results in significant improvements in signal/noise ratio and hence in limits of detection. The effect is noticeable at all wavelengths of measurement and especially over those regions of normal high flame background, e.g. OH region around 320 nm or the many CN band emission regions in the nitrous oxide–acetylene flame. (In the latter flame the reducing zone of the flame may also be increased in size.) Separation may be effected using a silica separator tube[116] or by blowing a stream of nitrogen or argon concentrically around the top of the burner stem[117–119].

Although flames are convenient in practice, they are not ideal atom reservoirs and especially so for fluorescence purposes. The flame background absorption and emission at the wavelength of measurement, or thermal emission from analyte or concomitant elements at this wavelength, often gives a high noise level which is unacceptable in atomic fluorescence spectrometry. Furthermore, the volume of sample solution available for analysis may be less than that required for conventional flame nebulisation and the analyte concentration may be too low to allow further dilution. Also, the sample for analysis may be solid and the presentation of such materials without dissolution is difficult using flames. More important, in many applications, e.g. analysis of radioactive materials, the use of flames is not permitted without excessive precautions. Other theoretical factors include the poor degree of atomisation for many elements in all flame types regardless of oxidant:fuel ratio, the relatively high concentration of quenching molecules in flames of greatest analytical utility and the dilution effect on the attainable atom concentration by the relatively high flow-rate of the flame gases. Vaporisation of the analyte solution and flame gas expansion (c. tenfold) also contribute to dilution of the atomic cloud. In consequence, many workers have devoted efforts towards the development of alternative methods of atomisation. At the present time the most successful of these are the various forms of electrically heated furnaces and filaments.

In general, the special advantages of these systems are as follows:

(i) An increased atomic concentration because the atomic vapour is contained in a smaller volume and little gas expansion occurs.

(ii) There is a decreased quenching of radiationally excited atoms because most non-flame systems are purged with an inert gas.

(iii) The noise contribution from a non-flame system, provided that it is properly screened from the detector, is much less than that associated with an analytically useful flame. In consequence, the limit of detection attainable is usually much lower with a non-flame cell.

(iv) Most non-flame systems can be used in the vacuum ultraviolet region of the spectrum.

Massmann[120] first demonstrated the usefulness of graphite cells for atomic fluorescence spectrometry (Figure 3.12). With a heated graphite cell (temperature of 2873 K) in an argon atmosphere he was able to detect down to 4×10^{-14} g of zinc; cadmium, silver, antimony, iron, thallium, lead, magnesium and copper were detected also although high intensity sources were not available for all elements. The sample size was between 30–100 µl with relative standard deviations between 12–4% respectively. As stated above, the transient nature of the detected signals means that either peak height or peak area may be measured; the latter measurement was found to be essentially independent of the matrix composition. Considerable

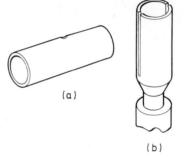

Figure 3.12 Graphite cell for (a) atomic absorption and (b) atomic fluorescence spectroscopy
(From Massmann[120], by courtesy of Microfilms International Marketing Corporation)

(a)

(b)

effort has been devoted recently to this type of furnace for atomic absorption spectrometry and a device similar to that originally described by Massmann[121] has been introduced as an accessory to an atomic absorption spectrometer[122].

West and Williams[123] more recently have reported the construction and use of a device known as the filament atom reservoir for use in both atomic absorption and fluorescence spectrometry. The solid carbon filament was originally within a chamber having quartz windows and the sample solution (c. 5 µl) was introduced into a depression in the filament. The chamber was purged with argon and the filament then heated electrically. Atomic fluorescence detection limits were reported for magnesium and silver of 10^{-15} and 3×10^{-11} g respectively using high intensity hollow cathode lamps. The device was simple and safe in operation although the precision was relatively poor. An improved design (Figure 3.13) eliminated the need for an enclosed chamber and considerably improved precision[49].

The effective lifetime of the free atoms produced by the filament atom reservoir is short compared with that of most furnace systems. The decay of the atomic population immediately above the hot filament is promoted by condensation of analyte atoms with like atoms or with atoms and molecules formed from concomitant elements present in the sample. The atomic vapour must be viewed therefore close to the surface of the filament which in turn gives rise to a very rapid transient signal as compared with the Massmann furnace techniques. There are many chemical and physical interference effects associated with both types of cell, but these depend to a large extent on the geometric design of the device as well as the matrix and analyte element.

A modified form of the filament atom reservoir[124] is available commercially as an accessory for atomic absorption purposes and consists of a hole drilled through the filament parallel to the radiation path. The sample solution

(0.5–1.0 µl) is introduced into the hole and sheathing is accomplished as before or with a hydrogen diffusion flame in place of the nitrogen or argon. These modifications appear to result in improved atomisation and they tend to minimise inter-element effects. It is worthy of note that this device is known as the 'Mini-Massmann' furnace.

Other non-flame cells used for atomic fluorescence purposes include the platinum furnace with continuous nebulisation[125] and the heated platinum

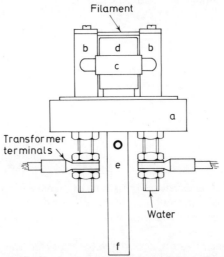

Figure 3.13 Filament atom reservoir modified for use in open atmosphere. (a) base, (b) water-cooled electrodes, (c) water link between electrodes, (d) laminar flow box, (e) inlet for shield gas, (f) support stem for reservoir
(From Alder and West[49], by courtesy of Elsevier)

loop atomiser[126, 127]. The former system is still in the process of development, but is meant to combine the advantages of continuous nebulisation and non-flame atomisation. The latter device is the most simple system to date, but is less precise than most of the other non-flame cells. The sample (down to 0.1 µl) is applied to the loop with a syringe or by dipping the loop into the analyte solution. As before the loop can be sheathed with an inert gas and atomisation is effected via electrical heating. Detection in this instance has been with d.c. amplification or preferably by photon counting systems because of the short atom residence time.

Table 3.3 summaries the limits of detection obtained using the various non-flame cells described above. A more detailed account of non-flame cells for use in atomic absorption and fluorescence spectrometry can be found in a review by Kirkbright[128].

3.5.4 Dispersion systems

Atomic fluorescence spectrometry is very closely related to atomic emission and atomic absorption spectrometry and it is not surprising therefore that almost all knowledge of the technique has been obtained using instruments primarily designed for emission and/or absorption measurements. In many instances such instruments have been unsuitable for atomic fluorescence studies, especially with respect to the small solid angle of measurement and the attainable limit of detection. On the other hand such instruments, which have generally employed the use of prism or grating monochromators, have restricted the amount of radiation received from the flame cell and hence have limited undesirable noise at the detector. This has been an important consideration even with apparently low background flames such as air–hydrogen or hydrogen diffusion flames which emit radiation at wavelengths greater than about 300 nm (e.g. OH emission). However, the recent introduction of solar-blind photomultipliers has resulted in atomic fluorescence studies using much less complex systems and, in fact, elimination of the monochromator altogether. The successful operation of this type of system depends on the fact that most of the radiation emitted by flames lies outside the sensitivity of the response curve of the photomultiplier (Figure 3.14) and consequently it produces little or no noise at the detector. In most instances radiation is allowed to fall directly on the solar-blind photomultiplier, although a multi-element analysis system, or alternatively the use of a con-

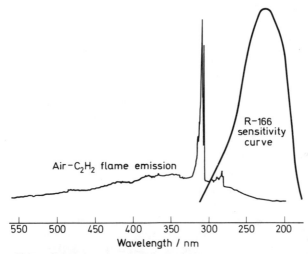

Figure 3.14 Spectral sensitivity curve of caesium telluride solar-blind photomultiplier R-166
(From Dagnall, R. M. and Kirbright, G. F. (1970). *Atomic Absorption Spectroscopy*, by courtesy of Butterworths)

tinuous source of excitation, would benefit from the use of narrow band-pass filters in conjunction with the detector. Another special advantage of the non-dispersive type of system is that it allows collection of atomic fluorescence radiation over a greater solid angle than is possible with a monochromator. It also is able to record the total atomic fluorescence signal resulting from several transitions providing that they fall within the sensitivity curve of the detector, thus enhancing the sensitivity of the method. This is quite different from atomic absorption analysis where maximum sensitivity can only be obtained by isolating the line possessing maximum oscillator strength.

Preliminary studies using solar-blind detectors were reported by Larkins *et al.*[129] for iron and by Vickers and Vaught[130] for cadmium, mercury and zinc. Since then the most complete studies have been due to Larkins[131], Larkins and Willis[132] and Mitchell and Johansson[133]. The work of Larkins involved the use of a HTV type R166 (Hamamatsu TV Company, Japan) solar-blind photomultiplier in conjunction with hollow cathode lamp excitation and a separated air acetylene flame (see Figure 3.15). The elements

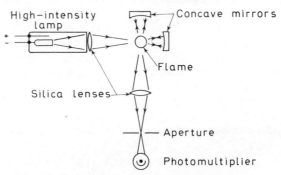

Figure 3.15 Schematic diagram of the optical system of a
solar-blind photomultiplier
(From Larkins[131], by courtesy of Pergamon Press)

studied were Au, Bi, Cd, Co, Fe, Ga, Hg, In, Ir, Mg, Mn, Pb, Pd, Pt, Sb, Se, Sn, Te, Tl and Zn. For Au, Cd, Hg, Mg, Ni, Sb and Zn the limits of detection obtained were significantly better than those obtained using a conventional atomic absorption spectrophotometer (Techtron AA–4). The application of more intense sources of radiation (e.g. vapour discharge lamps or high intensity hollow cathode lamps) gave even better limits of detection (from 6–400 fold) and in this instance even better than some of those quoted by Slavin[134] for atomic absorption. Certain elements possess major resonance lines or have their strongest fluorescence lines outside the range of the R166 photomultiplier. Some of these, e.g. Ag, Cr and Cu, were determined by replacing the R166 tube with one sensitive in the 300–400 nm region and a simple absorption filter. The use of a nitrogen sheathed nitrous oxide-acetylene flame[132] extended the range of elements to include Be, Mg, Al, Ge, Ti, Pd and Pt.

The special advantages of this type of system compared with that utilising a monochromator for atomic fluorescence purposes are as follows: (i) superior limits of detection for many elements; (ii) simple, compact, robust, readily aligned instrumentation free from wavelength drift problems; (iii) low overall cost of instrument; (iv) no adjustment of detector system required when changing from one analyte element to another.

The disadvantages include a restricted range of application (i.e. not all elements had their strongest fluorescence lines within the sensitivity range of the solar-blind photomultiplier) and the possibility of more spectral interferences than normally expected using a monochromator. In this connection the line source must give a 'pure' spectrum and not give a non-resonance line which lies too close to the resonance line of an extraneous element and so cause it to fluoresce. Larkins[131] observed this in one case, namely the excitation of iron (at 271.902 nm) by a platinum high intensity hollow cathode lamp radiating at 271.904 nm. Finally, the ratio of scatter to fluorescence was found to be higher than when using a monochromator, although again it was not a significant factor in this work.

The simplicity of the instrumentation used in the above studies has been developed further by Mitchell *et al.*[133, 135–137] and by Walsh[138] for the simultaneous analysis of several elements. Walsh used high intensity hollow cathode lamps grouped around a single solar-blind photomultiplier. The boosting discharges of the various light sources were switched on in rapid sequence, and by using a time-sharing amplification system the atomic fluorescence signals were recorded with one detection system.

The instrument suggested by Mitchell and Johansson[133] employs sequentially pulsed hollow cathode lamp sources and optical interference filters rotating in front of a single ultraviolet-visible response photomultiplier. This arrangement is the basis of the recently introduced Technicon AFS-6 atomic fluorescence instrument (Figure 3.16). This instrument utilises the Cassegrain mirror system described previously and is capable of measuring atomic fluorescence or thermally excited emission of up to six analyte elements via a multiplexed signal processing system with a digital read out. Detection limits, ranges of linearity and precision data have been presented[136] for 16 elements using both air and nitrous oxide–acetylene flames and it is concluded that the results are quite satisfactory for analytical purposes. In addition, the large dynamic range available minimised the need for

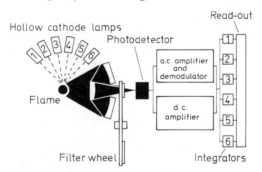

Figure 3.16 Six channel instrument for simultaneous multi-element analysis for a Technicon AFS-6 atomic fluorescence instrument
(From (1971). *Advances in Automated Analysis*, Technicon Int. Congress, Vol. II, p. 508, by courtesy of Thurman Associates, Miami)

multiple sample dilutions and the use of automated gain and zero settings and automated calibration procedures permitted full use of the instrument's multi-element analytical capability, without introducing additional complexity.

The most severe disadvantage of such multi-element instruments is the inability of low background emitting flames to provide a useful atomic reservoir for all elements or to eliminate chemical and physical interference effects encountered in real analyses. Frequently two or more separate runs are required in order to achieve maximum analytical sensitivity, one for readily atomised elements using an air-acetylene flame and another using a separated nitrous oxide-acetylene flame for difficultly atomised elements or for the elimination of a particular type of interference. However, the use of suppressing or releasing agents with the cooler flames and the relatively uncritical dependence of the atomic fluorescence intensity with height of observation for many elements makes the atomic fluorescence technique sufficiently adaptable for many multi-element applications.

3.6 APPLICATIONS

The volume of applications literature available for atomic fluorescence spectrometry is very small (at the time of writing) compared with that for atomic absorption spectrometry. The potential range of the technique is, however, at least equal to that of atomic absorption and the inherent sensitivity available may mean that direct trace analysis in the future can be accomplished without recourse to pre-concentration procedures such as solvent extraction or co-precipitation. As a consequence of the large dynamic range of atomic fluorescence (provided a sufficiently low limit of detection is obtained), the need for sample dilution may also be minimised or made more uniform for a wide range of analyte elements.

The applications in the scientific literature demonstrate the high sensitivity and selectivity of the technique and also provide evidence that scatter of source radiation is no more troublesome in atomic fluorescence than in atomic absorption. Table 3.4 summarises the determination of various ele-

Table 3.4   Some applications of atomic fluorescence spectrometry

Analyte element(s)	Matrix	Ref.
Ag, Cu, Fe, Mg, Ni, Pb	Jet lubricating oils	139
Cu, Cr, Ni, Fe, Ag, Mg	Lubricating oils	137
Ni	Gas oils	140
Cu, Fe, Pb	Hydrocarbon fuels	141
Ca, Cu, Mg, Mn, K Zn	Soil extracts	142
As, Bi, Pb, Se, Te	Nickel-base alloys	143
Pb	Steels	143
Pb	Blood, urine	145
Bi	Aluminium alloys	146
Si	Steels	147

ments in different matrices. The work of Smith, Stafford and Winefordner[139], Matousek and Sychra[140], Cotton and Jenkins[141], Mitchell, Demers et al.[137] and Dagnall, West et al.[142, 143] illustrates the analytical versatility of the technique for the direct determination of several analyte elements. The use of solvent extraction as a pre-concentration technique has been demonstrated for the determination of several low-melting-point elements in nickel-base alloys and steels[144]. The only biological estimation to date appears to be that due to Amos *et al.*[145] for the determination of lead in blood and urine. Two applications which demonstrate the freedom from interference problems due to scattered radiation are the determination of bismuth in aluminium alloys[146] and the determination of silicon in steels[147]. In the former study the non-resonance fluorescence signal at 302.46 nm was utilised, after excitation at 206.17 nm with an iodine electrodeless discharge lamp; in the latter instance a sheathed fuel-rich nitrous oxide–acetylene flame was used which efficiently atomised the matrix element and, therefore, eliminated any interference due to scatter.

The most recent application studies in atomic fluorescence spectrometry have been the simultaneous determination of several elements using prototype multi-channel instruments[137, 142, 144].

3.7 CONCLUSION

Atomic fluorescence flame spectrometry is a sensitive and selective technique which is complementary to other flame spectrometric methods. It may be expected that instrumentation specifically designed for atomic fluorescence purposes will provide extremely high sensitivity for trace analysis in a variety of matrices. However, atomic fluorescence is a relatively new technique and there are still necessary developments before it is likely to be regarded as a routine method or preferable to atomic absorption or atomic emission spectrometry. The primary advantage of atomic fluorescence is the inherent sensitivity, but in the absence of an intense, stable, source of excitation for all elements this cannot be realised. However, the current advances being made in laser technology and source pulsing techniques suggests that this aspect may not be too far off in reality. Similarly, the very recent interest in non-flame methods of atomisation for atomic absorption spectrometry has resulted in the development of highly efficient, low radiative background atomisers. Modification of these devices for atomic fluorescence purposes would again give considerable benefit with respect to sensitivity and freedom from many chemical and physical interference effects. The advantages of such developments are obvious in atomic absorption, but most benefit will undoubtedly occur in atomic fluorescence spectrometry.

The multi-channel aspects of atomic fluorescence spectrometry are also likely to receive considerably more attention in the future now that the possibilities of such instruments have been so clearly demonstrated. Furthermore, the range of applications using very simple single-channel instruments is likely to expand to such an extent that it is possible to envisage the introduction of inexpensive, portable instruments for the determination of a restricted range of elements.

Hopefully this chapter will encourage research and development in these areas and at the same time familiarise the reader with the technique of atomic fluorescence spectrometry, its potential and its advantages as a trace analytical technique.

References

1. Price, W. J. and Browning, D. R. (editors). (1969). *Spectroscopy*. (New York: McGraw Hill)
2. Ellis, D. W. and Demers, D. R. (1968). *Advan. Chem. Ser.*, **73**, 326
3. Smith, R. and Winefordner, J. D. (editors). *Spectrochemical Methods of Analysis*, in press. (New York: John Wiley)
4. Winefordner, J. D., Smith, R. and Mavrodineanu, (editors). *Analytical Flame Spectrometry Selected Topics*, in the press. (Eindhoven, Netherlands: Centrex)
5. Wood, R. W. (1904). *Proc. Amer. Acad.*, **X1**, 396
6. Bogros, A. (1926). *Compt. Rend.*, **183**, 124
7. Boeckner, C. (1930). *Bureau of Stand. J. Res.*, **5**, 13
8. Wood, R. W. (1913 and 1919). *Researches in Physical Optics*, I and II. (New York: Columbia University Press)
9. Terenin, A. (1925). *Z. Phys.*, **31**, 26
10. Ponomarev, N. and Terenin, A. (1926). *Z. Phys.*, **37**, 95
11. Fridrichson, J. (1930). *Z. Phys.*, **64**, 43
12. Ellett, A. (1925). *J. Opt. Soc. Amer.*, **10**, 427
13. Schüler, H. (1926). *Z. Phys.*, **35**, 323
14. Mitchell, A. C. G. and Zemansky, M. W. (1961). *Resonance Radiation and Excited Atoms*. (Cambridge: Cambridge University Press)
15. Nichols, E. L. and Howes, H. L. (1923). *Phys. Rev.*, **22**, 425
16. Badger, R. M. (1929). *Z. Phys.*, **55**, 56
17. Mankopff, R. (1933). *Verhandl. Deutsch. Phys. Ges.*, **14**, 16
18. Robinson, J. W. (1961). *Anal. Chim. Acta*, **24**, 254
19. Alkemade, C. Th. J. (1963). *Proc. 10th Colloquium Spectroscopium Internationale*. (Washington, D.C.: Spartan Books)
20. Winefordner, J. D. and Vickers, T. J. (1964). *Anal. Chem.*, **36**, 161
21. Winefordner, J. D. and Staab, R. A. (1964). *Anal. Chem.*, **36**, 165
22. Winefordner, J. D. and Staab. (1964). *Anal. Chem.*, **36**, 1367
23. Winefordner, J. D., Mansfield, J. M. and Veillon, C. (1965). *Anal. Chem.*, **37**, 1049
24. Dagnall, R. M., West, T. S. and Young, P. (1966). *Talanta*, **13**, 803
25. Weisskopf, V. (1933). *Phys. Z.*, **34**, 1
26. L'Vov and Plyushch, G. V. (Minsk, 1963). *Proceedings of the 15th Conference on Spectroscopy*. **2**, 159. (AN SSSR, Moscow, 1964)
27. Voigt, W. (1912). *Munch. Ber.*, 603
28. Davies, J. T. and Vaughan, J. M. (1963). *Astrophys. J.*, **137**, 1302
29. Orthmann, W. and Pringsheim, P. (1927). *Z. Phys.*, **43**, 9
30. Milne, E. A. (1930). *Thermodynamics of the Stars Handbuck der Astrophysik*, Vol. 3 (Berlin: Springer)
31. Unsold, A. (1955). *Phyzik der Sternatmosphere*. (Springer)
32. Hooymayers, H. P. (1968). *Spectrochim. Acta*, **23B**, 567
33. Van Trigt, C., Hollander, T. J. and Alkemade, C. Th. J. (1965). *J. Quant. Spectro Radiative Transfer*, **5**, 813
34. Chester, J. E., Dagnall, R. M. and Taylor, M. R. G. (1970). *Anal. Chim. Acta*, **51**, 95
35. L'Vov, B. (1970). *Atomic Absorption Spectrochemical Analysis*. (London: Adam Hilger)
36. Kreysig, E. (1967). *Advanced Engineering Mathematics*. (New York: John Wiley)
37. Seymour, S. and Jones, J. J. (1967). *Modern Communication Principles*, 17. (London: McGraw-Hill)
38. Dagnall, R. M., Sharp, B. L. and West, T. S. (1971). *Nature (London)*, **234**, 69
39. Aldous, K. M., Dagnall, R. M., Sharp, B. L. and West, T. S. (1971). *Anal. Chim. Acta*, **54**, 233
40. Alger, D. A., Dagnall, R. M., Sharp, B. L. and West, T. S. (1971). *Anal. Chim. Acta*, **57**, 1
41. Cooke, D., Dagnall, R. M., Sharp, B. L. and West, T. S. (1971). *Spectrosc. Letters.*, **4**, 91
42. Dagnall, R. M., Sharp, B. L. and West, T. S. (1972). *Nature (London)*, **235**, 56, 65
43. Walsh, A. (1969). *Proc. Int. Atomic Absorption Spectroscopy Conf.*, Sheffield, England
44. Walsh, A. (1958). *U.S. Patent*, 2, 847, 899
45. Mavrodineanu, R. and Hughes, R. C. (1968). *Applied Optics*, **7**, 1281
46. Mitchell, D. G. and Johannson, A. (1970). *Spectrochim. Acta*, **25B**, 175

47. Mitchell, D. G. and Johannson, A. (1971). *Spectrochim. Acta*, **26B**, 11, 677
48. Winefordner, J. D., Svoboda, V. and Cline, L. J. (1970). *CRC Crit. Rev. Anal. Chem.*, **1**, 233
49. Alder, J. F. and West, T. S. (1970). *Anal. Chim. Acta*, **51**, 365
50. Browner, R. F., Dagnall, R. M. and West, T. S. (1969). *Anal. Chim. Acta*, **46**, 207
51. Browner, R. F., Dagnall, R. M. and West, T. S. (1970). *Anal. Chim. Acta*, **50**, 375
52. Aldous, K. M., Dagnall, R. M. and West, T. S. (1969). *Analyst*. **94**, 347
53. Winefordner, J. D. and Staab, R. A. (1964). *Anal. Chem.*, **36**, 165
54. Winefordner, J. D. and Staab, R. A. (1964). *Anal. Chem.*, **36**, 1367
55. Mansfield, J. M., Winefordner, J. D. and Veillon, C. (1965). *Anal. Chem.*, **37**, 1049
56. Dagnall, R. M., West, T. S. and Young, P. (1966). *Talanta*, **13**, 803
57. Prugger, H., *Zeiss Information* No. 56, p.54. (1965). (work by R. Klaus)
58. Goodfellow, G. I. (1966). *Anal. Chim. Acta*, **36**, 132
59. Dagnall, R. M., Thompson, K. C. and West, T. S. (1966). *Anal. Chim. Acta*, **36**, 269
60. Paschen, F. (1916). *Ann. Physik*, **50**, 901
61. Schüler, H. (1926). *Z. Phys.*, **35**, 323
62. Schüler, H. (1930). *Z. Phys.*, **59**, 149
63. Russel, B. J., Shelton, J. P. and Walsh, A. (1957). *Spectrochim. Acta*, **8**, 317
64. Jones, W. G. and Walsh, A. (1960). *Spectrochim. Acta*, **16**, 249
65. Dinnin, J. I. (1967). *Anal. Chem.*, **39**, 1491
66. Dinnin, J. I. and Heltz, A. W. (1967). *Pittsburgh Conference on Analytical Chemistry and Applied Spectroscopy*, Pittsburgh, Pa., March, 1967
67. Dinnin, J. I. and Heltz, A. W. (1968). *Anal. Chem.*, **40**, 481
68. Rossi, G. and Omenetto, N. (1969). *Talanta*, **16**, 263
69. Armentrout, D. N. (1966). *Anal. Chem.*, **38**, 1235
70. Sullivan, J. V. and Walsh, A. (1965). *Spectrochim. Acta*, **21**, 721
71. Manning, D. C. and Vollmer, J. (1967). *Atom Abs. Newsletter*, **6**, 38
72. Vollmer, J. (1966). *Atom Abs. Newsletter*, **5**, 35
73. Gillies, W., Yamasaki, G. and Burger, J. C. (1966). Paper, *5th national meeting of the Society for Applied Spectroscopy*, Chicago, Ill., June 1966
74. Matousek, J. and Sychra, V. (1970). *Anal. Chim. Acta*, **49**, 175
75. Fleet, B., Liberty, K. V. and West, T. S. (1969). *Anal. Chim. Acta*, **45**, 205
76. West, T. S. and Williams, X. K. (1968). *Anal. Chim. Acta*, **42**, 29
77. Manning, D. C. and Heneage, P. (1967). *Atom. Abs. Newsletter*, **6**, 124
78. West, T. S. and Williams, X. K. (1968). *Anal. Chem.*, **40**, 335
79. Matousek, J. and Sychra, V. (1969). *Anal. Chem.*, **41**, 518
80. Sychra, V. and Matousek, J. (1969). *Chem. Listy.*, **63**, 177
81. Manning, D. C. and Heneage, P. (1968). *Atom. Abs. Newsletter*, **7**, 8
82. Veillon, C., Mansfield, J. M., Parsons, M. P. and Winefordner, J. D. (1966). *Anal. Chem.*, **38**, 204
83. Manning, D. C. and Heneage, P. (1968). *Atom. Abs. Newsletter*, **7**, 80
84. Ellis, D. W. and Demers, D. R. (1966). *Anal. Chem.*, **38**, 1943
85. Dagnall, R. M., Thompson, K. C. and West, T. S. (1966). *Anal. Chim. Acta*, **36**, 269
86. Dagnall, R. M., Thompson, K. C. and West, T. S. (1967). *Talanta*, **14**, 551
87. Dagnall, R. M., Thompson, K. C. and West, T. S. (1967). *Talanta*, **14**, 557
88. Dagnall, R. M., Thompson, K. C. and West, T. S. (1967). *Talanta*, **14**, 1151
89. Dagnall, R. M., Thompson, K. C. and West, T. S. (1967). *Talanta*, **14**, 1467
90. Dagnall, R. M., Thompson, K. C. and West, T. S. (1968). *Talanta*, **15**, 677
91. Dagnall, R. M., Taylor, M. R. G. and West, T. S. (1965). *Spectrosc. Letters*, **1**, 397
92. Mansfield, J. M., Bratzel, M. P., Norgordon, H. O., Knapp, D. O., Zacha, K. E. and Winefordner, J. D. (1968). *Spectrochim. Acta*, **23B**, 389
93. Zacha, K. E., Bratzel, M. P., Winefordner, J. D. and Mansfield, J. M. (1968). *Anal. Chem.*, **40**, 1733
94. Jackson, D. A. (1928). *Proc. Rov. Soc. (London)*, **A121**, 432
95. Meggers, W. F. (1948). *J. Opt. Soc. Amer.*, **38**, 7
96. Meggers, W. F. and Westfall, F. O. (1950). *J. Rest. Nat. Bur. Stand.*, **44**, 447
97. Meggers, W. F. and Kessler, K. G. (1950). *J. Opt. Soc. Amer.*, **40**, 737
98. Fehsenfeld, F. C., Evenson, K. M. and Broida, H. P. (1965). *Rev. Sci. Instr.*, **36**, 294
99. Fehsenfeld, F. C., Evenson, K. M. and Broida, H. P. (1964). *NBS Report 8701, U.S. Dept. of Commerce*

100. Ivanov, N. P., Minervina, L. V., Boranov, S. V. and Pofralidi, L. G. (1966). *Zh. Analit. Khim.*, **21**, 1129
101. Cresser, M. S. and West, T. S. (1970). *Spectrochim. Acta*, **25B**, 61
102. Mansfield, J. M. (1967). *Ph.D. Thesis*, University of Florida, U.S.A.
103. Browner, R. F., Dagnall, R. M. and West, T. S. (1969). *Anal. Chim. Acta*, **45**, 163
104. Browner, R. F., Dagnall, R. M. and West, T. S. (1969). *Talanta*, **16**, 75
105. Cooke, D. O., Dagnall, R. M. and West, T. S. (1971). *Anal. Chim. Acta*, **54**, 381
106. Silvester, M. D. and McCarthy, W. J. (1969). *Anal. Letters*, **2(5)**, 305
107. Silvester, M. D. and McCarthy, W. J. (1970). *Spectrochim. Acta*, **25B**, 229
108. Aldous, K. M., Alger, D., Dagnall, R. M. and West, T. S. (1970). *Lab. Prac*, **19**, 587
109. Browner, R. F., Dagnall, R. M. and West, T. S. (1969). *Anal. Chim. Acta*, **45**, 163
110. Dagnall, R. M., Silvester, M. D. and West, T. S. (1971). *Talanta*, **18**, 1103
111. Fraser, L. M. and Winefordner, J. D. (1971). *Anal. Chem.*, **43**, 1693
112. Bratzel, M. P. Jr., Dagnall, R. M. and Winefordner, J. D. (1969). *Anal. Chem.*, **41**, 713
113. Bratzel, M. P. Jr., Dagnall, R. M. and Winefordner, J. D. (1967). *Anal. Chem.*, **41**, 1527
114. Teclu, N. (1891). *J. Prakt. Chem.*, **44**, 246
115. Smithells, A. and Ingle, H. (1892). *J. Chem. Soc.*, **61**, 204
116. Kirkbright, G. F. and West, T. S. (1968). *Applied Optics*, **7**, 1305
117. Dagnall, R. M., Kirkbright, G. F., West, T. S. and Wood, R. (1969). *Anal. Chim. Acta*, **47**, 407
118. Ibid. (1970). *Analyst*, **95**, 425
119. Ibid. (1970). *Anal. Chem.*, **42**, 1029
120. Massmann, H. (1968). *Spectrochim. Acta*, **23B**, 215
121. Massmann, H. (1967). *Z. Anal. Chem.*, **225**, 203
122. Manning, D. C. and Fernandez, F. (1970). *Atomic Absorption Newsletter*, **9**, 65
123. West, T. S. and Williams, X. K. (1969). *Anal. Chim. Acta*, **45**, 27
124. Amos, M. D., Bennett, P. A., Brodie, K. G., Lung, P. W. Y. and Matousek, J. P. (1971). *Anal. Chem.*, **43**, 211
125. Black, M. S., Glenn, T. H., Bratzel, M. P. Jr. and Winefordner, J. D. (1971). *Anal. Chem.*, **43**, 1769
126. Bratzel, M. P. Jr., Dagnall, R. M. and Winefordner, J. D. (1969). *Anal. Chim. Acta*, **48**, 197
127. Bratzel, M. P. Jr., Dagnall, R. M. and Winefordner, J. D. (1970). *Appl. Spectr.*, **24**, 518
128. Kirkbright, G. F. (1971). *Analyst*, **96**, 609
129. Larkins, P. L., Lowe, R. M., Sullivan, J. V. and Walsh, A. (1969). *Spectrochim. Acta*, **24B**, 187
130. Vickers, T. J. and Vaught, R. M. (1969). *Anal. Chem.*, **41**, 1476
131. Larkins, P. L. (1971). *Spectrochim. Acta*, **26B**, 477
132. Larkins, P. L. and Willis, J. B. (1971). *Spectrochim. Acta*, **26B**, 491
133. Mitchell, D. G. and Johansson, A. (1970). *Spectrochim. Acta*, **25B**, 175
134. Slavin, W. (1968). *Atomic Absorption Spectroscopy*, (New York: Interscience)
135. Mitchell, D. G. (1970). *Advances in Automated Analysis*, Technicon International Congress, Vol. 2, 503, Industrial Analysis, Thurman Associates, Miami, U.S.A.
136. Demers, D. D. and Mitchell, D. G. (1970). Ref. 135, p.507
137. Gardels, M., Demers, D. D. and Mitchell, D. G. (1970). Ref. 135, p.513
138. Walsh, A. (1969). *Proceedings of the International Atomic Absorption Spectroscopy Conference*, Sheffield, England
139. Smith, R., Stafford, C. M. and Winefordner, J. D. (1969). *Canad. Spectrosc.*, **14**, 1
140. Matousek, J. and Sychra, V. (1970). *Anal. Chim. Acta*, **52**, 376
141. Cotton, D. H. and Jenkins, D. R. (1970). *Spectrochim. Acta*, **25B**, 283
142. Dagnall, R. M., Kirkbright, G. F., West, T. S. and Wood, R. (1971). *Anal. Chem.*, **43**, 1765
143. Dagnall, R. M., Taylor, M. R. G. and West, T. S. (1971). *Lab. Pract.*, **20**, 209
144. Dagnall, R. M., Taylor, M. R. G. and West, T. S. (1970). *Analyst*, in the press
145. Amos, M. D., Bennett, P. A., Brodie, K. G., Lung, P. W. Y. and Matousek, J. P. (1970). *21st Pittsburgh Conference in Analytical Chemistry and Applied Spectroscopy*, Cleveland, Ohio, March 1970
146. Hobbs, R. S., Kirkbright, G. F. and West, T. S. (1971). *Talanta*, **18**, 9, 859
147. Kirkbright, G. F., Rao, A. P. and West, T. S. (1969). *Anal. Letters.*, **2**, 465
148. Dagnall, R. M. and West, T. S. (1968). *Applied Optics*, **7**, 1287

4

Ion-selective electrode analysis

J. TENYGL

J. Heyrovský Institute of Polarography, Czechoslovak Academy of Sciences, Prague

4.1   ION-SELECTIVE ELECTRODES

4.1.1   Introduction

Ion-selective electrodes (ISEs) are electrochemical sensors which react by reversible change of the electrode potential to a change in the activity of one ionic species. The response of some ISEs is very selective for one particular ion and is not much affected by other ions. The ideal case is a specific electrode which reacts to a change in the activity of only one ionic species. Such an electrode does not at present exist, and the reading given by every ISE is influenced more or less by the presence of other ions, the electrode exhibiting higher or lower selectivity. For this reason the term ion-selective electrode is to be recommended over ion-specific electrode.

The sensing element is an ion-selective membrane which need not be thin. The first ISE was without doubt the glass electrode for determination of hydrogen ion activity. It was discovered at the beginning of this century by Cremer[1] and Haber and Klemensiewicz[2]. The specific response of this electrode to the hydrogen ion activity, which is not affected by many other ions and redox substances, can be taken as a typical example of high sensor selectivity. On the other hand, the 'sodium error' at high pH values illustrates the interference of sodium ions in the determination of hydrogen ion.

An enormous amount of research work and experimental ingenuity has been devoted to the development of electrodes for determining other substances and ionic species with a selectivity similar to that of the pH electrode for hydrogen ions. Perfect perm-selective membranes have been developed which are able to separate anions and cations but not particular types of cation or anion. The pH-sensitive glass electrode has been developed to its present perfect state and a few glass electrodes for determining alkali–metal ions have been introduced. The theory of membrane potentials has been developed by Donnan[3], Teorell[4], Marshall[5], Scatchard[6], Nicolsky[7] and others. A comprehensive review of this field of research has been published by Hills[8], Solner[9], Eisenman[10] and Lakshminarayanaiah[11].

ISEs have been the subject of numerous papers and reviews by Coving-ton[12], Pungor[13, 14], Rechnitz[15, 16] and Simon[17]. Theory and practice have been discussed by a number of authors in a monograph edited by Durst[18]. A compilation of theory has been made by Koryta[19].

4.1.2 The basic principles of ion-selective electrodes

The ISE contains a sensing element which is an ion-selective membrane. The sensing element is sealed in a body, usually made of a plastic tube, containing the internal reference electrode and internal filling solution. The measurement of activity of the ion of interest is in principle consistent with the measurement of the activity of hydrogen ions by means of a pH-sensitive glass electrode. Both electrodes are shown in Figure 4.1. Ion-selective membranes have been made of a special composition glass, a disc ground from a monocrystal, or have been manufactured by pressing a crystalline mixture from an inert matrix in which a sparingly soluble precipitate was dispersed; a layer of liquid ion exchanger has also been used. As the electrical

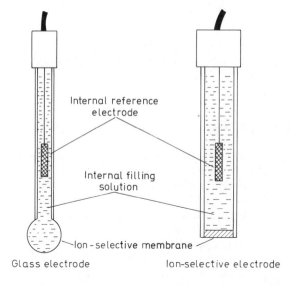

Internal reference
electrode

Internal filling
solution

Ion-selective membrane

Glass electrode   Ion-selective electrode

Figure 4.1 Comparison of an ion-selective electrode with the glass pH electrode

conductivity of a membrane material is usually high, the membrane can be made relatively thick and strong and thus resistant to mechanical stress.

The internal filling solution contains ions in reversible equilibrium with the internal reference electrode and is saturated with the active material in the ion-selective membrane. The potential of an ISE is the algebraic sum of the internal reference electrode potential and of the internal and external membrane surface potentials. The first two potentials are constant during the measurement and are included in the constant E'. The potential of an ion-selective electrode, E_{ISE}, is given by the equation

$$E_{ISE} = E' + 2.303 \frac{RT}{z_i F} \log a_i \tag{4.1}$$

where R is the gas constant, T the absolute temperature, z_i the charge of the ion and F the Faraday constant.

4.1.3   Origin of the membrane potential

Suppose we have a membrane separating two solutions. The solution' on the left-hand side of the membrane contains the ions i and j, with activities a_i' and a_j' respectively; the solution" on the right-hand side of the membrane contains the same ions with activities a_i'', a_j'', respectively. Suppose again that the membrane is ion-selective for the ion i, which can freely move across the membrane from the solution of higher activity, a_i', to the solution of lower activity, a_i''. The ion j (or any other ion) is prevented from passing through the membrane. In the membrane between the two solutions there will arise a potential difference, i.e. a membrane potential:

$$E_M = \frac{RT}{z_i F} \ln \frac{a_i'}{a_i''} \tag{4.2}$$

This equation is equivalent to the Nernst equation for a concentration cell.

If the activity a_i'' on one side of the membrane is kept constant, equation (4.2) can be converted into the form:

$$E_M = \text{const.} + \frac{RT}{z_i F} \ln a_i' \tag{4.3}$$

The membrane potential is thus a logarithmic function of the activity of the ion i.

As ion-selective membranes, glasses of special composition have been used which work as ion-exchangers and permit the transfer of only one ion. Some crystalline materials with defects in the crystal lattice have high ionic conductivity and exhibit high selectivity for the ions which form their crystal lattice. Liquid membranes have also been used. These consist of a liquid ion exchanger dissolved in an organic solvent.

The mechanism of ion selectivity is not fully understood and a general theory has not yet been given. Ion-selective glasses have been studied by Nicolsky et al.[7], Doremus[20], Karrelman[21], Conti[22] and Eisenman[23], ion-exchange resins by Bose[24], Basu[25] and others[8] and heterogeneous membranes by Buck[26, 27] and Pungor and co-workers[28, 29, 30].

The similarities between solid and liquid ion exchangers have been described by Eisenman[31]. The membrane potential of a thin glass membrane separating solutions of the same composition, as in the case given in the derivation of equation (4.2), is given[31] by equation (4.4).

$$E_M = \frac{RT}{F} \ln \frac{a_i' + \frac{u_j^*}{u_i^*} K_{ij} a_j'}{a_i'' + \frac{u_j^*}{u_i^*} K_{ij} a_j''} \tag{4.4}$$

where u_i^*, u_j^* are the mobility of ions i and j respectively within the membrane. The membrane selectivity to ion i in the presence of ion j is given by $(u_j^*/u_i^*)K_{ij}$, where K_{ij} is the equilibrium constant for the ion exchange reaction between the solution and the membrane:

$$j^+ \quad + \quad i^{+*} \quad \underset{\longleftarrow}{\overset{K_{ij}}{\longrightarrow}} \quad j^{+*} \quad + \quad i^+$$
(in solution) (in membrane) (in membrane) (in solution)

The selectivity depends on the value of both terms u_j^*/u_i^* and K_{ij} which tend to oppose one another. It was accepted and proved experimentally[22, 32] that the more strongly an ion is preferred by an ion-exchanger, the more slowly it moves within the membrane. This phenomenon is one reason for the selectivity limitations of solid ion exchangers[31].

For liquid ion exchangers the factors governing the membrane potential tend to be more complicated[22, 31, 33]. The selectivity of the membrane is given by the constant K_{ij}, and by the mobility of the 'trapped' ion within the liquid membrane. If the trapped ion is completely dissociated, the selectivity depends on the mobility ratio of u_i^* and u_j^* in the solvent forming the membrane. In the case of complete association, the selectivity depends on the mobility of the neutral ion pairs and the association equilibrium constant. If neutral carriers are used as a liquid membrane, the selectivity depends mainly on the specificity of the interaction of the carrier with neutral sequestering molecules[31]. Buck[26, 27] and Pungor[34] applied Eisenman's concept to heterogeneous membranes.

4.1.4 Measurement with ion-selective electrodes

The potential of an ISE is determinated by measuring the electromotive force of a cell consisting of an ISE and a reference electrode. The measurement can be carried out in one of two ways: in a cell with, or without, liquid junction.

4.1.4.1 Cell with liquid junction

The solution to be analysed is in contact with the solution of the reference electrode (Figure 4.2). At the boundary of the two solutions there arises a liquid-junction potential, E_l. The electromotive force, E, of the cell, which can be measured by a millivoltmeter with a high input resistance, is given by the equation:

$$E = E_{ISE} - (E_{ref} + E_l) \tag{4.5}$$

For the determination of E_{ISE} to be possible, both the remaining potentials, E_{ref} and E_l, should be of a constant value. The potential of the reference electrode is reasonably stable during the measurement and can be taken as constant. The source of uncertainty lies in the liquid-junction potential, E_l, which varies with changes in the compositions of the solutions, e.g. when the sample and standard solution differ greatly in ionic strength. An arrangement must therefore be made to reduce such changes to a minimum during the measurement. A schematic representation of a cell with a liquid junction may be written

reference	||	sample		ion-selective
electrode	||	solution		electrode

A single vertical line represents the electrode-liquid interface and a double vertical line, the boundary between two dissimilar solutions. A salt bridge is sometimes placed between the sample and reference electrode solutions.

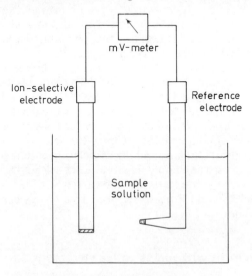

Figure 4.2 Schematic experimental arrangement for measurement with ion-selective electrodes

It usually consists of a concentrated potassium chloride solution. Schematic representation of such a cell is

reference	||	salt bridge	||	sample		ion-selective
electrode	||	solution	||	solution		electrode

4.1.4.2 Cell without liquid junction

Both the ISE and the reference electrode are in direct contact with the sample solution, which must contain an ionic species which maintains a constant reference electrode potential. Schematic representation of the cell is

reference		sample		ion-selective
electrode		solution		electrode

The electromotive force is given by the equation

$$E = E_{ISE} - E_{ref} \qquad (4.6)$$

The liquid-junction potential is thus eliminated. A source of uncertainty, however, lies in the potential of the reference electrode. The activity of the ionic species responsible for the reference electrode potential is not known as it is influenced by the other ionic species present in the sample. This activity can change, e.g. after adding a standard solution. However, these changes can be more easily avoided during measurements than can the changes of E_l, and, when working with ISEs, cells without liquid junction are recommended for use whenever possible. Unfortunately, this system cannot often be used since there is a lack of suitable electrode pairs. The reference electrode ionic species must not interfere with the ISE and *vice versa*.

A well-known example of a cell without liquid junction is the Harned cell[35], used for the direct determination of the mean activity coefficient of hydrochloric acid and the pK of weak acids[36]. In the Harned cell, hydrogen and silver/silver chloride electrodes were used.

$$\text{Pt, H}_2 \quad | \quad \text{H}^+, \text{Cl}^- \quad | \quad \text{AgCl, Ag}$$

Measurement of the activity of the solution components is possible because of the unique feature of this electrode pair. Hydrogen does not undergo electrochemical oxidation at the silver/silver chloride electrode and thus does not change its potential. The rate of reduction of Ag^+ on the Pt electrode is probably kinetically limited. The solubility of H_2 and AgCl in hydrochloric acid is low so they have little effect on the mean activity coefficient[27, 37, 38].

Cells without liquid junction were used for the determination of interferences in measurements with glass pH electrodes[39, 40, 41].

Manahan[42] used a fluoride ISE as a reference electrode in the determination of nitrate with a nitrate ISE. (see Section 4.2.6).

4.1.5 The liquid-junction potential

If two dissimilar solutions or two solutions of the same composition but different activity are in contact, a liquid junction potential, E_l, arises on the boundary between them. This potential difference is caused by diffusion of ions. In a cell with liquid junction, a potential difference as high as a few dozen millivolts between the sample and reference electrode solutions can occur. The absolute value of E_l is not of great importance but its stability during the measurement is. Fluctuations of the liquid-junction potential cause considerable error in the measurement of electromotive force. Only a brief outline, basic for understanding the problem, is given here. More detail can be found elsewhere[43, 44]. The influence of the liquid-junction potential on measurement with an ISE is discussed by Covington[45]. The liquid-junction potential between solutions I and II is given by the Planck equation[46-48].

$$-E_j = \frac{RT}{F} \int_I^{II} \sum_i \frac{t_i}{z_i} \, d \ln a_i \tag{4.7}$$

in which t_i, z_i and a_i are the transfer number, charge, and activity, respectively of the ion i. Equation (4.7) was solved by Planck[47, 48] and Henderson[49, 50].

Several simplifications concerning the physical constants and the structure of the liquid–liquid boundary were involved in the solution. The ionic mobilities were assumed to be constant and independent of ionic concentration, and the mean ion activity was used.

In a few instances, the solution of equation (4.7) can be expressed in a simple form. The liquid-junction potential of two uni–univalent electrolytes of the same composition but different concentration is given by the equation,

$$HCl \quad || \quad HCl$$

$$c_1 = 0.1 \text{ mol l}^{-1} \qquad c_2 = 0.1 \text{ mol l}^{-1}$$

$$E_1 = (2t^+ - 1)\frac{RT}{F}\ln\frac{c_1}{c_2} \qquad (4.8)$$

The liquid-junction potential, calculated by means of equation (4.8), is, for the example given, about 39 mV, where $T = 25\,°C$, $t^+ = 0.83$ and $a = c$. It does not depend on the manner in which the liquid–liquid boundary is formed.

The liquid-junction potential of two uni–univalent electrolytes of the same concentration with one common ion[46], is given by the Lewis–Sargent[51] equation

$$HCl \quad || \quad KCl$$

$$c_1 = 0.1 \text{ mol l}^{-1} \qquad c_2 = 0.1 \text{ mol l}^{-1}$$

$$E_1 = \frac{RT}{F}\ln\frac{U_{H^+} + U_{Cl^-}}{U_{K^+} + U_{Cl^-}} = \frac{RT}{F}\ln\frac{\Lambda_{HCl}}{\Lambda_{KCl}} \qquad (4.9)$$

where U and Λ are the mobility and the equivalent conductivity. For the example mentioned above, E_1 is about 29 mV and depends on the manner in which the liquid–liquid boundary is formed. If the two dissimilar solutions in contact have different concentrations, the equation for the liquid-junction potential does not reduce to the simple form. It was solved for a few cases graphically. As in the previous case, E_1 depends on the structure of the boundary.

4.1.5.1 Types of liquid junction

A detailed study of different types of liquid junctions was made by Guggenheim[52], who distinguished mixture, free diffusion, constrained diffusion, and flowing junction boundaries. The mixture boundary was considered by Henderson[49, 50] for the calculation of E_1 and it is assumed to be formed of two layers of solution, between which is a transition diffusion layer with a linear concentration gradient. This type of junction is difficult to realise and the nearest equivalent to it is a free diffusion boundary. This is formed by an initial sharp boundary between two solutions, which are then allowed to diffuse freely into each other, e.g. through a stopcock[53, 54] or by a sliding disc[55]. If cylindrical diffusion symmetry is maintained, the liquid-junction potential is stable. The constrained diffusion boundary was introduced by Planck[47] who considered the separation of two solutions by a porous wall. The concentration on both sides of the wall and the length of the diffusion path were assumed to be constant. The flowing junction[56-59] boundary has a stable and reproducible potential. The sharp boundary is formed by a continuous flow of two solutions differing in specific gravity. The liquid junction used in the commercially available reference electrodes is difficult to classify. Electrical connection of the solutions is made through a porous plug of glass, ceramic, asbestos or cellophane, or through a small diameter opening, e.g. a capillary[60, 55], a leakage in a ground glass joint, a leakage made by contraction of a palladium annulus sealed in glass, etc. There is usually a very slow flow of the reference electrode solution into the sample which prevents the contamination of the reference electrode and keeps the liquid-junction potential more stable.

The strong influence of suspensions and colloids on the magnitude of liquid-junction potentials, known as the 'Pallman effect'[43, 61-63] was observed during pH measurements in biological materials. The response of the sodium ion-selective glass electrode is also influenced by presence of a suspension[64, 65].

4.1.5.2   Reducing liquid-junction potentials

A salt bridge interposed between the sample and reference electrode solution has often been used for this purpose. It is formed by a concentrated solution of an electrolyte which contains an anion and a cation of similar mobilities. Guggenheim[52] suggested saturated potassium chloride. The transfer number of potassium, t_{K+}, is 0.49 and a concentrated solution can easily be prepared ($c_{sat} = 4.2M$). Guggenheim observed the appreciable reduction of E_l after interposing a salt bridge between the boundary $0.1M$ HCl $\|$ $0.1M$ KCl. On the basis of some contradictory results, it is not quite clear if this method is equally efficient in the case of the double liquid junction of a heteroionic solution[45, 66, 67]. Another method of reducing liquid-junction potentials consists of adding an indifferent electrolyte in a large excess to both the solutions forming the liquid boundary[45].

4.1.6   The selectivity ratio

The influence of interfering ions on the ISE potential is expressed by the selectivity ratio, K, which is sometimes called the selectivity constant. The latter term is incorrect since K is not really a constant and depends on the composition of the solution[68]. The potential of an ISE in the presence of an interfering ion, j, is given by the following empirical equation[69-73]

$$E = E' + \frac{RT}{z_i F} \ln\left(a_i + \sum_j K_{ij} a_j^{z_i/z_j}\right) \qquad (4.10)$$

where K_{ij} is the selectivity ratio of ion i over ion j, z_i is the charge of the ion i, being measured, z_j, is the charge of the interfering ion j, and E' is a constant. If only one interfering ion j is present in the solution, and both the interfering ion and the ion being measured are univalent, then equation (4.10) simplifies to

$$E = E' + \frac{RT}{F} \ln (a_i + K_{ij} a_j) \tag{4.11}$$

The interference of ion j is determined by the product, $K_{ij} a_j$. The lower the value of K_{ij}, the lower the interference. If $K_{ij} = 1$, the ISE has the same selectivity for both ions i and j and its selectivity is then very low. Some good ISEs have selectivity ratios lower than 5×10^{-3}.

The selectivity ratio may be determined by various methods. Eisenman[33] defines the potentiometric selectivity ratio for an ion exchange membrane as

$$K_{ij}^{Pot} = \frac{u_j^*}{u_i^*} K_{ij} \tag{4.12}$$

where u^* is the mobility of the ion in the membrane and K_{ij} is the equilibrium constant of the ion exchange reaction. The selectivity ratio is determined from the potential difference between two measurements[32]. The first value, E_1, is obtained by measuring the potential in the presence of the ion being measured at the concentration $c = 0.1\text{M}$ in the absence of interfering ions. The second value, E_2, is obtained with the same concentration of interfering ion in the absence of the ion being measured. The selectivity ratio is determined from the following equation

$$\log K_{ij} = \frac{E_1 - E_2}{2.3\ RT/F} \tag{4.13}$$

Rechnitz et al.[74, 75] defined the selectivity ratio as $K_{ij} = a_i/a_j$ which expresses the ratio of the activities of ions i and j, producing the same change of potential when present separately. In this method, the mutual influence of the ions is not taken into account. This is not the case in the methods used by Pungor et al.[14, 76, 77] in which the selectivity ratio was determined by direct and indirect methods.

4.1.6.1   The reduction of interferences

Besides the TISAB buffer (see Section 4.1.7.2), which works on the principle of adding a stronger complexing agent, formation of precipitates have also been used. Havas[78] reduced the influence of interfering ions by saturating the sample with the same sparingly soluble salt as is embedded in the heterogeneous membrane of the ISE. Solid silver chloride is added to the sample when an AgCl-based heterogeneous ISE is used. If an interfering ion, e.g. iodide, is present in the sample, precipitation of AgI occurs and the concentration of free iodide ions is considerably reduced. The number of chloride ions released by the precipitating reaction is negligible. According to Havas[78], iodide does not interfere even in concentrations 1/50–1/100 of that of chloride, while, if uncompensated, an iodide concentration 10^{-5} that of chloride interferes.

Interfering ions may be removed by chemical methods before the measurement. The use of ion exchangers is very popular.

4.1.7   Activity and concentration

Potentiometric measurements with ISEs give information about the *activity* of the ion being measured. The relation between this value and the concentration is given by equation

$$a_i = y_i c_i \tag{4.14}$$

where y_i is the activity coefficient and c_i is the concentration of ion i. The magnitude of the activity coefficient depends on the ionic strength of the solution and can be determined by means of the Debye–Hückel equation. Solutions of standard activity and performance tests for ISE were given by Bates and Alfenaar[79].

In some cases, it is more useful to know activity than concentration, e.g. biological processes such as nerve conduction, muscle contraction, and blood clotting are controlled by the activity of Ca^{2+} and not by its concentration[80].

4.1.7.1   Free ion concentration

The activity of free ions is decreased in the presence of substances which react with the ions being measured to form undissociated molecules. Typical reactions decreasing the concentration of free ions are the formation of un-dissociated weak acids or bases, precipitation, and complex formation. The activity of the free ion, i, is given by the following relation[81]

$$a_i = y_i x_i c_i \tag{4.15}$$

where y_i is the activity coefficient, x_i is the fraction of free ion and c_i is its total concentration.

The potential of an ISE in a solution containing an indifferent electrolyte, a complexing agent, and interfering ions is given by the equation

$$E = E' + 2.3 \frac{RT}{z_i F} \log \left(y_i x_i c_i + \sum_j K_{ij} a_j \right) \tag{4.16}$$

The activity coefficient and the fraction of free ions must be kept constant during the measurement. A decrease of the free ion concentration by the effects described above is a secondary form of interference, usually not included in lists of interferences given by manufacturers.

4.1.7.2   Maintenance of a constant ionic strength

The reading given by an ISE is dependent on the ionic strength of the solution because of its influence on the activity coefficient of the ion being measured. The ionic strength of the solution should be kept as constant as possible during the measurement and must not change on addition of the standard solution. If a sample of unknown ionic strength is analysed, the ionic strength fluctuation can be minimised by adding an indifferent electrolyte in a large excess to the sample. The contribution of the ionic strength of the sample is often negligible and can be ignored.

Frant and Ross[82] used a solution, known under the name TISAB (Total Ionic Strength Adjustment Buffer) for determining fluoride in water with a fluoride ISE. This consists of 1M sodium chloride, 0.25M acetic acid, 0.75M sodium acetate and 0.001M sodium citrate, pH = 5.0, ionic strength $I = 1.75$. Equal volumes of sample and buffer TISAB are mixed before the analysis. The buffer keeps the optimal pH and a stable ionic strength, and eliminates the influence of secondary interferences, i.e. Fe^{3+} and Al^{3+}. This is done by citrate ions, which bind the ferric and aluminium ions into strong complexes, thus releasing fluoride ions.

4.1.8 Properties of ion-selective electrodes

4.1.8.1 The accuracy of measurement

The potential of an ISE is a logarithmic function of the ion activity. The logarithmic relation causes the principal limitation on the accuracy. The change of electrode potential is 59.2 mV for a tenfold change in the activity of a univalent ion and 29.6 mV for that of a divalent ion. The relation between the overall electromotive force error ΔE and the concentration can be derived from the Nernst equation by deriving with respect to concentration[83].

$$\Delta E = \frac{0.2568}{z_i} \frac{\Delta c\,100}{c} \; (\text{mV, } 25\,°C) \tag{4.17}$$

$$\Delta E = \frac{0.2568}{z_i} \text{RE} \tag{4.18}$$

where RE is the percentage relative error in concentration. An error of 1 mV in the potential measurement corresponds to 3.9% RE in concentration for a univalent ion and 7.8% RE for a divalent ion. Since the fluctuation of the liquid-junction potential is included in the measurement of E, an accuracy of better than 0.5 mV can scarcely be obtained. According to Ross[84], under typical field conditions, a precision of 4.0 mV is more common. ISEs are, therefore, more suitable for the measurement of low concentrations, where even a high relative error corresponds to a smaller absolute error.

4.1.8.2 The dependence on temperature

The influence of the temperature on measurements with an ISE is very complex since it affects the potentials of the reference electrode, the liquid junction, the ion-selective membrane, and the internal reference electrode, as well as the activity of the solution. Accurate measurements must be made in thermostatted solutions. Covington[45] discusses the problem of temperature change compensation. Light[83, 85] differentiated the Nernst equation with respect to temperature and pointed out that the main source of error in pH and ISE measurements lies in the assumption that it is sufficient to compensate for the change in the Nernst slope with temperature.

4.1.8.3 The lower limit of detection

The lower limit of detection depends on the solubility of the active substance of the ISE. Pungor and co-workers[13, 86] have shown that, with precipitate-based membranes, this limit is determined by the solubility product of the appropriate precipitate. A simple formula for estimation of the lower limit of detection $c = (9.1 \times K_{sp})^{1/2}$ was given by Covington[87]. The values of c for AgI, AgBr, and AgCl are 3.8×10^{-8}M, 2.7×10^{-6}M and 3.8×10^{-5}M, respectively.

The minimum determinable concentration can be lowered by reducing the solubility of the precipitate, e.g. by cooling the sample[88] to 0 °C or by adding an organic solvent such as acetone, methanol, or dioxan[88, 89].

4.1.8.4 Response time

The dynamic response of ISEs depends on the type being used, the concentration of the ion being measured, and the experimental arrangement. A response time of 1–10 s is typical. At low concentrations and in the presence of interfering ions, the response time is longer. The dynamic response of heterogeneous membrane electrodes with a silicone rubber matrix was determined by Rechnitz et al.[74, 75] and more recently by Pungor and Toth[77, 90]. An equilibrium state is reached in a few tenths of a second[77, 90] and the transition curve can be described by a simple exponential function, in contrast to liquid membranes where a short-term overshooting of the potential was observed[91, 92]. The response time of solid electrodes has also been studied[93–95]. ISEs were used by Rechnitz et al.[91, 96–99] for a kinetic study of chemical reactions and Fleet and Rechnitz[100] used a continuous flow system with turbulent flow to measure the kinetics of complex formation of Ca^{2+}, Mg^{2+} and Be^{2+} with biologically important ligands.

4.2 REFERENCE ELECTRODES

The requirements for reference electrodes are well known. The electrode must keep a constant, reproducible potential which does not change with time.

Useful information on the reproducibility and stability of the reference electrode potential can be obtained by polarisation tests[38, 101–103]. The electrode is polarised by passage of current from an external source in both the anodic and cathodic directions. The polarisation curve, plotted as the dependence of potential $v.$ current, should be linear and exhibit no hysteresis.

Information on the theory and practical use of reference electrodes can be found in the monograph by Ives and Janz[38] and elsewhere[43, 45].

4.2.1   Internal reference electrodes

The internal reference electrode is in equilibrium with the internal filling solution of the ISE, which is saturated with the active material of the ion-selective membrane and contains the ions which keep the reference electrode potential constant. The ions of the reference electrode must not cause irreversible changes in the potential of the ion-selective membrane and *vice versa*. Silver/silver chloride or other silver/silver halide electrodes are used, e.g. in Pungor's heterogeneous membranes[14]. Information on the internal reference electrodes of commercially available ISEs is very limited. This is also true for the so-called 'solid state' electrodes, which contain no internal filling solution but instead a metal wire in direct contact with the ion-selective membrane[104–107].

4.2.2   The hydrogen electrode

The hydrogen electrode is a reference electrode of the first kind, with a very high exchange current density and an accurately reproducible potential. The electrode reaction

$$2H^+ + 2e \rightleftarrows H_2 \tag{4.19}$$

takes place on a platinum-black-covered platinum electrode. Platinum electrodes are very sensitive to the presence of sulphide, cyanide and arsenic compounds in trace amounts, and to some heavy metals such as mercury. The addition of a small amount of lead acetate to the chloroplatinic acid before deposition of the platinum black extends the life of the electrode. The hydrogen used for saturation of the solution must be of a high purity and even low concentrations of oxygen must be absent: a palladium purifier can be used[108]. An electrolytic source of ultra-pure hydrogen was constructed by Jansta[109] in which hydrogen was produced under high pressure by electrolysis at catalytic porous electrodes.

4.2.3   The silver/silver chloride electrode

After the hydrogen electrode, this electrode has the most stable and reproducible potential and has been extensively used. A stable potential is maintained by a constant silver ion concentration, given by the solubility product of AgCl. Any substances which form precipitates of lower solubility product, or a complex with a lower dissociation constant change the electrode potential and must be absent. The silver/silver chloride electrode is, therefore, very sensitive to the presence of bromide ions[110]. The effect of iodide is anomalously small. The oxygen effect is high, especially in acid solution and a potential shift as high as 1.5 mV was observed[111, 112] in the presence of oxygen.

Photoelectric effects have been observed and studied by many authors with varying results[38]. The silver/silver chloride electrode is prepared[103] by anodic oxidation of metallic silver or sintered silver powder in chloride media or by thermal decomposition of a mixture of silver and silver chlorate. Applications in electrochemistry[103] and biochemistry[113] have been recently reviewed.

4.2.4 The calomel electrode

The potential of this electrode is controlled by the activity of mercurous ions, given by the solubility product of mercurous chloride. A saturated or 3.5M solution of potassium chloride is often used, since concentrated electrolyte reduces the liquid-junction potential. The preparation, behaviour and stability of this electrode were exhaustively studied by Hills and Ives[38, 114]. They recommend mixing calomel and mercury in the dry state and spreading a thin film on the mercury surface; the electrolyte is then added. For work at higher temperatures, a thick layer of calomel paste has been used[14, 115]. The calomel electrode is very sensitive to dissolved oxygen in the electrolyte and traces of Zn, Cd, Sn, Pb and Cu in the mercury. After preparation, the electrode potential reaches equilibrium slowly (1–2 days), a fact which has been explained by a slow decomposition of Hg_2Cl_2 to $HgCl_2$, and by complex formation[116, 117]. Covington[108] does not recommend use of the calomel electrode because of the observed temperature hysteresis.

4.2.5 Other reference electrodes

The 'Thalimide electrode' is the commercial name of a reference electrode based on the system thallium amalgam–thallous chloride in a potassium chloride solution[108, 113, 118]. It is claimed to be free of temperature hysteresis and thus usable at higher temperatures. It is very sensitive to the presence of oxygen.

The mercury/mercurous sulphate electrode maintains a constant potential, but is inconvenient because of the high solubility of Hg_2SO_4 and its tendency to hydrolysis.

4.2.6 Ion-selective electrode reference electrodes

A glass pH electrode, selective to hydrogen ion activity, has been used as a reference electrode[119]. The potential is kept constant by adding a buffer to the measured solution. Measurement with conventional pH-meters causes difficulties, however, since they have only one high impedance input and the reference electrode input should be grounded. These problems have been discussed by Brand and Rechnitz[119], who have described a differential high impedance input solid-state amplifier. Manahan[42] has used the fluoride ISE as a reference electrode in the determination of nitrate ions by means of the nitrate ISE. A stable reference potential is maintained by adding NaF to the sample solution. The input impedance of the fluoride electrode is substantially lower (100–200 kΩ) than that of a glass electrode.

4.3 TECHNIQUES OF ANALYTICAL MEASUREMENT

4.3.1 Direct potentiometry

The activity of the ion being determined can be calculated from the Nernst equation

$$E = E_{ISE} - (E_{ref} + E_l) \qquad (4.20)$$

$$E = E' + 2.3\frac{RT}{z_iF} \log a_i - E_{ref} - E_l \qquad (4.21)$$

$$\log a_i = (E - E' + E_{ref} + E_l)/S \qquad (4.22)$$

where $S = 2.3 \, RT/z_iF = 59.2$ mV at 25 °C.

The calculation of the activity of ion a_i is difficult since E' and E_l are not constant. E' is the 'zero potential' of the ISE in the absence of the ion being measured and must be determined by calibration. The liquid-junction potential depends on the composition of the solutions.

4.3.2 The concentration cell technique

In this method, two identical ISEs are used. One electrode is in contact with the sample, the second with a reference solution. The solutions are connected by a salt bridge. The composition of one of the solutions is changed by adding a standard solution until the potential of the electrodes is identical or until each equals its respective residual potential. At this point, the activity of the ion being measured is identical in both solutions. This technique of titration to the null point was developed by de Brouckere[120], refined by Malmstadt and co-workers[121–124], and modified by Durst and Taylor[125]. These authors called it 'linear null-point potentiometry'; it has been used for the determination of silver[125, 126] and fluoride[127]. The Durst method[127] is especially suitable for the titration of very small volumes of sample, e.g. 10 μl. For this purpose a reference solution with a volume of 100 cm³ is titrated instead of the sample. This permits measurement of the titrant with much higher accuracy.

4.3.3 Calibration curves

The influence of non-ideal behaviour of the electrode can be suppressed by evaluation of the results by means of an empirical calibration curve. When this curve is constructed, the data are plotted semi-logarithmically as a dependance of E v. $\log c_i$. A linear calibration curve is obtained with a slope dependent on the behaviour of the electrode and on the ionic strength of solution[128]. The calibration curve should, therefore, be made in a solution containing the same concentration of ionic species as in the sample. An inert electrolyte in large excess is sometimes added to the sample to reduce the influence of fluctuations of the ionic strength of the sample. The calibration curve is made by adding small volumes of a concentrated standard solution to minimise the influence of dilution.

The calibration curve technique is suitable for samples with constant ionic strength and no interfering ions.

4.3.4 Standard addition

The method of standard addition is suitable for samples of variable ionic composition. It is also called the 'spike'[129] or 'known-increment'[130] method. The concentration of the sample, c_x, is determined from the difference of the electromotive force of the cell, before (E_1), and after (E_2), adding standard solution of concentration c_S and volume V_S. The volume of the sample V_x is known and it is supposed that V_S is sufficiently small and dilution of the sample is negligible. The concentration, c_x, can be calculated from the following equation[128, 131]

$$\Delta E = E_1 - E_2 = 2.3 \frac{RT}{z_i F} \log \left(\frac{c_x +}{c_x} \right) \tag{4.23}$$

where Δc is the increase in the concentration of the sample produced by a standard addition and equals $V_S c_S / V_x$. If the dilution of the sample is not negligible, the unknown concentration, c_x, can be calculated[128] as follows:

$$c_x = c_S \frac{V_S}{V_x + V_S} \left[10^{\Delta E/S} - \left(\frac{V_x + V_S}{V_x} \right) \right]^{-1} \tag{4.24}$$

The total ionic strength of the sample and the concentration of complexing agent (if it is present in a large excess) need not be known. It is supposed that the addition of standard solution does not change the ionic strength of the sample or the liquid-junction potential.

The factors determining the accuracy of the method are connected and work against each other, so that a compromise must be used in practice[128, 131–135]. If a small volume of a standard solution is added, the dilution of the sample is negligible but an error is caused by inaccuracies in metering. Large volumes of standard solution can be measured more accurately, but the dilution of the sample is no longer negligible and an error is caused by the change of ionic strength and liquid-junction potential. The double known addition method is recommended in some cases[134].

The known subtraction method[132] is based on adding a reagent, which reacts with the ion being determined and forms a precipitate or an undissociated complex. The concentration of the sample is calculated from the decrease of the concentration. In 'analate addition potentiometry'[136], the sample is added into a standard solution. Dilution of the sample by an indifferent electrolyte was suggested for diminishing the influence of interfering ions[135]. This method can be used when the interfering and measured ions differ in charge, as can be deduced from equation (4.10). The use of this method is limited.

4.3.5 Potentiometric titrations

Difficulties connected with the accurate measurement of electromotive force can be avoided by using a titration method. The concentration of the ion being measured is determined from the volume of titrant added, an ISE being used for the determination of the equivalence point only. This point is determined from the change in the electromotive force at the equivalence point. The accuracy of such measurements depends mainly on the accuracy of determining the equivalence point, that is, on the magnitude and sharpness of the potential break at the end-point. In the absence of interfering ions, the basic relation derived by Meites and co-workers[137–139] for acid–basc, precipitation, and chelometric titrations is valid. In the presence of interfering ions, a distortion of the titration curve and deformation in the vicinity of the equivalence point have been observed. The concentration of the ion being measured decreases during the course of the titration while the interfering ion concentration changes by dilution only. The calculation of the titration error was simultaneously performed by Carr[140] for precipitation titrations and by Schultz[141, 142] for precipitation and chelometric titrations. For estimation of the titration error, Carr[140] introduced two dimensionless factors, the first of which is a function of the solubility product and the original sample concentration, while the second depends on the interfering ion concentration and selectivity ratio of the ISE used. Schultz[141, 142] considers three possible cases, where the interfering ion is present (i) in the sample only, (ii) in the titrant only, (iii) in both the sample and the titrant. It is assumed that the liquid junction potential does not change and the selectivity ratio is constant during the titration. According to Schultz, the higher the original concentration of the ion being determined and the lower the solubility product, the sharper is the break in the titration curve after the equivalence point is reached. The most important factor on which the accuracy depends is the selectivity ratio, while dilution has a relatively small effect. The inflection point of the titration curve in the presence of interfering ions does not correspond to the equivalence point and the difference is the higher, the more the titration curve is flattened. The calculated error can be as large as several per cent.

The error mentioned above can be eliminated by using the graphical method originated by Gran[143, 144], which was used for measurement with ISEs by the Orion Research workers[145, 146]. The titration curve is plotted as a dependence of the concentration of ion measured v. the volume of the titrant, instead of the more common plot of potential v. volume of the titrant. The concentration of the ion being measured during the course of the titration is calculated from the measured potential of the ISE. To facilitate calculation, preprinted paper with a built-in correction for dilution can be used[145]. The linear titration curves obtained by this method are extrapolated to the equivalence point. A few points only, obtained at the beginning of the titration, are used for constructing the titration curve, since the influence of interferences and other factors is negligible in this region.

4.3.6 Continuous measurement

An ISE can be used for continuous monitoring of the composition of samples in industrial analysis. The sample is allowed to flow through an electrolytic cell and the signal is registered or used for control purposes.

If the sample does not contain interfering ions and its ionic strength does not change appreciably, it is possible to place the ISE and the reference electrode directly in the sample stream. Unfortunately, this seldom occurs in industrial analysis. The sample is, therefore, mixed with a reagent which maintains a constant ionic strength or pH or otherwise reduces the effect of interferences. This is done by metering of the sample and reagent in a constant ratio. To avoid interferences from the sample, a reference electrode of the second kind or an ion-selective reference electrode can be placed in the reagent stream. The liquid-junction potential, arising at the point where reagent and sample mix, cannot be eliminated in this case and is still a source of error in this system.

ISEs can be used as sensors in continuous titration or in a continuously performed method of known addition or known subtraction (see Section 4.3.4). In these methods, the sample and the reagent are continuously mixed in a constant ratio and the change in the potential is measured.

For long-term measurements, automatic calibration and zeroing of the electrode system is required. The higher the accuracy required, the more frequently the calibration curve must be checked. Besides the previously discussed factors, it is necessary to check the value of the 'zero potential' E' and the stability of the Nernst slope RT/z_iF. The 'constant' E' does not correspond to the standard electrode potential, E^0, and changes with time. A fluctuation of the order of a few tenths of a millivolt per day is typical, although larger changes have been observed. The Nernst slope is usually reasonably stable and changes only when the ISE is passivated.

Information about the lifetime of an ISE in continuous contact with solution in continuous measurement is very limited.

Reviews of the methods of continuous analysis are given by Light[83, 147], Riseman[148], Mallisa et al.[149] and in the manufacturers' literature[150]. ISEs have however been used for the continuous measurement of fluoride content in water[151, 152] and the concentration of sulphide in black digesting liquor in cellulose production[153], for checking the efficiency of ion exchange water purification[152], for measuring hydrogen chloride in gas[154, 155] and as sensors in the automatic titration of gaseous pollutants[156].

4.4 THE CLASSIFICATION OF ION-SELECTIVE ELECTRODES

The classification of ion-selective electrodes is not yet uniform. These electrodes can be classified as cation or anion selective, homogeneous or heterogeneous and on the basis of the physical state of the ion-selective membrane. The latter will be considered here as the main classification criterion.

4.4.1 Solid electrodes

4.4.1.1 Glass electrodes

The early discovery that the glass electrode sensitive to hydrogen ion activity also responds to the alkali metal ion[157], led to attempts to develop an electrode selective for these ions. It was found that selectivity for different ions is controlled by the composition of the glass[158, 159]. Eisenman and co-workers[69] carried out systematic research on the electrochemical properties of sodium aluminosilicate glasses in a large concentration region and have shown that selectivity to different cations is a continuous function of the glass composition. The literature of this field of research is very extensive and has been systematically evaluated by many authors in the book edited by Eisenman[10] where information about glass composition can be found. Rechnitz[15] evaluated the properties of glass selective electrodes and divided them into three main groups:

(i) pH electrodes with selectivity in the order

$$H^+ \gg Na^+ > K^+, Rb^+, Cs^+, \ldots \gg Ca^{2+}$$

The properties of glass pH electrodes have been exhaustively described elsewhere[10, 43, 160–163].

(ii) cation selective electrodes with selectivity in the order

$$H^+ > K^+ > Na^+ > NH_4^+, Li^+, \ldots \gg Ca^{2+}$$

The selectivity of these electrodes for potassium is about 10 times that for sodium. For reasons of selectivity, valinomycin electrodes are preferable for potassium determinations. Cationic glass electrodes have been used for the determination of NH_4^+, but with poor selectivity. Krull et al.[164] have described a potassium selective sensor with a biological material in the solid membrane. No details have been disclosed on these or on the ammonium ion electrode mentioned by Cosgrove et al.[165].

(iii) sodium ion-selective electrodes with selectivity in the order

$$Ag^+ > H^+ > Na^+ \gg K^+, Li^+, \ldots \gg Ca^{2+}$$

These electrodes are about 500–1000 times more sensitive to sodium then to potassium. Silver ions must be absent and the hydrogen ion concentration is controlled by addition of a buffer. Corning Glass Works NAS 11–18 sodium electrode[15] consists of 11% Na_2O, 18% Al_2O_3, 71% SiO_2, while Beckman Instruments have been using a glass consisting of 27% Na_2O, 4% Al_2O_3, 68% SiO_2. Štefanac and Simon[166] announced a glass with 10^4 times greater selectivity for sodium then for potassium, consisting of 69% SiO_2, 11% Li_2O, 12% Al_2O_3, 6% B_2O_3 and 2 mol % Ga_2O_3.

4.4.1.2 Crystalline electrodes

(a) *Fluoride electrode* — The first ion-selective electrode for the determination of fluorides was described by Frant and Ross[167]. This electrode contains a single crystal of LaF_3, with low electric resistivity, which can be further reduced by doping the crystal with a divalent cation such as Eu^{2+}. The LaF_3 electrode is highly selective for fluoride ions and only hydroxyl ions interfere. Their concentration must be at least ten times lower than the concentration of fluoride ions.

The potential of the LaF_3 electrode is given by the equation[80]

$$E = E'' - 2.3\frac{RT}{F}\log a_{F^-} \tag{4.25}$$

This electrode can also be used to measure the lanthanum ion activity in solutions which initially contain no fluoride. The fluoride activity at the electrode surface is then determined by the solubility product of lanthanum fluoride $K_{sp(LaF_3)} = a_{La^{3+}} \cdot a_{F^-}^3$ and the electrode potential is then given by the equation

$$E = E' + 2.3\frac{RT}{3F}\log a_{La^{3+}} \tag{4.26}$$

Before the measurement, finely divided LaF_3 must be separately added to the sample as the rate of equilibration of a solid single crystal with aqueous solution is slow[80].

The dependence of the potential on $\log a_{F^-}$ is linear in the concentration range $10^0 - 5 \times 10^{-5}$ M fluoride. A lower detection limit in the range 10^{-6}–10^{-7} M has been given by various authors. The limit depends on the solubility of the LaF_3 membrane in the sample solution. If substances which form strong complexes with La^{3+} or F^- are present, e.g. citrate (or EDTA), the concentration of fluoride at the electrode surface increases because of the reaction

$$LaF_3(s) + Cit^{3-} \longrightarrow LaCit^0 + 3F^- \tag{4.27}$$

The concentration of released fluoride ions, which depends on the citrate concentration, determines the lower limit of detection[80].

The mechanisms of hydroxyl-ion interference has not been explained in full. Frant and Ross[167] suggested that hydroxyl ions penetrate through the crystal lattice. A different interference mechanism, suggested by Butler[37], is the formation of $La(OH)_3$ at the electrode surface with subsequent release of F^-.

A secondary form of interference is caused by ions that form stable fluoride complexes and reduce the free fluoride ion concentration in the solution, e.g. Fe^{3+}, Al^{3+}, Be^{2+}. In acid solution, complexes such as HF_2^- and undissociated HF are formed[168] causing a lower electrode reading. On the other hand, fluorosilicate is practically completely hydrolysed and does not influence the potential of the electrode[93]. The effect of secondary interferences has been discussed in detail by Butler[37].

The behaviour of the lanthanum fluoride electrode 'CRYTUR'[169] has been evaluated by Weis[170b].

(b) *Silver halide crystalline electrodes.* Membranes made of fused AgCl and AgBr were proposed by Kolthoff and Sanders[171] for the determination of chloride and bromide ions.

Membranes ground from a single crystal of AgCl or AgBr, and a membrane of polycrystalline AgI, have been used in ISEs for the determination of halides in various electrodes, e.g. in the electrode 'CRYTUR'[169] and in other electrodes, the composition of which has not been disclosed[172-174]. These electrodes react to the silver ion activity which is determined in the presence of halides by the appropriate solubility product. The potential of the electrode is given by the equation

$$E = E'' - 2.3\frac{RT}{F} \log a_{x^-} \qquad (4.28)$$

where a_{x^-} is the activity of halide and E'' is a constant. The lower limit of detection is determined by the solubility of the crystal used in the membrane. When different halide ions are present simultaneously in the sample solution, the selectivity is determined by the ratio of the solubility product of the interfering ion and that being measured. If the interfering ion forms a sparingly soluble precipitate with silver, mixed crystals are fomed on the electrode surface and poisoning of the electrode occurs. The surface of a poisoned electrode can be regenerated by mechanical polishing. Complexing agents such as CN^-, $S_2O_3^{2-}$ and NH_3 interfere with the measurement but do not cause irreversible damage of the electrode surface. Sulphide must be absent because the very insoluble salt Ag_2S is formed. Common anions such as SO_4^{2-}, NO_3^-, PO_4^{3-}, F^-, HCO_3^- and oxidising agents such as Cu^{2+}, Fe^{3+}, MnO_4^- do not interfere. Strong reducing agents such as photographic developers reduce silver halides to metallic silver and ISEs cannot be used in these media. The maximum value of the ratio of the concentration of interfering ion to the concentration of the ion being measured is for:

(i) the chloride electrode: $Br^- - 3 \times 10^{-3}$, $I^- - 5 \times 10^{-7}$, $CN^- - 2 \times 10^{-7}$, $OH^- - 80$, $NH_3 - 0.12$, $S_2O_3^{2-} - 1 \times 10^{-2}$; S^{2-} must be absent.

(ii) the bromide electrode: $Cl^- - 4 \times 10^2$, $I^- - 2 \times 10^{-4}$, $CN^- - 8 \times 10^{-5}$, $OH^- - 3 \times 10^4$, $NH_3 - 2$, $S_2O_3^{2-} - 20$; S^{2-} must be absent.

(iii) the iodide electrode: $Cl^- - 1 \times 10^6$, $Br^- - 5 \times 10^3$, $CN^- - 0.4$, $S_2O_3^{2-} - 1 \times 10^5$; S^{2-} must be absent.

(c) *The Ag_2S–AgX halide electrode.* Ross[80] described ISEs in which silver halides are dispersed in a silver sulphide matrix. Because of its low solubility product, silver sulphide can be considered a chemically inert matrix with a high ionic conductivity. Ag_2S decreases the photo-electric effect and improves the mechanical properties of the membrane. This principle has been used by Orion Research and electrodes for chloride, bromide, iodide, cyanide and thiocyanate determination have been made commercially available. The parameters of these electrodes are similar to those of crystalline electrodes.

The formation of soluble silver complexes has been used in electrodes for the determination of cyanide and thiocyanate. The silver sulphide matrix serves as a diffusion barrier and a steady state at the electrode surface is quickly established. For the determination of complexing agents, iodide electrodes have been used. Silver iodide dissolves in the presence of cyanide ions[80].

$$AgI(s) + 2CN^- \longrightarrow Ag(CN)_2^- + I^- \qquad (4.29)$$

The electrode potential is determinated by the concentration of Ag^+ at the electrode surface via the solubility product of AgI, and the electrode reacts to the cyanide ion concentration in an almost Nernstian manner. The membrane of the electrode dissolves during the measurement and its life-time is, therefore, limited. The sample solution should be stirred at a constant rate. Iodide electrodes have been used[175] for the determination of mercuric ions. Iodide and silver ions must be absent. It is assumed that the following reaction takes place

$$Hg^{2+} + AgI \longrightarrow HgI^+ + Ag^+ \tag{4.30}$$

Concentrations as low as 10^{-8} M Hg^{2+} can be determined. The electrode surface must be regenerated from time to time by repolishing.

(d) *Sulphide ion-selective electrodes with Ag_2S.* The physical properties of silver sulphide enable formation of a rigid and mechanically-resistant membrane by a pressing technique. Silver sulphide has high ionic conductivity and a low solubility product ($K_{Sp} = 10^{-51}$). It is resistant to oxidising and reducing agents and equilibrates rapidly with the solution. Thus it is a suitable material for sulphide ISEs[80, 176]. This electrode reacts to a change in the silver ion activity according to the equation

$$E = E' + 2.3 \frac{RT}{F} \log a_{Ag^+} \tag{4.31}$$

where a_{Ag^+} is given by the solubility product

$$a_{Ag^+} = \left(\frac{K_{sp(Ag_2S)}}{a_{S^{2-}}} \right)^{\frac{1}{2}} \tag{4.32}$$

Substituting equation (4.32) in equation (4.31) results in equation (4.33) which expresses the dependence of the electrode potential on the sulphide ion activity

$$E = E'' - 2.3 \frac{RT}{2F} - \log a_{S^{2-}} \tag{4.33}$$

The potential of the electrode in the presence of sulphide is controlled according to equation (4.33) in the concentration range from saturated solutions to 10^{-8} M. The lower limit of detection is set by absorption and desorption on the walls of the vessel. In the presence of complexing agents, free sulphide ion activities as low as 10^{-19} M can be determined. The concentration of free sulphide ions decreases in acid solution by formation of HS^- and H_2S. Silver and mercury ions interfere, the latter poisoning the electrode surface by formation of mercury sulphide.

(e) *Mixed silver sulphide — metal sulphide electrodes.* — Membranes made of a mixture of Ag_2S and finely divided sulphide of an appropriate metal were used by Ross[80] for the determination of cations such as Cu^{2+}, Pb^{2+} and Cd^{2+}. The electrode detects the silver ion activity, which is determined at the sample-membrane interface by the solubility products of both sulphides[80];

$$K_{sp(Ag_2S)} = a_{Ag^+}^2 \cdot a_{S^{2-}}; \quad K_{sp(MS)} = a_{M^{2+}} \cdot a_{S^{2-}}$$

Eliminating $a_{S^{2-}}$ gives

$$a_{Ag^+} = \left(\frac{K_{sp(Ag_2S)}}{K_{sp(MS)}} a_{M^{2+}} \right)^{\frac{1}{2}} \tag{4.34}$$

Substituting a_{Ag^+} into the Nernst equation for silver ions we obtain

$$E = E'' + \frac{2.3RT}{2F} \log a_{M^{2+}} \tag{4.35}$$

The solubility product of the metallic sulphide MS must be higher than the solubility product of Ag_2S, but sufficiently small that the concentration of M^{2+} produced as a result of the solubility of MS is negligible compared to the concentration of M^{2+} in the sample solution. The rate of equilibration with the sample must be fast[80]. On this principle, ISEs for determining Cu^{2+}, Pb^{2+} and Cd^{2+} were constructed and have been manufactured[177]. The measurable concentration range is about $10^0 - 1 \times 10^{-7}$ M.

4.4.1.3   Heterogeneous membrane electrodes

In these electrodes, the active material, e.g. sparingly soluble precipitate or an ion exchanger, is dispersed in an inert matrix. The mechanical properties of the matrix, which should be rigid, elastic and free of pores and cracks, are important for the preparation of a good ion-selective membrane. A large number of materials have been used. Silicone rubber[28] was found to be an extremely good matrix material.

(a) *The silicone rubber matrix.* — Pungor[13, 14, 178] described the preparation of ion-selective membrane electrodes in which the active material, a sparingly soluble silver precipitate, is dispersed in polysiloxane. The catalyst and cross-linking agents are then added and after homogenisation, a membrane 0.3–0.5 mm thick is formed by calendering. Discs are cut and sealed to the end of a glass tube; the electrode is then filled with an internal filling solution and a reference electrode is added. Before use, the electrode must be activated by soaking for a few hours in solution.

The conductivity of the membrane and its properties depend on the ratio of silicone rubber to the active material (usually a 1:1 mixture is used), the grain size (5–15 μm), and crystalline form. Electrodes of Pungor's type are manufactured by Radelkis Instruments[179] for the determination of Cl^-, Br^-, I^-, S^{2-}, CN^-, SCN^- and Ag^+. These electrodes were evaluated by Rechnitz et al.[74, 75] and more recently by Pungor and Toth[14]; a review was given by Covington[87]. The properties of heterogeneous electrodes are similar to those of crystalline electrodes.

Silicone rubber membranes impregnated with AgCl, AgBr and AgI have been used for the determination of chloride[28, 180], bromide[180], and iodide[29].

A cyanide heterogeneous ISE has been used for the determination of free cyanide[90, 181, 182]. The cyanide electrode is an iodide electrode which reacts to the cyanide concentration by a mechanism similar to that described above. Toth and Pungor[182] examined the response of the cyanide ISE in solutions of the cyanide complexes of copper, nickel, mercury, cadmium, and zinc, which are used in plating baths. In accordance with theoretical considerations, it was found that the cyanide electrode does not respond to complexes with stability constants lower than the stability constant of $Ag(CN)_2^-$; e.g. with the cyanide complexes of copper, nickel and mercury.

$Zn(CN)_4^{2-}$ and $Cd(CN)_4^{2-}$ complexes have a small effect. Cyanide ions bound in undissociated HCN are also not determined.

Silver-sulphide-impregnated silicone rubber[183] has been used for the determination of sulphide and polysulphide[184, 185]. The selectivity of electrodes impregnated with $BaSO_4$ for sulphate determination[28-30] and with $BiPO_4$ for phosphate determination[186] is low. Buchanan and Seago[187] studied the behaviour of impregnated paraffin and silicone rubber membranes toward the transition metals. Hirata and Date[104, 105] obtained copper and lead selective electrodes by impregnating silicone rubber with Cu_2S and PbS. Heterogeneous membrane electrodes for fluoride determination have also been examined[188].

(b) *Miscellaneous matrices.* — A great number of different materials have been tested as matrices. The prepared membranes were initially perm-selective or only partially selective. Frequently used were paraffin[187, 189-195] high molecular substances[196], collodium[9, 197-199], and many others. More details about the materials and methods of preparation can be found elsewhere[8, 9, 11].

Dobbelstein and Diehl[200] prepared solid state electrodes with a limited selectivity for nitrate. Different types of membrane for calcium determination were investigated by Shatkay[201]. Růžička and co-workers[106, 107] used a hydrophobic porous graphite impregnated with a mixture of silver halides and silver sulphide for the determination of halides. Later results have shown that the active material need not be dispersed in the entire matrix. It is sufficient to rub in the active material as a thin layer on the surface of the electrode. Polyvinyl acetate impregnated with HgS [316] and a thermoplastic polymer[202, 203] impregnated with silver halide and metallic sulphides were used for the determination of Cl^-, Cu^{2+} and S^{2-}.

4.4.2 Liquid membrane electrodes

The sensing element of these electrodes is formed by a layer of an organic solvent, in which an ion exchanger is dissolved, and which separates the sample and internal filling solutions, as shown in Figure 4.1. The ion exchanger binds the ion being measured, forming charged or neutral particles, which can move freely within the liquid membrane, in contrast to solid ion-exchangers. As an active substance, neutral carriers, e.g. valinomycin, have also been used. The systems based on the extraction of associated ions[204, 205] or metallic chelates[206] do not show sufficient selectivity. The theory of liquid membrane potentials was discussed by Eisenman, Ciani and Szabo[31, 207, 208].

The organic solvent should have a low solubility in water, a high stability, a low vapour pressure and a high viscosity so that it is retained in the pores of the matrix[15]. The active substance must be sparingly soluble in water, exhibit high selectivity to the ion being measured, and the rate of the ion exchange reaction must be fast[80].

The main problem connected with the design of liquid membrane ion-selective electrodes is mechanical. The liquid membrane must form a well defined and stable boundary with the sample solution, not influenced, e.g. by stirring the solution. ISEs in which an organic phase separates two solutions in U-type cells or in cells shaped like the letter W (the organic solvent has a lower specific gravity than water), are not suitable for practical use. The organic solvent is, therefore, retained between two thin, permeable membranes, e.g. cellulose dialysis membranes[80] or sintered glass discs[209]. In the commercial forms of these electrodes, the organic solvent and ion exchanger are usually immobilised by soaking in a porous disc, which is sometimes connected with a container of organic solvent. The design of some electrodes enables simple exchange of the porous disc when necessary.

The calcium selective electrode, formed by a calcium salt of dodecyl phosphoric acid dissolved in di-n-acetyl phenyl phosphonate[12], was described by Ross[73, 80]. The measurable range is 10^{-1}–10^{-4} M Ca^{2+}. Ions such as H^+, Zn^{2+}, Fe^{2+}, Pb^{2+} and Cu^{2+} interfere with the measurement. The selectivity for calcium is about 100 times higher than for magnesium. This electrode was modified for the determination of the total concentration of divalent cations: the sensitivity for calcium and magnesium is nearly the same, and it is used for the determination of water hardness[177].

Shatkay et al.[201, 210] have used a membrane composed of a tributyl phosphate solution of theonyltrifluoroacetone dispersed in polyvinyl chloride as a calcium selective electrode with a similar selectivity ratio.

Rechnitz and Hseu[211] evaluated Beckman Instrument's calcium selective electrode. No details on the composition have been disclosed.

Anion selective electrodes for the determination of NO_3^-, ClO_4^- and BF_4^- are based on a liquid anion exchanger, containing the o-phenanthroline chelating group[80].

The nitrate selective electrode is insensitive to cations and measures NO_3^- in the concentration range 10^{-1}–10^{-6} M. Principal interferences are ClO_4^-, I^-, ClO_3^- and Br^-. Common anions such as Cl^-, SO_4^{2-}, PO_4^{3-} and CO_3^{2-} have little effect.

The nitrate selective electrode can be used for the determination of fluoroborate ions in the absence of nitrate if the liquid membrane is first converted to the fluoroborate form[212].

Perchlorate electrodes can be used for the determination of ClO_4^- concentrations from 10^{-1}–10^{-5} M in the pH range 4–10; hydroxyl ions interfere. Grekovich et al.[213] have used a thick layer of a toluene solution of tetraoctylammonium perchlorate as an ion exchanger for perchlorate selective electrodes.

A liquid membrane chloride selective electrode is based on the dimethyl-distearylammonium cation[80]. This electrode can be used for chloride determination in the presence of strong reducing agents and sulphide, which must be absent when the AgCl-based electrode is used. Anions such as NO_3^-, ClO_4^-, Br^- and I^- interfere.

The potassium selective valinomycin electrode was designed by Pioda, Stankova and Simon[17, 214]. The active substance is a cyclic polypeptide, the antibiotic valinomycin, which strongly binds potassium ions. The membrane is formed by a 'Millipore' filter, which is saturated by a solution of valinomycin in diphenyl ether. The valinomycin electrode has about 4000 times greater selectivity for potassium than for sodium, with selectivity ratios $K_{K^+/Me}$: $Na^+ - 2.5 \times 10^{-4}$; $H^+ - 5 \times 10^{-5}$; $Li^+ - 2 \times 10^{-4}$; $NH_4^+ - 1 \times 10^{-2}$; $Ag^+ - 2 \times 10^{-9}$; $Cs^+ - 4 \times 10^{-1}$; $Rb^+ - 1.9$. Divalent cations and common anions such as Cl^-, Br^-, I^-, NO_3^- and SO_4^{2-} do not interfere.

The selectivity ratio of other antibiotics of the same group is smaller[215, 216], e.g. for nonactin and monactin, $K_{K^+/Na^+} = 1 \times 10^{-2}$ and 8×10^{-3}, respectively.

Valinomycin dissolved in various solvents was used by Frant and Ross[217] in a potassium electrode[177] which was evaluated by Lal and Christian[218]. The selectivity ratio of different substances was given[219, 220].

A barium ISE was described by Levins[221]. The selective liquid membrane is formed by a polyethylene glycol derivative, dissolved in p-nitroethylbenzene. The selectivity for Ba^{2+} over Ca^{2+} and Mg^{2+} is better than 10^4:1. The selectivity ratios $K_{Ba^{2+}/M}$ are : $Cu^{2+} - 2 \times 10^{-3}$; $Sr^{2+} - 1 \times 10^{-3}$; $K^+ - 1.10^{-3}$; and smaller than 1×10^{-4} for Ca^{2+}, Mg^{2+}, Ni^{2+}, Co^{2+}, Zn^{2+}, Fe^{2+}, Na^+, NH_4^+, Li^+, and H^+. Common anions do not interfere. The electrode was used for the potentiometric titration of sulphate with $BaCl_2$.

Accounts of a few other electrodes with limited selectivity have been published, e.g. electrodes for the determination of polyvalent cations[222], anions with long hydrocarbon chains[223], metallic cations[224], metallic and tetraalkyl ammonium ions[203] and for acetylcholine[225].

4.4.3   Membrane-covered electrodes

The selectivity of ISEs can be increased in some cases by covering the sensing element with an auxiliary layer, which prevents the interfering ions from reaching the electrode surface, or which reacts with the ions of interest forming another ionic species, which is detected by the electrode. These sensors do not yet have a generally accepted name but 'membrane-covered electrodes', 'diffusion electrodes' and others have been used.

The location of the reference electrode in the sensor depends on the electric conductivity of the auxiliary layer. If the conductivity is sufficiently high, the reference electrode is placed in the sample solution. If a non-conductive layer is used, the reference electrode is in contact with a thin layer of electrolyte within the separating membrane.

4.4.3.1   Permeable membranes

A permeable layer through which all ionic species present in the sample solution can diffuse has been used in enzyme electrodes. The sensing element is covered with a layer of immobilised polymeric gel, in which the appropriate enzyme is dispersed. The substance to be determined (substrate) diffuses through the gel layer and reacts in the following way

$$\text{substrate} \xrightarrow{\text{enzyme}} I$$

The concentration of the product, I, is measured by an ISE. Guilbault and co-workers[226-229] determined the concentration of urea by covering a cationic glass electrode sensitive to NH_4^+ with the immobilised enzyme, urease. The concentration of NH_4^+, which is produced by enzymatic hydrolysis, is proportional to the concentration of urea in the sample. In a similar way, amino acids were determined by using L-amino acid oxidase[230, 231]. The selectivity of enzyme electrodes is, at present, limited by the selectivity of glass electrodes, which is not high. Another problem involved in this method is the decrease of the enzyme concentration in the immobilised layer by dissolution in the sample or by chemical attack. It is advantageous that the enzyme-containing layer can be easily replaced when necessary.

Rechnitz and Llenado[232] used a cyanide ISE, which was covered with immobilised β-glucosidase, for the determination of amygdaline.

Uppdike and Hicks[233] constructed an enzyme sensor for the determination of glucose. The decrease in oxygen concentration caused by the enzymatic oxidation of glucose is measured by a sensor working on the polarographic principle (see below). Similar principles have been used in other sensors[234, 235].

4.4.3.2   Semipermeable membranes

The sensing element is covered with a thin membrane, made e.g. of rubber, polyethylene, PTFE, or other plastic material, which separates the sensing element from the solution. High molecular weight substances, colloids, ions, and, to a certain degree, the solvent, cannot penetrate through the membrane, while gases and some volatile substances can diffuse through to the surface of the sensing element. Between the sensing element and the membrane, there is a thin layer of electrolyte in which the gases dissolve and are detected by the sensor.

A high degree of selectivity can be obtained by this method. Unfortunately, it cannot be univerally used.

The membrane material is chosen according to its mechanical properties and permeability to different gases. The response time depends on the diffusion resistance of the layer formed by the semipermeable membrane and the film of electrolyte on the surface of the sensor and on the rate of absorption and desorption of the gas therein. A thin membrane and a thin electrolyte film are, therefore, used. A great number of sensors have been described which differ mainly in the way in which the above mentioned requirements are achieved.

A sensor for the determination of the partial pressure of carbon dioxide was described by Stow[236]. A glass pH electrode was covered by a thin rubber membrane through which CO_2 diffused from the sample to the thin layer of water on the electrode surface. The change in pH caused by the H_2CO_3 formed was proportional to the partial pressure of CO_2. The process is reversible and equilibrium partial pressures on both sides of the membrane are maintained. The sensitivity of the determination can be increased by adding bicarbonate to the electrolyte[237, 238]. Gases which penetrate through the membrane and dissolve in the electrolyte with a subsequent change in pH, e.g. SO_2, interfere. This electrode has been used especially for biological measurement[237–240].

An ammonia electrode recently developed by Orion Research works on a similar principle (Figure 4.3). The sensing element is a flat pH electrode, covered by a gas-permeable membrane. The ammonia in the sample diffuses through the membrane into the electrolyte film until reversible a equilibrium is reached. Ammonia dissolves in the electrolyte and the concentration

$$NH_3 + H_2O \rightleftharpoons NH_4^+ + OH^- \qquad (4.36)$$

of hydroxyl ions is measured. The electrode potential obeys the Nernst equation and is proportional to the ammonia concentration in the range 10^0–10^{-6} M. The only interferences are volatile amines; cations, anions and gases such as CO_2, SO_2, H_2S and HCN do not interfere, because the pH of the sample must be at least 11. It is possible to measure other substances

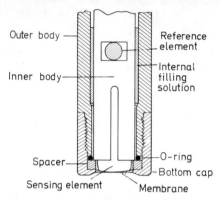

Figure 4.3 A membrane-covered sensor for the determination of ammonia (By courtesy of Orion Research, Inc.)

after conversion into NH_3, e.g. NH_4^+ (by adding hydroxide), NO_3^- (after reducing to NH_3) and organic nitrogen (after Kjeldal digestion).

An oxygen sensor, working on the polarographic principle, was described by Clark et al.[241] and many others brought it to its present perfection[242–246]. The sensing element is a silver cathode, covered by a thin polyethylene membrane and an electrolyte film. Oxygen from the sample diffuses through the membrane and is reduced on the cathode according to the equation

$$O_2 + 2H_2O + 4e \rightleftharpoons 4OH^- \qquad (4.37)$$

The oxygen concentration is proportional to the current passing through the external circuit. This sensor is suitable for the determination of oxygen concentrations above 1% vol.

The above mentioned polarographic sensor is included among potentiometric ISEs to illustrate various measuring possibilities. It can be expected that ISEs will follow the development of other electroanalytical methods and sensors based on a combination of several principles will be constructed. This is necessary for the measurement of high concentrations, where the obtainable accuracy of potentiometric methods is limited by the electrode's logarithmic response. A membrane-covered sensor, working on both the polarographic and potentiometric principles, is not difficult to visualise. The development of selective electrodes for the determination of chlorine, iodine, nitrogen oxides and sulphur dioxide can be expected. The construction of a potentiometric sensor for the determination of traces of oxygen is hampered since the oxygen electrode reacts reversibly to the oxygen concentration at temperatures of a few hundred degrees centigrade. Progress in the field of fuel-cells suggests, however, that the temperature necessary for reversible function could be substantially reduced. The development of a potentiometric oxygen sensor can, therefore, also be expected.

4.5   ANALYTICAL APPLICATIONS OF ION-SELECTIVE ELECTRODES AND COMMERCIALLY AVAILABLE MODELS

Fluoride electrodes have found extensive practical use since they simplify and facilitate analyses. They have been used for the determination of fluoride in water[247–250], plant tissue and vegetation[251, 252], chromium plating baths[253], minerals[170, 254–259], fluorocarbons[260, 261], toothpaste[262], urine and biological materials[263], silicate rocks without separation of aluminium[254], plants and air[264]. Fluoride electrodes have served as indicators in potentiometric titrations[260, 265–267], and as detectors in the continuous titration of gaseous pollutants[156]. Filter paper impregnated with K_2CO_3 [268], CaO[269] and sodium formate[156] has been used for accumulation of traces of fluoride in the atmosphere. Some applications were reviewed by Bazzelle[270].

The calcium selective liquid membrane electrode has been used for the direct determination of calcium in clays and soils[271, 272], milk[273], sea water[274], sera and biological fluids[275–277], and as an indicator in potentiometric titrations of water hardness with EDTA[278] and of high concentrations of calcium[279, 280]. Some biological applications have been reviewed[281].

The chloride ISE was used for the determination of chloride in perspiration[282, 283], cheese[284], milk[285], plating baths[253], plant tissue[286], biological fluids[287], sea water[288], in the presence of interfering ions[265, 289] and for the determination of HCl in gases[154, 155].

The iodide ISE has been used for the determination of iodide in organic materials[290, 291], selenium[258], minerals waters[13], for the indirect determination of sulphur dioxide after reaction with iodine[13], and as an indicator in potentiometric titrations[292].

The bromide ISE was used for the determination of bromide in water[238a], in sera, and as an indicator in potentiometric titrations[288].

The sulphide ISE has been used for the determination of traces of silver[293], of sulphide and disulphide groups in organic compound and proteins[294, 295], in thiols[296], and in digesting liquors used in the production of cellulose[83]. Analytical applications of heterogeneous ISEs have also been described[297].

The cyanide ISE was used for the determination of cyanide in plant tissues[298] and in plating baths[253]. The direct determination of cyanide in the presence of cyanide complexes with heavy metals has been described[182].

The nitrate ISE has been used for the determination of nitrate in limestone[299, 300], soils[300–304], natural waters[305], vegetables[306], plants tissues[302, 307, 308], sodium nitrite[309], and for the determination of nitrogen oxide and nitrogen dioxide in gases[310], and, after modification, for the determination of fluoroborate[212].

The perchlorate ISE has been used as an indicator electrode in the potentiometric titration of perchlorate with tetraphenyl arsonium chloride[311].

The sodium ISE was used for the determination of sodium in biochemical measurements[312], and as an indicator in potentiometric titrations[313]. The potassium electrode was used for the determination of potassium ions in sera[217] and for the measurement of the formation constant of potassium–adenosine triphosphate complex[314].

ISEs have also found extensive use in biochemistry for measurements *in vivo* and *in vitro*. A review of biological application was given by Khuri[312].

Microelectrodes with a liquid ion-exchanger were reviewed by Walker[315] and biological applications of the calcium ISE were given by Moore[281].

Some commercially available ISEs are given in Table 4.1.

Table 4.1   Commercially available ISE

Species sensed	Type	Principal interferences
Ammonia	M	volatile amines
Bromide	S, H	S^{2-}, I^-, CN^-
Cadmium	L	Ag^+, Hg^{2+}, Cu^{2+}
Calcium	L	$Zn^{2+}, Fe^{2+}, Pb^{2+}$
Chloride	S, H	$S^{2-}, CN^-, I^-, Br^-, SCN^-$
	L	$ClO_4^-, I^-, Br^-, NO_3^-; S^{2-}$ can be present
Cupric	S	Ag^+, Hg^{2+}
Cyanide	S, H	S^{2-}, I^-
Fluoride	S	OH^-
Fluoroborate	L	I^-, NO_3^-
Iodide	S, H	S^{2-}, CN^-
Lead	S	Ag^+, Hg^{2+}, Cu^{2+} and high level of Cd^{2+} and Fe^{3+}
Nitrate	L	$ClO_4^-, I^-, ClO_3^-, Br^-$
Perchlorate	L	OH^-
Potassium	L	Cs^+
Sulphide (silver)	S	Hg^{2+}
Sodium	G	Ag^+
Thiocyanate	S	S^{2-}, I^-, CN^-, Br^-
Water hardness (divalent cation)	L	$Zn^{2+}, Fe^{2+}, Cu^{2+}, Ni^{2+}$

G, glass; H, heterogeneous; L, liquid; M, membrane covered; S, solid

References

1. Cremer, M. (1906). *Z. Biol.,* **47,** 562
2. Haber, F. and Klemensiewicz, Z. (1909). *Z. Physik. Chem.,* **67,** 385
3. Donnan, F. G. (1925). *Chem. Revs.,* **1,** 73
4. Teorell, T. (1935). *Proc. Nat. Acad. Sci.,* **21,** 152
5. Marshall, C. E. (1948). *J. Phys. Chem.,* **52,** 1284
6. Scatchard, G. (1953). *J. Amer. Chem. Soc.,* **75,** 2883
7. Nicolsky, B. P., Schultz, M. M., Belijustin, A. A. and Lev, A. A. *Recent Developments in the Ion-Exchange Theory and its Application in the Chemistry of Glass,* Chapter 6, see Ref. 10
8. Hills, G. J. (1961). *Membrane Electrodes,* Chapter 6, see Ref. 38
9. Solner, K. (1968). *Ann. N. Y. Acad. Sci.,* **148,** 154
10. Eisenman, G., Ed. (1967). *Glass Electrodes for Hydrogen and Other Ions,* (New York: Marcel Dekker)

11. Lakshminarayanaiah, N. (1965). *Chem. Rev.*, **65**, 491
12. Covington, A. K. (1969). *Chem. Brit.*, **5**, 388
13. Pungor, E. (1967). *Anal. Chem.*, **39**, 29A
14. Pungor, E. and Toth, K. (1970). *Analyst*, **95**, 625
15. Rechnitz, G. A. (1967). *Chem. Eng. News*, **45**, No. 25, 146
16. Rechnitz, G. A. (1969). *Anal. Chem.*, **41**, No. 12, 109A
17. Simon, W. Wuhrman, H. R., Vašak, M., Pioda, L. A. R., Dohner, R. and Štefanac, Z. (1970). *Angew. Chem.*, Internat. Ed. **9**, 445
18. Durst, R. A. Ed. (1969). *Ion-Selective Electrodes*, NBS Special Publ. No. 314, (Washington: U. S. Government Printing Office)
19. Koryta, J. (1972). *Ion-Selective Membrane Electrodes*, (Prague: Academia) (in Czech, English edition in prep.)
20. Doremus, R. H., Chapter 4, *Diffusion Potential in Glass*, See ref. 10
21. Karrelman, G. and Eisenman, G. (1962). *Bull. Math. Biophys.*, **24**, 413
22. Conti, F. and Eisenman, G. (1965). *Biophys. J.*, **5**, 247
23. Eisenman, G., *The Origin of the Glass Electrode Potential*, Chapter 5, See Ref. 10
24. Bose, S. K. (1960). *J. Indian Chem. Soc.*, **37**, 465
25. Basu, A. S. (1958). *J. Indian Chem. Soc.*, **39**, 619
26. Buck, R. P. (1968). *Anal. Chem.*, **40**, 1432
27. Buck, R. P. (1968). *Anal. Chem.*, **40**, 1439
28. Pungor, E., Havas, J. and Toth, K. (1964). *Acta Chim. Hung.*, **41**, 239
29. Pungor, E., Toth, K. and Havas, J. (1965). *J. Hung. Sci. Instr.*, **3**, 2
30. Pungor, E. and Havas, J. (1966). *Acta Chim. Hung.*, **50**, 77
31. Eisenman, G., *Theory of Membrane Potentials*, Chapter 1, see Ref. 18
32. Eisenman, G. (1968). *Ann. N. Y. Acad. Sci.*, **148**, 5
33. Eisenman, G. (1968). *Anal. Chem.*, **40**, 310
34. Pungor, E., Havas, J. and Toth, K. (1966). *Acta Chim. Hung.*, **48**, 17
35. Harned, H. S. and Ehlers, R. W. J. (1932). *J. Amer. Chem. Soc.*, **54**, 1350
36. Harned, H. S. and Owen, B. B. (1958). *The Physical Chemistry of Electrolytic Solutions*, (New York: Reinhold Publ. Corp.)
37. Butler, J. N., *Thermodynamic Studies*, Chapter 5, see Ref. 18
38. Ives, D. J. G. and Janz, G. J. (1961). *Reference Electrode, Theory and Practice*, (New York: Academic Press)
39. Beck, W. H., Bottom, A. E. and Covington, A. K. (1968). *Anal. Chem.*, **40**, 501
40. Bottom, A. E. and Covington, A. K., Chapter 3, see Ref. 18
41. Covington, A. K. and Lilley, T. H. (1967). *Phys. Chem. Glasses*, **8**, 88
42. Manahan, S. E. (1970). *Anal. Chem.*, **42**, 128
43. Bates, R. G. (1964). *Determination of pH, Theory and Practice*, (New York: Wiley)
44. Mattock, G. and Band, B. M. (1967). *Glass Electrodes for Hydrogen and Other Cations*, (G. Eisenman, editor). Chapter 1, (New York: M. Dekker)
45. Covington, A. K., *Reference Electrodes*, Chapter 4, see Ref. 18
46. MacInnes, D. A. (1961). *Principles of Electrochemistry*, Chapter 13, (New York: Dover Publ.)
47. Planck, M. (1890). *Ann. Physik.*, **39**, 161
48. Planck, M. (1890). *Ann. Physik.*, **40**, 561
49. Henderson, P. (1908). *Z. Physik. Chem.*, **60**, 325
50. Henderson, P. (1907). *Z. Physik. Chem.*, **59**, 118
51. Lewis, G. N. and Sargent, L. W. (1909). *J. Amer. Chem. Soc.*, **31**, 363
52. Guggenheim, E. A. (1930). *J. Amer. Chem. Soc.*, **52**, 1315
53. Alner, D. J. and Greczek, J. J. (1965). *Lab. Practice*, **14**, 721
54. Wynn, V. and Ludbrook, J. (1957). *Lancet*, **272**, 1068
55. Wright, M. P. (1959). *Symposium on pH and Blood Gas Measurements*, (R. F. Woolmer, editor). (London: Churchill)
56. Lamb, A. B. and Larson, A. T. (1920). *J. Amer. Chem. Soc.*, **42**, 229
57. Lekhani, J. V. (1932). *J. Chem. Soc.*, 179
58. Maclagan, N. F. (1929). *Biochem. J.*, **23**, 309
59. Roberts, E. J. and Fenwick, F. (1927). *J. Amer. Chem. Soc.*, **49**, 2787
60. Siggaard, O. and Anderson, O. (1961). *J. Clin. Lab. Invest.*, **13**, 205
61. Jenny, H., Nielson, T. R., Coleman, N. T. and Williams, D. E. (1950). *Science*, **112**, 164
62. Kater, J. A. R., Leonard, J. E. and Matsuyama, G. (1968). *Ann. N. Y. Acad. Sci.*, **148**, 54

63. Pallman, H. (1930). *Koll. Chem. Ber.*, **30**, 334
64. Bower, C. A. (1961). *Soil Sci. Soc. Proc.*, **25**, 18
65. Merril, C. R. (1961). *Nature (London)*, **192**, 1087
66. Covington, A. K. (1966). *Electrochim. Acta*, **11**, 959
67. Finkelstein, N. P. and Werdier, E. T. (1957). *Trans. Faraday Soc.*, **53**, 1618
68. Ross, J. W., Chapter 2, p. 72. see Ref. 18
69. Eisenman, G., Rudin, D. O. and Casby, J. (1957). *Science*, **126**, 831
70. Garrels, R. M., Sato, M., Thompson, M. E. and Truesdell, A. H. (1962). *Science*, **135**, 1045
71. Isard, J. O., see Ref. 38, p. 64
72. Nicolsky, B. P. (1937). *Acta Physikochim. USSR*, **7**, 597
73. Ross, J. W. (1967). *Science*, **156**, 1378
74. Rechnitz, G. A., Kresz, M. R. and Zamochnik, S. B. (1966). *Anal. Chem.*, **38**, 973
75. Rechnitz, G. A. and Kresz, M. R. (1966). *Anal. Chem.*, **38**, 1786
76. Pungor, E. and Toth, K. (1969). *Anal. Chim. Acta*, **47**, 291
77. Pungor, E. and Toth, K. (1970). *Hung. Sci. Instr.*, No. 18, 1
78. Havas, J. (1971, August 2–5). *Proceedings 2nd Conference on Applied Physical Chemistry*, 631. Veszprém (Hungary)
79. Bates, R. G. and Alfenaar, M., *Activity Standards for Ion-Selective Electrodes*, Chapter 6, see Ref. 18
80. Ross, J. W., *Solid State and Liquid Membrane Ion-Selective Electrodes*, Chapter 2, see Ref. 18
81. Orion Research, Inc. (1970). *Newsletter*, **2**, 5
82. Frant, M. S. and Ross, J. W. (1968). *Anal. Chem.*, **40**, 1169
83. Light, T. S., *Industrial Analysis and Control*, Chapter 6, see Ref. 18
84. Ross, J. W., Chapter 2, p. 60. see Ref. 18
85. de Bethune, A. J., Light, T. S. and Swendeman, N. (1959). *J. Electrochem. Soc.*, **106**, 616
86. Pungor, E., Toth, K. and Havas, J. (1966). *Microchim. Acta*, 689
87. Covington, A. K., *Heterogenous Membrane Electrodes*, Chapter 3, see Ref. 18
88. Orion Research, Inc. (1971). *Newsletter*, **3**, 30
89. Anderson, K. P., Butler, E. A. and Wolley, E. M. (1971). *J. Phys. Chem.*, **75**, 93
90. Toth, K. and Pungor, E. (1968). *Proc. IMECO Symposium on Electrochemical Sensors*, Veszprém (Hungary), p. 35.
91. Rechnitz, G. A., *Analytical Studies on Ion-Selective Membrane Electrodes*, Chapter 9, see Ref. 18
92. Rechnitz, G. A. and Kugler, G. C. (1967). *Anal. Chem.*, **39**, 1682
93. Bock, H. and Strecker, S. (1968). *Z. Anal. Chem.*, **235**, 322
94. Hseu, T. M. and Rechnitz, G. A. (1968). *Anal. Chem.*, **40**, 1054
95. Light, T. S. and Swartz, J. L. (1968). *Anal. Lett.*, **1**, 825
96. McClure, J. E. and Rechnitz, G. A. (1966). *Anal. Chem.*, **38**, 136
97. Rechnitz, G. A. and Lin, Z. F. (1967). *Anal. Chem.*, **39**, 1406
98. Rechnitz, G. A. and Lin, Z. F. (1968). *Anal. Chem.*, **40**, 696
99. Srinivasan, K. and Rechnitz, G. A. (1968). *Anal. Chem.*, **40**, 1818
100. Fleet, B. and Rechnitz, G. (1970). *Anal. Chem.*, **42**, 690
101. Consens, R. H., Ives, D. J. G. and Pittman, R. W. (1953). *J. Chem. Soc.*, 3972
102. Hammett, L. P. and Lorch, A. E. (1933). *J. Amer. Chem. Soc.*, **55**, 70
103. Janz, G. J. and Ives, D. J. G. (1968). *Ann. N. Y. Acad. Sci.*, **148**, 210
104. Hirata, H. and Date, K. (1970). *Talanta*, **17**, 883
105. Hirata, H. and Date, K. (1971). *Anal. Chem.*, **43**, 279
106. Růžička, J. and Lamm, G. G. (1971). *Anal. Chim. Acta*, **53**, 206
107. Růžička, J. and Lamm, G. G. (1971). *Anal. Chim. Acta*, **54**, 1
108. Covington, A. K., p. 112, see Ref. 18
109. Jansta, J., *Chem. Listy*, in press
110. Pinching, G. T. and Bates, R. G. J. (1970). *J. Res. Nat. Bur. Stand.*, **37**, 311
111. Mackfarlane, A. J. (1931). *J. Chem. Soc.*, 3212
112. Marov, G. G., De Lollis, N. J. and Acree, S. F. (1945). *Res. Nat. Bur. Stand.*, **34**, 115
113. Bates, R. G., see Ref. 43, p. 286
114. Hills, G. J. and Ives, D. J. C. (1951). *J. Chem. Soc.*, 311
115. Alner, D. J., Greczek, J. J. and Smeeth, A. G. (1967). *J. Chem. Soc.*, 1205
116. Armstrong, R. D., Fleischmann, M. and Thirsk, H. R. (1965). *Trans. Faraday Soc.*, **61**, 2238

117. Covington, A. K., Dobson, J. V. and Lord Wynne-Jones (1967). *Electrochim. Acta*, **12**, 525
118. Fricke, H. K. (1960). *Beitrage zur angewandte Glassforschung*, 175 (Schott E., editor). (Stuttgart: Wiss. Verlagsgesellschaft)
119. Brand, M. D. J. and Rechnitz, G. A. (1970). *Anal. Chem.*, **42**, 616
120. de Brouckere, L. (1928). *Bull. Soc. Chem. Belg.*, **37**, 103
121. Malmstadt, H. V. and Winefordner, J. D. (1959). *Anal. Chim. Acta*, **20**, 283
122. Malmstadt, H. V. and Pardue, H. L. (1960). *Anal. Chem.*, **32**, 1034
123. Malmstadt, H. V., Hadjiioanou, T. P. and Pardue, H. L. (1960). *Anal. Chem.*, **32**, 1039
124. Malmstadt, H. V. and Winefordner, J. D. (1961). *Anal. Chim. Acta*, **24**, 91
125. Durst, R. A. and Taylor, J. K. (1967). *Anal. Chem.*, **39**, 1374
126. Durst, R. A., May, E. L. and Taylor, J. K. (1968). *Anal. Chem.*, **40**, 977
127. Durst, R. A. (1968). *Anal. Chem.*, **40**, 931
128. Durst, R. A., *Analytical Techniques and Applications of Ion-Selective Electrodes*, Chapter 11, see Ref. 18
129. Garrels, R. M., p. 355–361, see Ref. 10
130. Orion Research, Inc., Model 94–16 Instruction Manual, Cambridge, Mass.
131. Orion Research, Inc. (1969). *Newsletter*, **1**, 10
132. Orion Research, Inc. (1970). *Newsletter*, **2**, 23
133. Orion Research, Inc. (1970). *Newsletter*, **2**, 7
134. Orion Research, Inc. (1970). *Newsletter*, **2**, 33
135. Orion Research, Inc. (1970). *Newsletter*, **2**, 43
136. Durst, R. A. (1969). *Microchim. Acta*, **3**, 611
137. Meites, L. and Goldman, J. A. (1963). *Anal. Chim. Acta*, **29**, 472
138. Meites, L. and Goldman, J. A. (1964). *Anal. Chim. Acta*, **30**, 18
139. Meites, L. and Meites, T. (1967). *Anal. Chim. Acta*, **37**, 1
140. Carr, P. W. (1971). *Anal. Chem.*, **43**, 425
141. Schultz, F. A. (1971). *Anal. Chem.*, **43**, 502
142. Schultz, F. A. (1971). *Anal. Chem.*, **43**, 1523
143. Gran, G. (1950). *Acta Chem. Scand.*, **4**, 559
144. Gran, G. (1952). *Analyst*, **77**, 661
145. Orion Research, Inc. (1970). *Newsletter*, **2**, 49
146. Orion Research, Inc. (1971). *Newsletter*, **3**, 1
147. Light, T. S. (1969). *Industrial Water Engineering*, **6**, No. 9, 33
148. Riseman, J. M. (1970, May 25–27) paper presented at *16th National Symposium, Instrument Society of America*, Pittsburg, Pa.
149. Malissa, H. and Jellinek, G. (1969). *Z. Anal. Chem.*, **245**, 70
150. Orion Research, Inc. (1970). *Newsletter*, **2**, 21
151. Babcock, R. H. and Johnson, K. A. (1968). *J. Amer. Water Works Ass.*, **60**, 953
152. Light, T. S. and Mannion, R. F., see Ref. 83
153. Light, T. S. and Swartz, J. L., see Ref. 18
154. Brittan, M. I., Hanf, N. W. and Liebenberg, R. R. (1970). *Anal. Chem.*, **42**, 1306
155. Lee, T. G. (1969). *Anal. Chem.*, **41**, 391
156. Elfers, L. A. and Decker, C. E. (1968). *Anal. Chem.*, **40**, 1658
157. Hughes, W. S. (1922). *J. Amer. Chem. Soc.*, **44**, 2860
158. Lengyel, B. and Blum, E. (1934). *Trans. Faraday Soc.*, **30**, 461
159. Nicolsky, B. P. and Tolmacheva, T. A. (1937). *Zh. Fiz. Khim.*, **10**, 513
160. Dole, M. (1941). *The Glass Electrode*, (New York: Wiley)
161. Eisenman, G. (1965). *Advances in Analytical Chemistry and Instrumentation*, Vol. 4, (New York: Wiley—Interscience)
162. Mattock, G. (1961). *pH Measurement and Titration*, (London: Heywood)
163. Mattock, G. (1963). *Advances in Analytical Chemistry and Instrumentation*, Vol. 2, (C. N. Reilley, editor). (New York: Wiley—Interscience)
164. Krull, I. H., Mask, C. A. and Cosgrove, R. E. (1970). *Anal. Lett.*, **3**, 43
165. Cosgrove, R. E., Mask, C. A. and Krull, I. H. (1970). *Anal. Lett.*, **3**, 457
166. Štefanac, Z. and Simon, W. (1967). *Anal. Lett.*, **1**, No. 2, 1
167. Frant, M. S. and Ross, J. W. (1966). *Science*, **154**, 1553
168. Srinivasan, K. and Rechnitz, G. A. (1968). *Anal. Chem.*, **40**, 509
169. Research Institute of Monocrystals, Turnov, (Czechoslovakia)
170. Weis, D. (1969). *Chem. Listy*, **63**, 1152

170a. Weis, D. (1971). *Chem. Listy,* **65,** 306
170b. Weis, D. (1971). *Chem. Listy,* **65,** 1091
171. Kolthoff, I. M. and Sanders, H. L. (1937). *J. Amer. Chem. Soc.,* **29,** 416
172. Beckman Instruments, Inc., Fullerton, Cal., Bulletin 7145—A
173. Corning Glass Works, Medfield, Mass., Bulletin ISE—1
174. Philips, Eindhoven, The Netherlands, Bulletin ISE
175. Orion Research, Inc. (1970). *Newsletter,* **2,** 41
176. Frant, M. S. and Ross, J. W. (1966). Paper presented at the *Eastern Analytical Symposium,* New York
177. Orion Research, Inc. Guide to Specific Ion Electrode, CAT/961
178. Havas, J. and Pungor, E. (1968). Proc. *IMECO Symposium on Electrochemical Sensors,* Veszprém (Hungary)
179. Radelkis, *Electrochemical Instruments,* Budapest, Hungary
180. Havas, J., Papp, E. and Pungor, E. (1967). *Magy. Kem. Foly,* **73,** 292
181. Pungor, E. and Toth, K. (1964). *Microchim. Acta,* 565
182. Toth, K. and Pungor, E. (1970). *Anal. Chim. Acta,* **51,** 221
183. Pungor, E., Smidt, E. and Toth, K. (1968). Proc. *IMECO Symposium on Electrochemical Sensors,* p. 121. Veszprém (Hungary)
184. Papp, J. and Havas, J. (1968). Proc. *IMECO Symposium on Electrochemical Sensors,* Veszprém (Hungary)
185. Papp, J. and Havas, J. (1970). *Hung. Sci. Instr.* No. 17, 17
186. Rechnitz, G. A., Lin, Z. F. and Zamochnik, S. B. (1967). *Anal. Lett.,* **1,** 29
187. Buchanan, E. B. and Seago, J. L. (1968). *Anal. Chem.,* **40,** 517
188. MacDonald, A. M. C. and Tóth, K. (1968). *Anal. Chim. Acta,* **41,** 99
189. Cloos, P. and Fripiat, J. J. (1960). *Bull. Soc. Chim. Fr.,* **86,** 423, 2105
190. Fischer, R. B. and Babcock, R. F. (1958). *Anal. Chem.,* **30,** 1732
191. Pungor, E. and Hollós-Rokosiayi, E. (1961). *Acta Chim. Hung.,* **27,** 63
192. Tendeloo, H. J. C. and Krips, A. (1957). *Rec. Trav. Chim.,* **76,** 703
193. Tendeloo, H. J. C. and Krips, A. (1957). *Rec. Trav. Chim.,* **76,** 946
194. Tendeloo, H. J. C. and Krips, A. (1959). *Rec. Trav. Chim.,* **78,** 177
195. Tendeloo, H. J. C. and van der Voort, F. H. (1960). *Rec. Trav. Chim.,* **79,** 639
196. Geyer, R. and Syring, W. Z. (1966). *Z. Chem.,* **6,** 92
197. Gragor, H. P. and Solner, K. (1954). *J. Phys. Chem.,* **58,** 409
198. Carr, C. W. and Solner, K. (1944). *J. Gen. Physiol.,* **28,** 119
199. Carr, C. W., Gregor, H. P. and Solner, K. (1945). *J. Gen. Physiol.,* **28,** 179
200. Dobbelstein, T. N. and Diehl, H. (1969). *Talanta,* **16,** 1341
201. Shatkay, A. (1967). *Anal. Chem.,* **39,** 1056
202. Růžička, J. and Tjell, J. Chr. (1969). *Anal. Chim. Acta,* **47,** 475
203. Sharp, M. and Johansson, G. (1971). *Anal. Chim. Acta,* **54,** 13
204. Coetzee, C. J. and Frieser, A. (1968). *Anal. Chem.,* **40,** 2071
205. Baum, G. (1970). *Anal. Lett.,* **3,** No. 3, 105
206. Růžička, J. and Tjell, J. Chr. (1970). *Anal. Chim. Acta,* **49,** 346
207. Ciani, S., Eisenman, G. and Szabó, G. (1969). *Membrane Biol.,* **1,** 1
208. Szabó, G., Eisenman, G. and Ciani, S. (1969). *Membrane Biol.,* **1,** 346
209. Covington, A. K. (1971). *Anal. Chim. Acta,* **55,** 453
210. Bloch, R., Shatkay, A. and Saroff, H. A. (1967). *Biophys. J.,* **7,** 865
211. Rechnitz, G. A. and Hseu, T. M. (1969). *Anal. Chem.,* **41,** 111
212. Carlson, R. M. and Paul, J. L. (1968). *Anal. Chem.,* **40,** 1292
213. Grekovich, A. L., Materova, E. A. and Belinskaya, F. A. (1971). *Elektrochimia,* **8,** 1275
214. Pioda, L. A. R., Stankova, V. and Simon, W. (1969). *Anal. Lett.,* **2,** No. 12, 665
215. Shemyakin, M. M., Ovchinikov, Yu. A., Ivanov, V. T., Antonov, V. K., Shkrob, A. M., Mikhaleva, T. T., Eustratov, A. V. and Malenkov, G. G. (1967). *Biochem. Biophys. Res. Comm.,* **29,** 834
216. Wipf, H. K., Pioda, L. A. R., Štefanac, Z. and Simon, W. (1968). *Helv. Chim. Acta,* **51,** 377
217. Frant, M. S. and Ross, J. W. (1970). *Science,* **167,** 987
218. Lal, S. and Christian, G. D. (1970). *Anal. Lett.,* **3,** 11
219. Orion Research, Inc. (1970). *Newsletter,* **2,** 14
220. Pedersen, C. J. (1967). *J. Amer. Chem. Soc.,* **89,** 7017
221. Levins, R. J. (1971). *Anal. Chem.,* **43,** 1045

222. Harrell, J. B., Jones, A. D. and Chopin, G. R. (1969). *Anal. Chem.,* **41,** 1459
223. Gavach, C. and Bertrand, C. (1971). *Anal. Chim. Acta,* **55,** 385
224. Scibona, G., Mantella, L. and Danesi, P. R. (1970). *Anal. Chem.,* **42,** 844
225. Corning Glass Works, Medfield, Mass.
225a. Corning Glass Works, Technical Information, Catalog No. 47126
226. Guilbault, G. G. and Montalvo, J. (1969). *Anal. Lett.,* **2,** 283
227. Guilbault, G. G. and Montalvo, J. (1969). *J. Amer. Chem. Soc.,* **91,** 2164
228. Guilbault, G. G. and Montalvo, J. (1970). *J. Amer. Chem. Soc.,* **92,** 2533
229. Guilbault, G. G. and Hrabánková, E. (1970). *Anal. Chim. Acta,* **52,** 287
230. Guilbault, G. G. and Hrabánková, E. (1970). *Anal. Chem.,* **42,** 1779
231. Guilbault, G. G. and Hrabánková, E. (1970). *Anal. Lett.,* **3,** 53
232. Rechnitz, G. A. and Llenado, R. (1971). *Anal. Chem.,* **43,** 283
233. Updike, S. J. and Hicks, G. P. (1967). *Nature (London),* **214,** 986
234. Williams, D. L., Doig, A. R. and Rokosi, A. (1970). *Anal. Chem.,* **42,** 118
235. anonymous (1971). *Chem. Eng. News,* **49,** No. 19, 23
236. Stow, R. W., Baer, R. F. and Randall, B. F. (1957). *Arch. Phys. Med. Rehabil.,* **38,** 646
237. Severinghaus, J. W. and Bradley, A. F. Jr. (1958). *J. Appl. Physiol.,* **13,** 515
238. Severinghaus, J. W. (1968). *Ann. N. Y. Acad. Sci.,* **148,** 115
238a. Harriss, R. C. and Williams, H. H. (1969). *J. Appl. Meteorol.,* **8,** 299
239. Dubowski, K. M. (1966). *Clin. Pathol.,* 124
240. Rispens, P. and Hoek, W. (1968). *Clin. Chim. Acta,* **22,** 291
241. Clark, L. C., Jr., Weld, R. G. and Taylor, Z. (1953). *J. Appl. Physiol.,* **6,** 189
242. Barendrecht, E. (1965). *Chem. Weekblad,* **61,** 555
243. Carritt, D. E. and Kanwisher, J. W. (1959). *Anal. Chem.,* **31,** 5
244. Fatt, I. (1968). *Ann. N. Y. Acad. Sci.,* **148,** 81
245. Mancy, K. H., Okun, D. A. and Reilley, C. N. (1962). *J. Electroanal. Chem.,* **4,** 65
246. Sawyer, D. T., George, R. S. and Rhodes, R. C. (1959). *Anal. Chem.,* **31,** 2
247. Crosby, N. T., Dennis, A. L. and Stevens, J. G. (1968). *Analyst,* **93,** 643
248. Mesner, R. E. (1968). *Anal. Chem.,* **40,** 443
249. Harwood, J. E. (1969). *Water Res.,* **3,** 273
250. Warner, T. B. (1969). *Anal. Chem.,* **41,** 527
251. Gyoerkoes, T., White, D. A. and Luthy, R. S. (1970). *Journal of Metals,* **22,** 29A
252. Levaggi, D. A., Oyung, W. and Feldstein, M. (1971). *J. Air Pollution Ass.,* **21,** 277
253. Frant, M. S. (1967). *Plating,* **54,** 702
254. Ingram, B. L. (1970). *Anal. Chem.,* **42,** 1825
255. van Loon, J. C. (1968). *Anal. Lett.,* **1,** 393
256. Oliver, R. T. and Clayton, A. G. (1970). *Anal. Chim. Acta,* **51,** 409
257. Rinaldo, P. and Montesi, P. (1971). *La Chimica e L'Industria,* **53,** 26
258. Westerlund-Helmerson, U. (1971). *Anal. Chem.,* **43,** 1121
259. Duff, E. J. and Stuart, J. L. (1970). *Anal. Chim. Acta,* **52,** 155
260. Light, T. S. and Mannion, R. F. (1969). *Anal. Chem.,* **41,** 107
261. Selig, W. (1970). *Z. Anal. Chem.,* **249,** 30
262. Shane, M. and Miele, D. (1968). *J. Pharm. Sci.,* **57,** 1260
263. Kakabadse, G. J., Manohin, B., Bather, J. M., Weller, E. C. and Woodridge, P. (1971). *Nature (London),* **229,** 626
264. Buck, M. and Reusman, G. (1971). *Fluoride,* **4,** 5
265. Havas, J., Papp, E. and Pungor, E. (1968). *Acta Chim. Hung.,* **58,** 9
266. Lingane, J. J. (1967). *Anal. Chem.,* **39,** 881
267. Lingane, J. J. (1968). *Anal. Chem.,* **40,** 935
268. Mukai, K. and Ishida, H. (1970). *Newsletter,* **2,** 17
269. Levaggi, D. A. Oyuhg, W. and Feldstein, M. (1970). *Newsletter,* **2,** 17
270. Bazzelle, W. E. (1971). *Anal. Chim. Acta,* **54,** 29
271. Fertl, W. and Jessen, F. W. (1969). *Clays and Clay Minerals,* 17
272. Woolson, E. A., Axley, J. H. and Kearney, P. C. (1970). *Soil Sci.,* **109,** 279
273. Muldoon, P. J. and Liska, B. J. (1969). *J. Diary Sci.,* **52,** 460
274. Thompson, M. E. and Ross, J. W. (1966). *Science,* **154,** 1643
275. Hattner, R. S., Johnson, J. W., Bernstein, D. S. Wachman, A. and Brackman, J. (1970). *Clin. Chim. Acta,* **28,** 67
276. Moore, E. W. (1970). *J. Clin. Invest.,* **49,** 318
277. Moore, E. W. and Makhlouf, G. M. (1968). *Gastroenterology,* **55,** 465

278. Oliver, R. T. and Mannion, R. F. (1970). *ISA Anal. Instr.*, (pre-print), VIII-3
279. Huston, R. and Butler, J. N. (1969). *Anal. Chem.*, **41**, 200
280. Li, T. K. and Piechocki, J. T. (1971). *Clin. Chem.*, **17**, 411
281. Moore, E. W., *Studies with Ion-Exchange Calcium Electrode in Biological Fluids: Some Applications in Biomedical Research and Clinical Medicine*, Chapter 7, see Ref. 18
282. Hansen, L. Buechele, M., Koroshec, J. and Warwick, W. J. (1968). *Amer. J. Clin. Path.*, **49**, 834
283. Kopito, L. and Schwachman, H. (1969). *Pediatrics*, **43**, 794
284. Holsinger, V. H., Posati, L. P. and Pallansch, M. J. (1967). *J. Diary Sci.*, **50**, 1189
285. Muldoon, P. J. and Liska, B. J. (1971). *J. Diary Sci.*, **54**, 117
286. LaCroix, R. L., Keeney, D. R. and Walsh, L. M. (1970). *Soil Sci. & Plant Anal.*, **1**, 1
287. Kriigsman, W., Mansveld, J. F. and Griepink, B. (1970). *Clin. Chim. Acta*, **29**, 575
288. Jagner, D. and Årén, K. (1970). *Anal. Chim. Acta*, **52**, 491
289. van Loon, J. C. (1968). *Analyst*, **93**, 1068
289a. van Loon, J. C. (1971). *Anal. Chim. Acta*, **54**, 23
290. Paletta, B. and Panzenbeck, K. (1969). *Clin. Chim. Acta*, **26**, 11
291. Paletta, B. (1969). *Microchim. Acta*, **6**, 1210
292. Czákvári, B. and Mészáros, K. (1968). *Hung. Sci. Instrum.*, **11**, 9
293. Müller, D. C., West, P. W. and Müller, R. H. (1969) *Anal. Chem.*, **41**, 2038
294. Harrap, B. S. and Gruen, L. C. (1971). *Anal. Biochem.*, **42**, 398
295. Selig, W. (1970). *Mikrochim. Acta*, 168
296. Gruen, L. C. and Harrap, B. S. (1971). *Anal. Biochem.*, **42**, 377
297. Mascini, M. and Liberti, A. (1970). *Anal. Chim. Acta*, **51**, 231
298. Gillingham, J. T., Shirer, M. M. and Page, N. R. (1969). *Agron. J.*, **61**, 717
299. Chalk, P. M. and Keeney, D. R. (1971). *Nature (London)*, **229**, 42
299a. Chatterjee, B. and Marshall, C. E. (1950). *J. Phys. Chem.*, **54**, 671
300. Mahendrappa, M. K. (1969). *Soil Sci.*, **108**, 132
301. Bremner, J. M., Bundy, L. G. and Agarwal, A. S. (1968). *Anal. Lett.*, **1**, 837
302. Mack, A. R. and Sanderson, R. B. (1971). *Can. J. Soil Sci.*, **51**, 95
303. Myers, R. J. K. and Paul, E. A. (1968). *Can. J. Soil Sci.*, **48**, 369
304. Øien, A. and Selmer-Olsen, A. R. (1969). *Analyst*, **94**, 888
305. Shaw, E. C. and Willey, P. (1969). *Calif. Agr.*, **5**, 11
306. Baker, A. V., Peck, N. H. and MacDonald, G. E. (1971). *Agron. J.*, **63**, 126
307. Baker, A. S. and Smith, R. (1969). *J. Agr. Food Chem.*, **17**, 1284
308. Paul, J. L. and Carlson, R. M. (1968). *J. Agr. Food Chem.*, **16**, 766
309. Gehring, D. G., Dippel, W. A. and Boucher, R. S. (1970). *Anal. Chem.*, **42**, 1686
310. Di Martini, R. (1970). *Anal. Chem.*, **42**, 1102
311. Baczuk, R. J. and Du Bois, R. J. (1968). *Anal. Chem.*, **40**, 685
312. Khuri, N. R., *Ion-Selective Electrodes in Biochemical Research*, Chapter 8, see Ref. 18
313. Carr, J. D. and Schwartzfoger, D. G. (1970). *Anal. Chem.*, **42**, 1238
314. Rechnitz, G. A. and Mohan, M. S. (1970). *Science*, **168**, 1460
315. Walker, J. L., Jr. (1971). *Anal. Chem.*, **43**, No. 3, 89A
316. Araki, S., Suzuki, S. and Tomita, Y. (1970). *Symposium on Analytical Chemistry*, Fukuoka, (Japan)

5
Kinetic Methods of Analysis

G. G. GUILBAULT
Louisiana State University in New Orleans

Abbreviations used in the text

ATP   = Adenosine triphosphate
FAD   = Flavine adenine dinucleotide
NAD   = Nicotinamide adenine dinucleotide = DPN
NADH   = Reduced NAD = DPNH
NADP   = Nicotinamide adenine dinucleotide phosphate
NADPH   = Reduced NADP
PMS   = Phenazine methosulphate

5.1   INTRODUCTION

5.1.1   General

In the last 10 years chemical analysis based on reaction-rate methods has gained considerable popularity. Because the rate of a chemical reaction is dependent on the concentration of the reactants, or in some cases on the concentration of the catalyst, it may be used as a method of analysis. Such techniques, called kinetic methods of analysis, may be used for the determination of either high or low concentrations of reactants. Kinetic methods of analysis are extremely sensitive when applied to catalytic reactions, since as little as 10^{-12} g cm^{-3} of many substances are capable of catalysing a reaction. In many cases these methods are highly specific, thus providing the two chief factors most analysts are seeking in a quantitative method.

Since, generally, the initial rate of reaction is measured kinetically, these techniques avoid undesirable side reactions that frequently occur and they are rapid compared to the usual analytical methods. For example, the quantitative reaction of phenols with chloroquinonimines to form indophenols is slow, requiring *c.* 15–30 min for equilibrium to be attained[1]. However, one can determine phenols using this same reaction[2] by a kinetic method of analysis, and obtain stable, reproducible results in 2–3 min.

The rate of chemical reaction is dependent on the temperature of the system and the ionic strength, so both of these factors must be controlled for good results. In general, the errors in kinetic methods are of the order of 1–3%, which are quite acceptable particularly at ultramicro-concentrations.

In this review we shall discuss some of the uses of kinetics in analysis. In particular, we will consider determinations of substrates, activators and inhibitors using enzymes as the active reagent, but will not include methods of assay of enzymes (these are more catalytic in nature). Also to be considered are methods for the assay of inorganic and organic substances which directly participate as reactants in a chemical system, but not those cases in which the substance determined is a catalyst for the reaction. The latter methods will be discussed in the review by Yatsimirskii on Catalytic Methods (this volume, Chapter 6).

5.1.2 Kinetic parameters affecting rates of reaction

5.1.2.1 Reactants

If substance A reacts with B to give products C and D, the rate of reaction is described by:

$$V = -d\frac{A \text{ or } B}{dt} = d\frac{C \text{ or } D}{dt} = K[A]^n[B]^m$$

where V = velocity or rate of the chemical reaction and $-dA/dt$ or $-dB/dt$ is the rate of disappearance of A or B and dC/dt or dD/dt is the rate of appearance of C or D. K is a proportionality constant which is independent of the reactant concentrations and is called the reaction rate constant. $[A]$ and $[B]$ are the concentrations of the reactants A and B.

The number of molecules that take part in each step of a reaction defines the molecularity of the reaction. The equation for a monomolecular reaction, according to the law of mass action, takes the form:

$$V = K[A]^n$$

and for a bimolecular reaction:

$$V = K[A]^n[B]^m$$

The kinetic equation, however, does not always indicate the molecularity of the reaction. This can only be determined from the actual mechanism of the reaction. This rate equation itself can only be determined by experiment. The coefficients n and m are determined by noting how the rate of the reaction is affected by changing the concentrations of the reactants A and B. These coefficients may be fractions and may even be negative.

The kinetic equation for a first-order reaction would be:

$$V = K[A]^1[B]^0 = K[A]$$

The rate is proportional to the concentration of one reactant, and linearly increases with increasing concentrations of that substance. It is independent of the amount of the second material. The actual concentrations of A and of the product C at any time can be calculated from the equation

$$dC/dt = K(A-C) \text{ which upon integration gives}$$

$$\ln[A/(A-C)] = Kt$$

The rate of a second-order reaction is dependent upon the concentrations of two reactants and can be expressed by the equation:

$$V = K[A][B] \quad \text{or} \quad dC/dt = K(A-C)(B-C)$$

Integration of this equation gives:

$$\frac{1}{B-A} = \ln \frac{A\,(B-C)}{B\,(A-C)} = Kt$$

5.1.2.2 Catalysts

A catalyst may be broadly defined as an agent which alters the rate of a chemical reaction without shifting the equilibrium of that reaction. The catalyst enters into the reaction mechanism at a critical state, but in a cyclic manner, such that it does not undergo a permanent change. A catalyst, therefore, only affects the rate of the reaction, but does not change the position of equilibrium[3].

A catalyst, then, is any agent which speeds up a reaction by offering an alternative path of reaction, which has a considerably lower activation energy or kinetic order, without changing the point of equilibrium.

The rate of a reaction, expressed schematically by the equation $A+B \rightarrow C+D$ can be sharply increased by the addition of a catalyst X. The simplest type of mechanism for such a reaction involves the catalyst and one reactant (substrate) to form a 'complex' which then decomposes to form the product. The catalyst is thus regenerated and combines once more with a substrate molecule. A mechanism for this type of reaction can be expressed as[4]:

$$A+X \xrightarrow{\text{fast}} AX \tag{5.1}$$

$$AX+B \xrightarrow{\text{fast}} C+D+X \tag{5.2}$$

$$A+B \xrightarrow{\text{slow}} C+D \tag{5.3}$$

The rate of this catalytic reaction can then be expressed by the equation:

$$V = K'[A]^n[B]^m[X] \tag{5.4}$$

where the coefficients m and n are determined by the actual mechanism. If the reaction mechanism were as given by equations (5.1) and (5.2) with (5.1) being the rate-limiting step, then the expression for the reaction rate would be:

$$V = K'[A][X] \tag{5.5}$$

and the catalyst X could be determined based on the effect it exerts on the reaction of A and B.

Enzymes, which are biochemical protein catalysts for chemical reactions occurring in living cells, are among the best known catalysts. The general mechanism for an enzyme reaction is given by

$$E+S \underset{k_{-1}}{\overset{k_1}{\rightleftharpoons}} ES \xrightarrow{k_2} P+E \qquad (5.6)$$

where E is the enzyme, S the substrate, $[E-S]$ is an addition complex and P is the product[5]. An equilibrium treatment[6] and a steady-state treatment[7] have been developed for this mechanism and a simplified rate equation is obtained which applies for most enzymatic reactions. This rate expression has the form

$$V_i = K_2[E]_0[S]_0/(K_m+[S]_0) \qquad (5.7)$$

where K_m is known as the Michaelis constant and

$$[E]_0 = [E]+[E–S] \quad \text{and} \quad [S]_0 = [S]+[P]$$

The initial rate of reaction, V_i, is a function of the maximum rate, V_{max} (which is proportional to the enzyme concentration, and is expressed by $K_2[E]_0$). The rate equation can then be written:

$$V_i = V_{max}[S]_0/(K_m+[S]_0) \qquad (5.8)$$

In Figure 5.1 the effect of substrate and enzyme concentration on the initial rate is shown, where the initial rate of the enzyme reaction is plotted as a

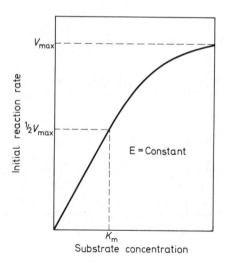

Figure 5.1 The initial rate of an enzyme-catalysed reaction, obeying equation (5.8), as a function of the initial substrate concentration

function of the substrate concentration. From this figure and equation (5.8) a physical interpretation of K_m is obtainable, that is, K_m is that substrate concentration which gives an initial rate equal to $\frac{1}{2}V_{max}$, where V_{max} is defined as the maximum attainable initial rate.

5.1.2.3 Activators

The rate of a catalytic reaction can often be significantly increased by the addition of very small amounts of certain substances, called activators. Specifically, an enzyme activator is a substance whose presence in the reaction mixture is required for the enzyme to be an active catalyst[9].

The mechanism generally proposed to explain the activity of an activator involves the formation of an activation–catalyst complex, although the activator may serve as a catalyst for some step earlier in the reaction.

$$E \text{ (inactive)} + \text{Activator} \rightleftharpoons E \text{ (active)}$$

Figure 5.2 shows the effect of an activator on the initial rate of an enzyme

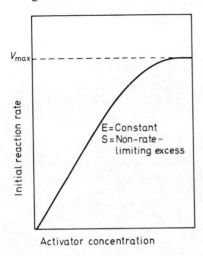

E = Constant
S = Non-rate-
limiting excess

Figure 5.2 The dependence of initial reaction rate upon the activator concentration

reaction of a constant enzyme and substrate concentration. The rate of reaction is seen to increase until all the enzyme becomes activated (V_{max}).

5.1.2.4 Inhibitors

An inhibitor is defined as any substance which slows down the rate of an enzyme-catalysed reaction[9]. One way that an inhibitor can slow down an enzyme-catalysed reaction is by forming a non-reactive complex with either the free enzyme or with the enzyme–substrate intermediate complex as follows:

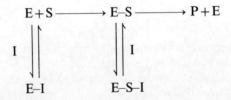

The inhibitor, I, reduces the amount of enzyme or complex available for catalysing the reaction and the initial rate of reaction will decrease with increasing inhibitor concentration. Figure 5.3 illustrates this effect of an inhibitor on the initial reaction rate.

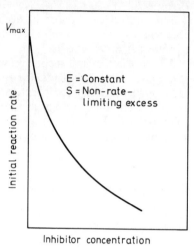

Figure 5.3 The dependence of initial reaction rate upon inhibitor concentration

5.1.2.5 Temperature

The rate of a chemical reaction generally increases with an increase in temperature. This dependence can be expressed quantitatively by the equation

$$\log K = \text{constant} - E/4.5\ T$$

where K is the reaction rate constant. This temperature dependence is generally true no matter whether the reaction is exothermic or endothermic.

As a rule of thumb, a $10\ °C$ rise in temperature of an enzyme-catalysed reaction approximately doubles to triples the reaction rate. This makes stringent temperature control of a reaction mixture necessary.

5.1.2.6 Foreign salts

The presence of foreign salts of other elements in enzyme-catalysed reaction mixtures can also affect the rate of reaction. One way these salts affect the rate is by changing the equilibrium of the formation of the activated complex. This is usually called a 'salt effect of the first kind'. The magnitude of this effect can be calculated using the equation:

$$\log K = \log K_0 + Z_A Z_B \sqrt{\mu} + (b_A + b_B + b^*)\mu$$

where K and K_0 are the rate constants with and without foreign salts present, Z_A and Z_B are the charges on the reacting substances A and B, μ is the ionic strength and b_A, b_B and b^* are empirical coefficients for A, B, and the activated complex respectively.

The equation shows that foreign salts can lead to either an increased or decreased rate of chemical reaction (like or unlike charges on particles A and B respectively).

The coefficients b_A and b_B are not equal to zero for electrically neutral compounds and thus a salt effect can occur due to these substances; however, the effect is much smaller.

A salt effect of the second kind is described as a displacement of weak bases and very weak acids due to the presence of foreign salts. The foreign salts serve to tie up the effective concentration of one of the reactants by forming a complex ion, a precipitate, or by shifting the ionisation equilibrium.

These salt effects are severe enough that care must be taken to eliminate unwanted foreign ions and to control the ionic strength of the reaction media.

5.1.3 Reaction rate calculations

5.1.3.1 General

In order to determine the amount of a reactant in solution by kinetic methods, the rate of the reaction must be calculated. To effect this calculation, a pseudo-first-order reaction condition must be established. This can be done by making all the reactants in excess except the one to be determined. If small changes in the concentration of a reactant, product, or indicator reagent can be measured accurately by some sensitive method, then the initial rate of the reaction can be measured, and such a kinetic method of analysis employed for the determination. By using non-rate-limiting concentrations of all substances except the one to be assayed, the initial rate of reaction is measured and is proportional to the concentration of the substance being determined. If

$$A + B \longrightarrow C + D$$

with excess B then $dC/dt = K'[A]$.

There are several initial reaction-rate methods by which the concentration of reactant A can be assayed. These methods include the initial slope, fixed concentration and fixed time methods, all of which will be discussed separately in the succeeding sections.

All of the methods mentioned share the three major advantages of initial rate methods; that is (a) since the reaction has proceeded to only a small fraction of completion, there is only a small concentration of the product present, hence the back-reaction makes no appreciable contribution to the overall rate of reaction; (b) complications resulting from any side reactions should be minimal during the initial reaction period; and (c) the concentration of the other reactants will not change appreciably during the initial period and the reaction will proceed via pseudo first-order kinetics. Thus, the reproducibility of the measurements is improved and the interpretation of the experimental rate data in terms of concentrations is simplified.

5.1.3.2 Initial-slope method

The initial-slope method or method of tangents is a method of determining the concentration of an unknown from the slope of a linear curve. In this method the concentration of a reactant, product, or indicator substance is plotted v. time as a reaction proceeds. The initial slope of the resulting curve is obtained by extrapolating to zero time. The slope is related directly to the

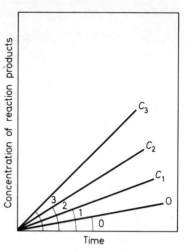

Figure 5.4 Dependence of concentration of reaction product on time at different concentrations of the substance being determined

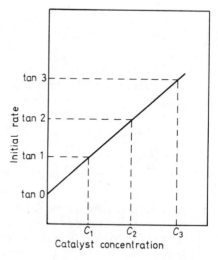

Figure 5.5 Calibration curve for determination of concentration of reactants by the tangent method

concentration of the reactant being determined (Figure 5.4). This method is generally employed by measuring the change in some physicochemical parameter of the reactant, product, or indicator substance. The initial slope method is illustrated in Figure 5.5 which shows a typical calibration curve obtained when the initial reaction rate is plotted v. catalyst concentration.

5.1.3.3 Fixed-concentration method

The fixed-concentration method involves the measurement of the time re-quired for the concentration of one of the reactants to reach a specified value. Any property of the substance denoting its concentration can be measured.

In this method of fixed concentration a linear relationship is always ob-

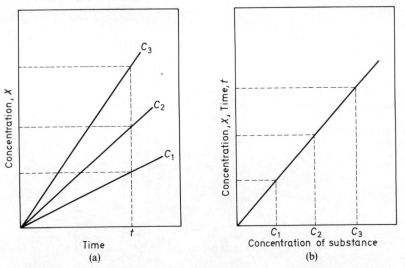

Figure 5.6 (a) Original graph for the determination of concentration by the fixed-concentration method. (b) Calibration graph for the determination of concentration by the fixed-concentration method

served between the experimentally determined concentration and the recipro-cal of the time required to reach a specified concentration of the substance (or the time for the reaction to proceed to a specified extent).

Figures 5.6a and 5.6b illustrate the typical curves obtained in this method.

5.1.3.4 Fixed-time method

In the fixed-time method the reaction is allowed to proceed for a predeter-mined time interval and after this time the concentration of the reactants or products is determined. This measurement can be done either by stopping the reaction at this time or by measuring some physiochemical property of the solution at the specified time.

Using this method, it is possible in some cases to make determinations using a calibration plot where the concentration of the substance being determined is plotted $v.$ some physicochemical parameter at a constant time. The time used, however, should be chosen to yield a linear curve (as at time 1 in Figure 5.7), which represents a typical calibration curve.

5.1.3.5 Conclusions

Ingle and Crouch[10] pre-formed a theoretical treatment of various rate equations which indicated that the choice of rate measurement approach depends on the characteristics of the reaction used for analysis, the kinetic role of the sought-for species, and the signal–concentration relationship of the transducer used to follow the reaction. The fixed time approach is superior for first- or pseudo-first-order reactions and determination of substrate concentrations in enzyme reactions. In contrast, rate analysis

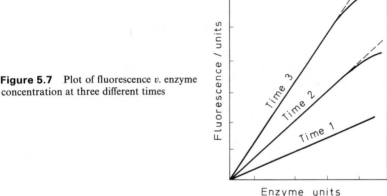

Figure 5.7   Plot of fluorescence v. enzyme concentration at three different times

of enzyme activity or other catalysts is best suited for the variable-time procedure. The variable-time approach is also best for non-linear response curves.

5.1.4   Monitoring the reaction rate

5.1.4.1   Chemical methods

To monitor the rate of a reaction, the change with time of the concentration of one of the reactants or products must be measured. Using a chemical method this is accomplished by periodically withdrawing a sample from the reaction mixture and measuring the desired reactant or product by a titration. To accomplish this measurement the reaction in the withdrawn sample must generally be stopped. The stopping of the reaction is accomplished by one of several methods: by adding a substance which either combines with one of the reactants or inhibits the reaction; by cooling the reaction suddenly; or, if the reaction is pH dependent, by rapidly adding acid or alkali to change the pH.

5.1.4.2 Instrumental methods

Instrumental methods of monitoring the reaction rate make it possible to follow a chemical reaction without stopping the reaction or withdrawing samples to be analysed. The reaction is followed instead by monitoring the appearance or disappearance of some species in the reaction. Usually a direct measurement is made of some physiochemical property of the species being followed. Coupled reaction sequences may also be used to follow the reaction.

In the direct method some property such as absorbance, pH, or fluorescence of either a reactant or product is followed. In a coupled reaction sequence a second reaction is used as an 'indicator reaction'. This reaction is chosen so as to provide an easily followed, measurable parameter.

5.1.5 General aspects of enzyme catalysed reactions

Enzymes, which are biochemical catalysts, are frequently used for chemical analyses. Enzyme-catalysed reactions have been used to determine the enzyme itself, substrates, activators, and inhibitors of the enzyme. Enzymes possess two outstanding advantages as analytical reagents, that is specificity and sensitivity. An enzyme is capable of catalysing a specific reaction of a particular substrate even though other reactions of that substrate may be just as

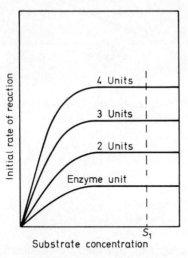

Figure 5.8 Plot of initial rate v. substrate concentration at various enzyme concentrations. S_1 = rate-limiting substrate concentration

favourable and even though other closely related homologues of that substrate may be present[9].

As catalysts, enzymes are always more effective than non-biological catalysts, and because of this catalytic effect small amounts of enzymes found in living systems can effect transformations of small amounts of their substrates. Thus, enzymes make exceptionally good analytical reagents and provide not only specificity but great sensitivity as well.

When combined with kinetic methods of analysis, especially instrumental methods, enzyme-catalysed reactions can provide rapid and sensitive methods of analysis. The high sensitivity and specificity of fluorescence measurements make this an excellent technique for following enzymatic reactions.

The use of kinetic measurements in enzymic analysis is illustrated by Figure 5.8. The initial rate of the reaction is a function of the enzyme concentration in a first-order fashion as is predicted from the Michaelis–Menten equation. The initial rate is also proportional to the substrate concentration at low concentrations. The analytically useful portion of the curve of a substrate generally lies in the region below $0.5\,K_m$ with best results below $0.1\,K_m$ (Ref. 4). Calibration plots of the initial rate v. enzyme concentration at non-rate-limiting excess of the substrate (S_1 in Figure 5.8), or of the initial rate versus substrate concentration at constant enzyme concentration, should be linear, allowing one to assay for either substrate or enzyme.

5.1.6   Reference publications

Guilbault[11] and Bergmeyer[12] have written books on Enzymic Methods of Analysis which cover many known methods for the assay of substrates, activators and inhibitors. Guilbault[11] furthermore, has included a discussion of the immobilisation and source of enzymes and of the automation of enzymic reactions. Reviews on the use of Enzymes as Analytical Reagents were prepared by Guilbault for *Analytical Chemistry* every 2 years[13] and for the *CRC Critical Reviews in Analytical Chemistry*[14].

Excellent reviews of the non-enzymic field of kinetics can be found in books by Mark and Rechnitz[15] and Yatsimirskii[16]. Good review articles on the use of kinetics in analysis have been prepared by Rechnitz[17] and Guilbault[18].

5.2   ANALYTICAL APPLICATIONS

5.2.1   Enzyme catalysed reactions

5.2.1.1   Assay of substrates

(a) *General* — At a fixed enzyme concentration, the initial rate of an enzymatic reaction increases with increasing substrate concentration until a non-limiting excess of substrate is reached, after which additional substrate causes no increase in rate. The region in which linearity is achieved, and in which an analytical determination of substrate concentration can be made based on the rate of reaction, lies below $0.1\,K_m$. The most important advantage of an enzymatic assay is its specificity. Frequently, only one member of a homologous series is active in the enzyme-catalysed reaction; other members are totally inactive or react at much slower rates. Most enzymes are also specific for one optical isomer of the substrate. Thus, in the enzymatic assay of amino acids, bacterial amino acid decarboxylase is specific for L-amino acid only. Another advantage in the use of enzymes for substrate analysis lies in the great sensitivity obtained. Glucose, for example, is oxidised at the rate of a few per cent per minute, regardless of concentration. Thus, a 10^{-7} mol l^{-1} solution can be analysed as easily as a 10^{-4} M one.

A complete compilation of enzymic methods for the assay of carbohydrates, amines, amino acids, organic acids, hydroxy compounds, esters, aldehydes, and inorganic substances has been prepared by Guilbault[11, 13] and Bergmeyer[12]. In this review only some of the more recent and unusual examples of the use of enzymes for analysis of substrates will be discussed.

(b) *Carbohydrates* — Enzymes offer advantages of specificity and sensitivity over other non-enzymic methods for the determination of carbohydrates. Furthermore, by the use of several enzymes, each selective for one carbohydrate, a complex mixture of carbohydrates could be assayed. Guilbault, Sadar and Peres[19] described fluorometric methods for the assay of mixtures of the carbohydrates glucose, fructose, maltose, cellobiose, lactose, glycogen, and salicin. Glucose and fructose are determined fluorimetrically using hexokinase and the resazurin–resorufin indicator reaction.

$$\text{Glucose} + \text{ATP} + \text{MgCl}_2 \xrightarrow{\text{hexokinase}} \text{glucose-6-phosphate}$$

$$\text{Glucose-6-phosphate} + \text{NADP} \xrightarrow{\text{Glucose-6-phosphate dehydrogenase}}$$
$$\text{NADPH} + \text{gluconophosphate}$$

$$\text{NADPH} + \text{resazurin} \xrightarrow{\text{Phenazine methosulphate}} \text{NADP} + \text{resorufin}$$
$$\text{(non-fluorescent)} \qquad\qquad \text{(fluorescent)}$$

The rate of formation of resorufin is measured and is proportional to the glucose present (or to fructose using phosphohexose isomerase).

$$\text{Fructose} + \text{ATP} + \text{MgCl}_2 \xrightarrow{\text{Hexokinase}} \text{fructose-6-phosphate}$$

Maltose, cellobiose, lactose, glycogen, and salicin are enzymatically hydrolysed to glucose, which is then determined fluorometrically using the respective enzyme, glucose oxidase, together with *p*-hydroxyphenylacetic acid and peroxidase:

$$\text{Glycogen} \xrightarrow{\text{Amyloglucosidase}} \text{glucose}$$

$$\text{Carbohydrate} \xrightarrow{\text{E}} \text{glucose}$$

$$\text{Glucose} \xrightarrow{\text{Glucose oxidase}} \text{H}_2\text{O}_2$$

$$\text{H}_2\text{O}_2 + p\text{-hydroxyphenyl acetic acid} \xrightarrow{\text{Peroxidase}} \text{Fluorescence}$$

Some of the enzyme systems, together with the range of concentrations and interferences in the determination of various sugars, are listed in Ref. 19. All the enzyme systems are highly selective: only cellobiose and salicin, two substrates seldom found together, are substrates for β-glucosidase; maltase acts specifically on malose; β-galactosidase (lactase) acts specifically on lactose and amyloglucosidase specifically on glycogen.

The analysis of a mixture of lactose, maltose, fructose, and glycogen was attempted using the four enzyme systems β-galactosidase, maltase, hexokinase, and amyloglucosidase, with no prior separation. The sample was split into four aliquots, and each was assayed for one of the four components. Excellent results are obtained for the simultaneous determination of these four carbohydrates.

(c) *Organic acids* — This same philosophy outlined above could be applied to the assay of any complex mixture of organic substances, provided an enzyme was available for the specific assay of each component of the mixture. Guilbault, Sadar and McQueen[20] described an assay procedure using six enzyme systems for the determination of mixtures of 21 organic acids. Lactate (Types II and IV), malate, glutamate, isocitrate, and β-hydroxybutyrate dehydrogenases are used, coupled with NAD, phenazine methosulphate, and resazurin, in a fluorometric procedure for the determination of acetic, adipic, benzilic, butyric, D-α- and D-β-hydroxybutyric, chloroacetic, DL-citric, formic, L-glutamic, glutaric, glycolic, threo-D-isocitric, L-lactic, L-malic, malonic, oxalic, phthalic, DL-succinic, and L-tartaric acids in the range of c. 0.1–500 µg with an accuracy and precision of c. 2%. The rate of production of the highly fluorescent resorufin is proportional to the concentration of the acid,

$$\text{Acid} + \text{NAD} \xrightarrow{\text{Dehydrogenase}} \text{Keto-acid} + \text{NADH}$$

$$\text{NADH} + \text{Resazurin} \xrightarrow[\text{or PMS}]{\text{Diaphorase}} \text{NAD} + \text{Resorufin}$$
$$\text{(Non-fluorescent)} \qquad\qquad \text{(Fluorescent)}$$

Some typical results for the analysis of acids are shown in Ref. 20. With β-hydroxybutyrate dehydrogenase, 1–75 µg cm^{-3} of D-β-hydroxybutyrate were analysed in the presence of L-glutamic, L-lactic, L-malic, and D-α-hydroxybutyric acids at concentrations of 1 mg cm^{-3} with an accuracy of $\pm 1.3\%$ and a precision of 2%. DL-Citric acid is the only interference determinable in the range 10–110 µg cm^{-3} with a precision and accuracy of c. 2%.

A four-component acid mixture of DL-citric acid, D-isocitric acid, L-lactic acid, and L-glutamic acid has been analysed. The sample was divided into four equal parts, and analysis of each acid performed with a different dehydrogenase: β-hydroxybutyrate dehydrogenase for DL-citric acid, isocitric dehydrogenase for D-isocitric acid, lactate dehydrogenase Type II for L-lactic acid, and glutamate dehydrogenase for L-glutamic acid. A phosphate buffer (pH 6.5) was used for ICDH, a glycine–hydrazine buffer (pH 9.5) for β-OH–BuDH, and tris buffer (pH 9.5) for LDH and GDH. Essentially, the same precisions and accuracies were obtained for the analysis of the mixture as were obtained in the determination of each acid alone.

(d) *Amines and amino acids* — (i) *General electrochemical methods.* Guilbault and co-workers have recently described several simple, direct electrochemical probes for the assay of amines and amino acids.

Guilbault, Smith and Montalvo[21] described the use of a cationic electrode that responds to NH_4^+ for the electrochemical assay of amines (urea, glutamine, asparagine, etc.) and amino acids:

$$\text{Urea} \xrightarrow{\text{Urease}} NH_4^+$$
$$\text{Glutamine} \xrightarrow{\text{Glutaminase}} NH_4^+$$
$$\text{Asparagine} \xrightarrow{\text{Asparaginase}} NH_4^+$$
$$\text{Amino acid} \xrightarrow{\text{Oxidase}} NH_4^+$$
$$\text{Amine} \xrightarrow{\text{Oxidase}} NH_4^+$$
$$\text{Glutamic acid} \xrightarrow{\text{Dehydrogenase}} NH_4^+$$

Some results obtained in the use of this method for the analysis of various enzymes and substrates are indicated in Ref. 21. The urease, glutaminase, asparaginase and D- and L-amino acid oxidase systems worked well. Attempts to get the glutamic acid dehydrogenase system to work failed, probably because of the presence of ions in the enzyme. This was indicated by the large potentials observed before initiation of the enzymic reaction and the serious drift problems encountered. All attempts to remove the ionic interferences failed. Similar problems were sometimes encountered in assay of the other enzyme systems, but such problems were generally eliminated by pre-treatment of the enzyme and/or substrate with an ion exchange resin.

(ii) *Enzyme electrode probes for the assay of substrates.* Carrying the concept of an electrochemical probe one step further, Guilbault and Montalvo[22] described a urea enzyme electrode suitable for the continuous assay of urea in body fluids.

The urea transducer is called a urease electrode because it is made by polymerising a gelatinous membrane of immobilised enzyme over a Beckman cationic glass electrode which is responsive to ammonium ions. Specificity for urea is obtained by immobilising the enzyme urease in a layer of acrylamide gel 60–350 µm thick on the surface of the glass electrode. When the urease electrode is placed in contact with a solution containing urea, the substrate diffuses into the gel layer of immobilised enzyme. The enzyme catalyses the decomposition of urea to ammonium ion as shown in the following equation.

$$Urea + H_2O \xrightarrow{Urease} 2NH_3 + CO_2$$

The ammonium ion produced at the surface of the electrode is sensed by the specially formulated glass which measures the activity of this monovalent cation in a manner analogous to pH determination with a glass electrode.

The potential of this electrode is measured after allowing sufficient time for the diffusion process to reach the steady state. This interval varies from *c.* 25 to 60 s for 98 % of the steady-state response, depending on the thickness of the gel membrane.

When the urea concentration is below the apparent K_m for the immobilised enzyme, but *c.* 0.6 mg of urea per 100 ml of solution, the potential of the electrode varies linearly with the logarithm of the urea concentration. Also, the response curve goes from first order at low urea concentrations to zero order at high substrate concentrations.

In order to improve the stability of the urea electrode a thin film of Cellophane was placed around the enzyme gel layer to prevent leaching of the urease into the surrounding solution[23]. In this way an electrode could be used continuously for more than 21 days at 25 °C with no loss of activity.

In a later paper, Guilbault and Montalvo[24] discussed the preparation of various types of urease enzyme electrodes, the immobilisation parameters that affect the response of the electrodes, factors that affect the stability of the immobilised enzyme, and the effect of foreign monovalent cations on the enzyme response.

Guilbault and Hrabankova[25] described a specific electrode for the direct assay of urea in body fluids (blood and urine). It was found that, by using a glass electrode (Beckman 39 137) as the reference electrode and an ion exchange resin to remove interferences from Na^+ and K^+, urea could be assayed in blood or urine. The precision obtainable is about 1% averaging three or more samples. The difference between the results obtained with the urea electrode and the standard spectrophotometric method is $c.$ 2–3%, but it is believed that the electrode method is more accurate due to the low reliability factory of the spectrophotometric method.

The stability of the urea electrode is excellent. With many electrodes no loss in activity occurred over the first 21 days regardless of the urea concentration; excellent reproducible (within 1–2%) voltages were obtained over a three-week period. With other electrodes a small drift in potential from day to day was observed. The same standard urea solution used above 5×10^{-4} mol l^{-1} was also used to control the electrode stability. If the potential drifted more than 1 mV, a new calibration plot was made.

The variation between electrodes was also compared, and it was found that of ten urea electrodes prepared as described above (polyacrylamide layer of urease over a Beckman cation electrode covered with a Cellophane sheet), at least nine had electrode responses that agreed within 3–5% of each other. This indicates that it is not difficult to make an urea electrode. For maximum precision, however, a separate calibration plot was made for each electrode, as is usually done with any selective ion electrode.

Guilbault and Hrabankova[26, 27] described the preparation of an amino acid electrode and described its electrode properties.

The L-amino acid electrode was made by placing a thin layer of L-amino acid oxidase (L-AAO) over a Beckman monovalent cation electrode. The enzyme catalyses the decomposition of amino acid to NH_4^+ ions by the reaction:

$$RCH \cdot NH_3^+ CO_2^- + H_2O + O_2 \xrightarrow{\text{L-AAO}} RCO \cdot CO_2^- + NH_4^+ + H_2O_2$$

The H_2O_2 formed reacts non-enzymatically with the α-keto acid product:

$$RCO \cdot CO_2^- + H_2O_2 \longrightarrow RCO_2^- + CO_2 + H_2O$$

If H_2O_2 is destroyed by catalase, the overall reaction is described by the following equation:

$$2RCH \cdot NH_3^+ CO_2^- + O_2 \longrightarrow 2RCO \cdot CO_2^- + 2NH_4^+$$

Ammonium ions formed are sensed by the electrode described, the steady state potential of which is proportional to the activity of NH_4^+ ions in the enzyme layer, i.e. to the concentration of amino acid in the solution.

The stability of electrodes containing 20 and 100 mg of L-AAO per cm^3 of solution was compared. Electrodes with the higher enzyme concentration were more stable.

Liquid membrane electrodes were stable for two weeks; after that time more rapid decrease of response was observed.

Next, let us consider the D-amino acid electrode. FAD is a coenzyme attached to the enzyme, necessary for activity. The D-amino acid electrodes had poor stability in solution since FAD diffuses out from the concentrated enzyme gel into the solution by dialysis. To get around this, the electrode must be stored in high concentrations of FAD: the FAD diffuses back into the membrane and the electrode now is stable for long periods of time. If no FAD is used in storage of the electrode the electrode response goes to zero after one day. The electrode is stable for c. 28 days if FAD is added to the solution used for storing the enzyme electrode between use.

The D-amino acid oxidase is a little less active than the L-enzyme so more enzyme must be used in the active layer, but a Nernstian type of response is achieved. The lower limit is only 2×10^{-4} M compared to 10^{-4} M with the L-amino acid electrode[27]. A pH of 8.5 is optimum for all concentration measurements.

Nernstian response curves are obtained for D-phenylalanine, alanine, and other D-amino acids[28]. D-Proline, which is cyclic, does not respond because it does not give ammonia.

Consider next the asparagine electrode. Asparaginase is a very unstable enzyme, but can be stabilised very nicely by physical entrapment in a poly-acrylamide gel and a layer of dialysis membrane paper over the outside. This electrode can be used for measurement of asparagine directly[28]. Stability data show that the initial potential reading was almost the same as that after 25 days showing the electrode is indeed quite stable.

A glutamine electrode was prepared using glutaminase[29] in the same manner as before. This is a Type II electrode like the asparagine. Finally, a L-glutamic acid electrode was made using NAD in solution and glutamate dehydro-genase in a gel. This was not followed up because of the problem of having to add NAD in solution, which is somewhat expensive.

(e) *Hydroxy compounds* – Guilbault and Sadar[30] have developed enzymic methods for the assay of mixtures of hydroxy compounds. Four enzyme systems were used, each for the assay of one alcohol in the mixture: alcohol dehydrogenase, alcohol oxidase, carbohydrate oxidase, and sorbitol de-hydrogenase. In each case a fluorometric monitoring system was used to follow the reaction.

(i) Alcohol or carbohydrate $\xrightarrow{\text{Oxidase}}$ H_2O_2

$$p\text{-Hydroxyphenylacetic acid} + H_2O_2 \xrightarrow{\text{Peroxidase}} \text{Fluorophor}$$
$$(\lambda_{ex} = 317 \text{ nm}; \lambda_{em} = 414 \text{ nm})$$

The rate of formation of the fluorescent product is followed and is pro-portional to the alcohol of carbohydrate content.

(ii) Alcohol + DPN $\xrightarrow{\text{Dehydrogenase}}$ DPNH + α-keto acid

or

$$\text{D-Sorbitol} + DPN^+ \xrightarrow[\text{Dehydrogenase}]{\text{Sorbitol}} DPNH + \text{D-fructose} + H^+$$

$$DPNH + \text{resazurin} \xrightarrow{\text{PMS}} DPN + \text{resorufin}$$
(Non-fluorescent)   (Fluorescent)

The formation of resorufin is measured and is proportional to the concen-tration of sorbitol or alcohol present. Sorbitol, methanol, ethanol, allyl alcohol, cyclohexanol, and s-butyl alcohol did not interfere in the deter-mination of glucose, xylose, or sorbose using carbohydrate oxidase. Likewise, sugars like glucose, xylose, or sorbose and alcohols such as methanol, ethanol, allyl alcohol, etc. did not interfere in the assay of sorbitol using sorbitol dehydrogenase, nor did sorbitol or any sugars interfere in the determination of alcohols using alcohol oxidase.

Some typical results of the analysis of a mixture of glucose and sorbitol using carbohydrate oxidase and sorbitol dehydrogenase, respectively, are shown in Ref. 30. An average error of about 1.4% and a precision of c. 1.5% were obtained.

Finally, a three-component mixture of methanol, xylose, and sorbitol was assayed using three different enzymes: alcohol oxidase, carbohydrate oxidase, and sorbitol dehydrogenase. The results indicated a precision of c. 1.5% and an average error of 1–1.5%.

(f) *Inorganic substances* — Although enzymic methods have been described for the assay of many inorganic substances, two very sensitive procedures that are worthy of special note are those for phosphate and ammonium ions. The enzymatic determination of ammonia in tissue body fluids was described by Faway and Dahl[31], in body fluids by Mondzac, Ehrlich and Seegmiller[32], in blood and tissue by Reichelt, Kvamme, and Tveit[33], and in blood by Kirsten, Gerez and Kirsten[34]. All used the oxoglutarate–glutamic dehydrogenase-NADH system, measuring the change in absorbance at 340 nm.

$$NH_4^+ + H^+ + NADH + oxoglutarate \xrightarrow{dehydrogenase} glutamate + NAD^+ + H_2O$$

The enzymatic procedure eliminates all the disadvantages of previous methods and is specific for ammonia in the presence of amines. Roch-Ramel[35] and Rubin and Knott[36] described fluorometric methods for ammonia using this reaction sequence and following the disappearance of the fluorescent NADH ($\lambda_{ex} = 340$ nm; $\lambda_{em} = 460$ nm). From 4×10^{-11} to 2×10^{-10} equivalents of NH_4^+ were determinable.

Schulz, Passonneau, and Lowry[37] have described a fluorometric enzymic method for the measurement of inorganic phosphate based on the following enzyme sequence.

$$Glycogen + phosphate \xrightarrow{Phosphorylase\ a} Glucose\text{-}1\text{-}phosphate$$

$$Glucose\text{-}1\text{-}phosphate \xrightarrow{Phosphoglutomutase} Glucose\text{-}6\text{-}phosphate$$

$$Glucose\text{-}6\text{-}phosphate + NADP \xrightarrow{Dehydrogenase} 6\text{-}Phosphogluconolactone + NADPH$$

NADPH is fluorescent, and the rate of its formation indicates the phosphate present at concentrations of $0.02–5 \times 10^{-10}$ M. Faway, Roth and Faway[38] assayed inorganic phosphate in tissue and serum using this same reaction sequence except for a spectrophotometric measurement of NADPH.

5.2.1.2   Determination of activators

An enzyme activator is a substance which is required for an enzyme to be an active catalyst:

$$E(inactive) + Activator \rightleftharpoons E(active)$$

The activity of the enzyme will increase until enough activator has been used to activate the enzyme fully. The initial rate of the enzyme reaction is proportional to the activator concentration at low concentrations, thus providing a method for its determination. Very little has been done on the analytical determination of activators. A method for magnesium in plasma is described by Baum and Czok[39], based on the activation of isocitric dehydrogenase. With constant amounts of enzyme, the rate is dependent on magnesium concentration down to 10^{-6} M. A thorough study of this reaction was made by Adler, Gunther, and Plass[40], and by Blaedel and Hicks[41], who found that only Mg^{2+} and Mn^{2+} efficiently activate this enzyme. The useful analytical range extends up to about 100 parts per billion for Mn^{2+} and 2×10^{-4} M for Mg^{2+}; Hg^{2+} and Ag^+ at 10^{-5} M, and Ca^{2+} at 10^{-4} M inhibited the Mn^{2+} activation completely.

A number of enzymes require for their activity a specific co-enzyme which participates in the enzymic reaction. By measuring the amount of activation of such an enzyme by the co-enzyme, a plot of initial rate of reaction v. co-enzyme concentration may be constructed. At low concentrations of co-enzyme, the degree of activation will be proportional to the concentration of co-enzyme added. For example, flavine adenine dinucleotide (FAD), which is the co-enzyme of D-amino acid oxidase, can be determined by this method:

$$\text{D-Alanine} + \text{FAD} \xrightarrow{\text{D-Amino acid oxidase}} \text{Pyruvic acid} + \text{FADH} + \text{NH}_3$$

$$\text{FADH} + \text{O}_2 \longrightarrow \text{FAD} + \text{H}_2\text{O}_2$$

Warburg and Christian[42] and Straub[43] were the first to describe this method. Unknown samples of FAD containing 0–0.3 μg of FAD can be determined by measuring the increase in the rate of oxidation of alanine by D-amino acid oxidase[44].

The firefly reaction has been shown to require ATP and Mg^{2+} in addition to luciferin, luciferase, and oxygen[45]:

$$\text{Luciferin} + \text{O}_2 + \text{ATP} \xrightarrow[\text{Mg}^{+2}]{\text{Luciferase}} \text{Oxyluciferin} + \text{ADP} + \text{Pi}$$

This reaction has been used as the basis for the most sensitive method for ATP that is known[46]. This reaction may also be used to assay oxygen at partial pressures below 10^{-3} mm, when the gas is passed through a bacterial emulsion containing all requirements for the luminescent reaction except oxygen[47].

5.2.1.3 *Determination of inhibitors*

(a) *General* — As mentioned previously an inhibitor is a compound that causes a decrease in the rate of enzyme reaction, either by reacting with the enzyme to form an enzyme–inhibitor complex or by reacting with the enzyme–substrate intermediate to form a complex:

$$\text{E} + \text{I} \rightleftharpoons \text{EI}$$

$$\text{E} + \text{S} \rightleftharpoons \text{E–S} \longrightarrow \text{P} + \text{E}$$
$$\downarrow \text{I}$$
$$\text{E} \text{——} \text{S} \text{——} \text{I}$$

In general, the initial rate of an enzymic reaction will decrease with increasing inhibitor concentration, linearly at low inhibitor concentrations, then will gradually approach zero.

Analytical working curves for inhibitor assay are generally constructed by plotting % inhibition v. concentration of inhibitor. The % inhibition is calculated as follows:

$$\% \text{ Inhibition} = \frac{(\text{Rate})_{\text{No Inhibitor}} - (\text{Rate})_{\text{Inhibitor}}}{(\text{Rate})_{\text{No Inhibitor}}} \times 100$$

A control rate is recorded with no inhibitor present, but with the same volume of the solvent used to contain the inhibitor added. This is especially critical in studies of inhibitors added in non-aqueous solution since most non-aqueous solvents will inhibit the enzyme in concentrations greater than 3%. In addition to the control (non-inhibited) rate, the rate of spontaneous (non-enzymic) hydrolysis of oxidation of the substrate should be measured and all rates corrected for such non-enzymic effects, if necessary.

Generally, a plot of % inhibition against concentration is a typical exponential type curve with a linear range extending from 0 to 60 or 70% inhibition. This linear region is the most analytically useful range. The concentration (M) of inhibitor that causes a 50% inhibition of enzymic activity is the I_{50} and is a measure of the strength of an inhibitor.

(b) *Assay of inorganic substances* — Enzymic methods have been described for the determination of many common inorganic cations and anions: Ag^+, Al^{3+}, Be^{2+}, Bi^{3+}, Ce^{3+}, Cd^{2+}, Co^{2+}, Cu^{2+}, Fe^{2+}, Hg^{2+}, In^{3+}, Mn^{2+}, Ni^{2+}, Pb^{2+}, Zn^{2+}, CN^-, $Cr_2O_7^{2-}$, F^-, and S^{2-}. A complete listing of these procedures was compiled by Guilbault[10].

(c) *Assay of pesticides* — One of the most interesting newer aspects of enzymic methods of analysis has been the selective assay of pesticides.

Several analytical methods have been proposed based on the inhibition of an enzyme reaction. Keller[48] has reported a fairly specific method for the determination of microgram concentrations of DDT, based on the inhibition of carbonic anhydrase. DDT inhibits this enzyme at concentrations at which other inhibitors (except the sulphonamides) are inactive. The most specific and sensitive method for some organophosphorus compounds is based on the inhibition of cholinesterase, and numerous papers have been published on this system. Kitz[49] has published a review with 158 references on the chemistry of anticholinesterase compounds. Giang and Hall[50] assayed TEPP, Paraoxan, and other insecticides that inhibit cholinesterase *in vitro*, and Kramer and Gamson[51] have developed a colorimetric procedure with compounds related to the indophenyl acetates for the determination of 1–10 µg of various organophosphorus compounds.

Guilbault, Kramer and Cannon[52] described an electrochemical method for the analysis of Sarin, Systox, parathion and malathion. The decrease in the rate of the cholinesterase catalysed hydrolysis of butyrylthiocholine iodide, as measured by dual-polarised platinum indicator electrodes, is linearly related to concentrations of the organophosphorus compounds.

Guilbault and Sadar[53] have developed sensitive methods, based on the use of enzyme systems for the determinations of chlorinated insecticides, carbamate insecticides and herbicides. Fluorescence methods were used because whenever these have been tried they have been shown to allow the determination of much lower concentrations of enzymic inhibitors (since lower enzymic activities can be measured). The substrate 4-methylumbelliferone heptanoate was used; it is cleaved by as little as 10^{-5} units of lipase to 4-methylumbelliferone. The chlorinated insecticides aldrin, Sevin, and the herbicide 2,4-D, were found to inhibit lipolytic activity and could be determined in the concentration range 0.1–30 µl ml^{-1}.

The results reported indicate that the method of lipase inhibition is the most sensitive enzymatic method yet reported for heptachlor, aldrin, lindane, and 2,4-D. The proposed inhibition scheme is likewise a good one for DDT. Although carbonic anhydrase is inhibited by lower concentrations of DDT the lipase procedure described is an easier and more convenient one to carry out.

A fluorometric procedure for the assay of aldrin, heptachlor, and methyl parathion was developed by Guilbault, Sadar, and Zimmer[54]. The method was based on the inhibition of acid and alkaline phosphatase by these pesticides.

The use of cholinesterase for the detection of pesticides has received the most interest from analysts. Reviews on cholinesterase and cholinesterase inhibitors have been prepared by Cohen et al.[55] and Delga and Foulhoux[56].

Because various animals and insects are known to be adversely affected by low concentrations of different pesticides, cholinesterases from these sources have been isolated and used for assay of pesticides. Guilbault, Kuan and Sadar[57] purified cholinesterases from honey bees and boll weevils and studied the effect of 12 different pesticides on these enzymes. Boll Weevil was used for the specific assay of Vapona (DDVP) and bee cholinesterase for Vapona and Paraoxan. Zhuravskaya and Bobyreva[58] found the esterase of the cotton aphid was most sensitive to methylmercaptophos while Laphygma moth was most sensitive to Sevin.

Soliman[59] and Foszezynska and Stycyznska[60] studied the inhibition of housefly-head cholinesterase by organophosphorus compounds, and Voss[61] found peacock plasma to be a useful cholinesterase source for the analysis of insecticidal carbamates.

The effect of 12 different pesticides on liver enzymes isolated from rabbit, pigeon, chicken, sheep and pig was studied by Guilbault, Sadar and Kuan[62]. The cholinesterases from these sources were found to be inhibited at very low concentrations by the organophosphorus pesticides DDVP, parathion, and methyl parathion. None of these enzymes was inhibited by any of the chlorinated pesticides, and pigeon and sheep liver cholinesterase were not inhibited by Sevin.

Guilbault et al.[63] studied the effect of 15 different pesticides including carbamates, chlorinated hydrocarbons, and organophosphorus compounds on cholinesterases from insects such as the housefly, sugar boll weevil, fire ant and German cockroach.

Fire ant cholinesterase is specifically inhibited by DDVP and methyl parathion. Even parathion had little effect on this enzyme. None of the other twelve inhibitors, Paraoxan, Aldrin, Captan, Dalapon, DDT, Dieldrin, 2,4-D-Acid, Heptachlor, Lindane, Mirex, Methoxychlor and Sevin, inhibits it at all. Cholinesterase from both types of housefly (NAIDM- and DDT-resistant strains) are specific for DDVP and Sevin, and others have little or no effect up to 10^{-3} M concentration of the inhibitor. Although not as sensitive as other cholinesterases, the sugar boll weevil is totally specific for DDVP, below 10^{-6} M concentration. Parathion interferes at concentrations greater than 10^{-3} M; although paraoxan was not tried, it is expected to interfere also. The enzyme from American cockroaches is the most sensitive of all but lacks specificity. Although it is most sensitive to DDVP, Paraoxan, parathion, and methyl parathion, a good number of chlorinated insecticides, carbamates, and herbicides inhibit it also.

From the results of this study it can be concluded that in addition to the sensitivity a further specificity and selectivity in the assay of pesticides using enzymic methods has been achieved.

Guilbault et al.[64] have discussed a specific enzymic method for the assay of chlorinated pesticides in the presence of herbicides, organophosphorus pesticides, carbamates and fungicides. The method is based on the selective inhibition of hexokinase by Aldrin, Chlorodane, DDT and Heptachlor. Parts per million concentrations of these four pesticides have been assayed in the presence of large amounts of other chlorinated pesticides (DDD, DDE, Dieldrin, Lindane, Methoxychlor, etc.), as well as other types of pesticides, with an accuracy of about 2%.

5.2.2   Non-enzymic reactions

5.2.2.1   General

Kinetic methods have increased to a point where they have become a common technique accepted by many analytical chemists for the assay of inorganic and organic substances. The accuracy and precision have been improved to a point where they are comparable to the equilibrium methods and many completely automated systems have been designed and described for kinetic systems.

With the growing popularity of kinetic methods it is not surprising to find that considerable effort has been expended in the development of new measuring techniques. Much of this effort has been exerted towards the automation of rate measurements to achieve greater reliability and improved convenience.

Pardue[65] has described an automatic method for measuring the slopes of rate curves using an electronic slope-machining device. This technique was applied to the quantitative assay of cysteine and glucose[66] with a 1–2% relative accuracy in the parts per million concentration range. Later this technique was modified to provide digital read-out in concentration units[67].

Mark et al.[68, 69] have described a semi-automatic modular instrument system for chemical analysis by either kinetic-based or equilibrium methods. The system includes a high-stability low-noise spectrophotometer that is designed to measure continuously the rate of chemical reactions. A coefficient multiplier is built in so that concentration in appropriate units can be read out directly. Pardue and Deming[70] have described a high stability, low-noise precision spectrophotometer using optical feed-back that can be used in kinetic measurements.

Stewart and Lum[71] have described a stopped-flow temperature-jump spectrophotometer that extends the time limitation down to about 20 μs. Chen, Schechter and Berger[72] conducted stopped-flow fluorescence experiments in the 0.05–10 s range with conventional instrumentation.

Sajo[73] described a method for the determination of concentrations $< 10^{-4}$ M by applying catalytic reactions and using a thermometric end-point determination. A differential chromatographic method for studying the kinetics of catalytic reactions was proposed by Roginskii et al.[74].

Klatt and Blaedel[75] derived, and verified experimentally, current-flow rate equations for the mass-transfer limited regeneration of a reactant at a tubular electrode. This steady-state, hydrodynamic, electrochemical system was used to study reactions with pseudo first order rate constants greater than 0.1 s^{-1}.

Cordos, Crouch and Malmstadt[76] described a new integration technique for obtaining direct readout of rates that have high noise immunity. The measurement cycle could be varied from milliseconds to hundreds of seconds for maximum performance, and results from synthetic signals showed that a relative error of c. 0.2% was achievable.

Many researchers have described the use of computers in kinetic analysis. Bard and Lapidus[77] used a computer to construct the kinetic models of experimental data in both homogeneous and heterogeneous systems. James and Pardue[78] described an instrument for the automatic measurement and computation of results in kinetic systems. The output used was a non-transient analogue signal that could be calibrated directly in desired concentration units. Reaction times of $6 \times 10^{-3} – 10^3$ s could be studied with errors of 0.5–2%. James and Pardue[79] used a small, general purpose, digital computer for the on-line processing of reaction-rate data for quantitative analyses. A wide range of analyses could be performed with this system with no changes in hardware. A precision of c. 1% was obtained with analysis of osmium and alkaline phosphatase. Parker, Pardue and Willis[80] described a method for determination of concentration in reaction-rate analyses. The instrument utilised digital circuitry in the computer, with a direct readout in units of concentration or reciprocal time. Data for determination of alkaline phosphatase showed deviations of $< 1\%$.

Toren and Davis[81] developed a simple inexpensive computer for the solution of the two simultaneous equations for two unknowns that occur in differential chemical analysis. The accuracy is c. 0.3% and the precision 0.5% for typical applications. A new method for computing the reciprocal of time for use in reaction-rate methods was described by Crouch[82]. The instrument is all electronic and can be used to measure the rates of both fast and slow reactions.

The use of a small computer in the clinical chemistry laboratory[83] and the micro-chemistry laboratory[84] has been proposed. Toren et al.[85] discussed the interfacing of spectrophotometric rate measurements with the computer. The LABCOM Version 4 on-line system was used.

5.2.2.2 Analysis of inorganic ions

Tanaka, Funahashi and Shirai[86] described a general method for the determination of mixtures of some heavy metal ions (CO–Ni, Co–Fe Cu–Pb,

Ni–Fe, Zn–Cd) by means of the differential reaction rates of ligand substitution reactions. Margerum *et al.*[87] achieved qualitative and quantitative analysis of mixtures of metal ions by means of the reaction of the metal–CDTA complexes with acid or with an exchanging metal ion. Analysis of mixtures of lanthanides, transition metals, Group II and III metals is possible, in concentrations as low as 10^{-6} M, with an accuracy better than 10%.

Pausch and Margerum[88] described a differential kinetic analysis of alkaline earth ions using stopped-flow spectrophotometry. The exchange reactions between Pb^{II} and the alkaline earth complexes of CDTA proceed at sufficiently different velocities to permit the kinetic determination of mixtures of Mg, Ca, Sr and Ba.

A fast photokinetic method for the determination of Fe^{III} was reported by Lukawiewicz and Fitzgerald[89] which employs the photoreduction of Fe^{III}, *in situ* complexation of the Fe^{II} formed, and continuous recording of the coloured complex.

Guilbault and McQueen[90] described an ultrasensitive catalytic method for metal ions and cyanide-containing organic molecules. The cyanide specific reaction of Guilbault and Kramer[91] is used; metal ions such as Hg^{2+}, Ag^+, Cu^{2+}, Co^{2+}, Zn^{2+}, Ni^{2+}, Cd^{2+}, which complex cyanide, decrease the rate of formation of the cyanhydrin and subsequent reduction of *o*-dinitrobenzene. As little as 3 p.p.b. of Ca^{2+}, 20 p.p.b. of Ag^+ or Hg^{2+}, or 60 p.p.b. of Co^{2+} can be determined.

Cordos, Crouch and Malmstadt[92] assayed phosphate with an automatic digital readout system for reaction rate studies. Crouch[93] used the assay of phosphate by a spectrophotometric rate method to demonstrate the application of a reciprocal time computer to routine analysis. Crouch and Malmstadt[94] conducted a mechanistic investigation of the molybdenum blue method for determination of phosphate, and described a method for the assay of phosphate[95] based on the formation of molybdenum blue from phosphate, Mo^{VI}, and ascorbic acid. Phosphate concentrations in the 1–12 p.p.m. range are determined with a 1–2 % relative error in aqueous solution; inorganic phosphate concentrations in blood serum can also be measured with similar errors. An automated fast reaction-rate system for quantitative phosphate determinations in the millisecond range was described by Javier, Crouch and Malmstadt[96]. The method utilises a completely automated stopped-flow system with direct digital readout of initial reaction rates to measure the initial rate of formation of the 12-molybdophosphoric acid. A kinetic method for the assay of phosphate in blood in the presence of organophosphates was described by Lorentz[97] and Crouch and Malmstadt[118] have proposed a kinetic method for assay of inorganic phosphate in the presence of ATP.

5.2.2.3 *Assay of organic compounds*

(a) *Single-component analysis* — Several analytical methods have been described for the assay of organic compounds in solution[98].

Mottola, Haro and Freiser[99] described a method for the assay of Cu^{II} complexing ligands such as cysteine, EDTA, 2-aminoethanethiol, salicylic acid, 1,10-phenanthroline and ethylenediamine, based on their interaction.

Kinetic methods for the assay of *o*-chloronitrobenzene or chlorodinitrobenzene in the presence of chloronitrobenzene were described by Leyrodi[100]. A kinetic method for the assay of α- and β-naphthol was described by Babkin[101] and a kinetic method for phenols was proposed by Rodziewicz et al.[102]. Phenol, o- and p-cresol, p-ethylphenol, o-phenylphenol, 2,4-, 2,6- and 3,5-xylenol and 4-chloro-3,5-xylenol can be determined with an error of 5–8%.

A kinetic method for the determination of 2-furaldehyde in the presence of acetone was described by Tikhonova[103]. The method was based on the reaction with bisulphite. Blaedel and Petitjean[104] reported a kinetic method for the assay of acetylacetone based on the reaction with hydroxylamine hydrochloride.

Another situation where a kinetic method can be more advantageous than a non-kinetic method occurs where specialised apparatus required for an analysis of a single species by a non-kinetic method is eliminated by a kinetic method. Burgess and Latham[105], for example, described a kinetic bromination reaction for either phenol or o- or p-cresol.

Finally, methods have been developed for the assay of components in a mixture based on a variation of reaction temperature. Bonting[106] described a method for the analysis of fructose–glucose mixtures; fructose reacts with anthrone at room temperature within ten minutes to give a highly coloured species, while glucose does not react appreciably even over a long period. Then,

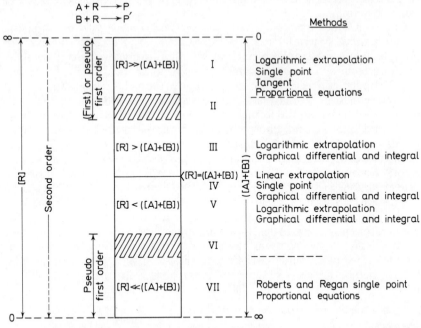

Figure 5.9 General analytical techniques applicable to second-order reactions

after fructose has reacted, the temperature is raised to 100 °C, and the glucose then reacts to completion in a few minutes.

(b) *Assay of mixtures of organic compounds* – The first study of the use of the kinetic method for the determination of closely related mixtures, based on the difference in reaction rates of the components with a common reagent and the mathematical interpretation of the results, was given by Lee and Kolthoff[107]. Many analytical techniques for the determination of closely related mixtures are based on the difference in reaction rates of the components with a common reagent[108–110]. Some of the general analytical techniques applicable to second-order reactions are indicated in Figure 5.9.

A method for the analysis of multicomponent ester mixtures by dual temperature differential kinetics was proposed by Nunnely[111]. A second-order extrapolation approach extended differential reaction analysis into the area of four- and five-component solutions. Lohman and Mulligan[112] described a method for assay of mixtures of ethanolamides by differential saponification rates. A kinetic analysis of mixtures of glycols was proposed by Bensone and Fletcher[113] based on a difference in specific rate constants for the cleavage of glycols with Pb^{IV}.

A method for the determination of binary ketone mixtures was described by Toren and Gnuse[114] based on the differential reaction rates of ketones with a large excess of hydroxylamine hydrochloride. For mixtures containing 10^{-5} mole of ketone, a relative error and precision of 1 % was obtained if the rate constants differed by a factor of 5.

A differential reaction-rate method for the determination of carbonyl mixtures was described by Mark and Greinke[115] and used as a kinetic experiment for the undergraduate laboratory.

Greinke and Mark carefully studied synergistic effects and other sources

Table 5.1   Typical organic compounds assayed by differential kinetic methods

Organic molecule	Reactant	Reference
Alcohols	Acetic anhydride	123
Aldehydes, ketones	Bisulphite	124, 125
Acetals, ketals	Hydrolysis	126, 127
Acids	Esterification–	
	diphenyldiazomethane	128
Acid chlorides	Esterification	129
Esters	Saponification	123
Amides	Hydrolysis	130
Nitriles	Hydrolysis	131
Amines	Phenylisocyanate	132
	Methyl iodide	133, 134
	Carbonyl compounds	135
Anilines	Methyl iodide	136
Double bonds	Perbenzoic acid	137, 138
	Bisulphite addition	122

of error in the rapid analysis of binary amine mixtures by differential reaction rates[116] and the effect of relative concentrations in the analysis of closely related carbonyl mixtures by differential reaction rates using the method of proportional equations[117].

Mark, Greinke and Papa[118] evaluated three major differential kinetic methods currently in use, e.g. proportional equations, graphical extrapolation and single point, and performed an error analysis in terms of optimum concentrations, magnitude of rate constants and rate ratios.

A modification of the differential kinetic method to complex reaction systems having mixed stoichiometries was described by Bond, Scullion and Conduit[119] who illustrated their method by resolving the hydrolysis reactions of a mixture containing cyclotrimethylenetriamine and cyclotetraethyl-enetriamine.

Hanna and Siggia[120] extended the usefulness of differential kinetic measurements to rapid reactions having reaction times in the millisecond range through the use of a continuous-flow technique, which provides rapid mixing of the reagents, but leisurely observation of the extent of reaction. Siggia et al.[121] also demonstrated the application of dialysis methods to the kinetic analysis of mixtures; two and three component mixtures of sugars and amino acids could be resolved by dialysis through thin films on the basis of their differing diffusion rates. Amides and nitriles were analysed with good accuracy by measuring the rate of evolution of ammonia generated by the alkaline hydrolysis of their compounds.

Some of the typical organic mixtures that have been determined are summarised in Table 5.1.

5.3 CONCLUSIONS

Kinetic methods have increased to a point where they have become a technique accepted by many analytical chemists. The accuracy and precision of rate methods have been improved to a point where they are comparable to equilibrium methods, and many completely automated systems have been designed and described for kinetic systems. Furthermore kinetic methods have a number of advantages over nonkinetic or equilibrium methods: (i) mixtures of very closely related compounds can be analysed; (ii) analysis is much faster; (iii) a larger number of chemical reactions can be used since side reactions are avoided and slow reactions can now be analytically useful; and (iv) such methods are easily adaptable to automation.

References

1. Gibbs, H. D. (1927). *J. Biol. Chem.*, **72**, 649
2. Guilbault, G. G., Kramer, D. N. and Hackley, E. (1966). *Anal. Chem.*, **38**, 1897
3. Frost, A. and Pearson, R. (1961). *Kinetics and Mechanics*, 2nd Edn. Ch. 3. (New York: Wiley)
4. Guilbault, G. G. (1967). *Flourescence. Theory, Instrumentation and Practice*, Ch. 8. (New York: Marcel Dekker)
5. Mark, H. and Rechnitz, G. (1968). *Kinetics in Analytical Chemistry*, Ch. 3. (New York: Interscience)
6. Michaelis, L. and Mentin, M. (1913). *Biochem. Z.*, **49**, 333
7. Briggs, G. and Holdane, J. (1925). *Biochem. J.*, **19**, 338
8. Blanchard, M., Green, D., Nocito-Carroll, V. and Ratner, S. (1946). *J. Biol. Chem.*, **163**, 137
9. Blaedel, W. and Hicks, G. (1964). *Advances in Analytical Chemistry and Instrumentation*, (Reilly, C. editor). (New York: Interscience)
10. Ingle, J. D. and Crouch, S. R. (1971). *Anal. Chem.*, **43**, 697
11. Guilbault, G. G. (1970). *Enzymatic Methods of Analysis*, (Oxford: Pergamon)
12. Bergmeyer, H. U. (1965). *Methods of Enzymatic Analysis*, 2nd Edn. (Mannheim: Verlag Chemie)

13. Guilbault, G. G. (1970). *Anal. Chem.*, (1966). **38**, 527R; (1968). **40**, 459R; (1970). **42**, 334R (1968). **40**, 459R; (1970). **42**, 334R
14. Guilbault, G. G. (1970). *Critical Reviews in Analytical Chemistry*, **1**, 377
15. Mark, H. and Rechnitz, G. (1968). *Kinetics in Analytical Chemistry*, (New York: Interscience)
16. Yatsimirskii, K. B. (1966). *Kinetic Methods of Analysis*, (Oxford: Pergamon Press)
17. Rechnitz, G. (1966). *Anal. Chem.*, **38**, 513R; (1968). **40**, 455R
18. Guilbault, G. G. (1970). *Anal. Chem.*, **42**, 334R·
19. Guilbault, G. G., Sadar, M. and Peres, K. (1969). *Anal. Biochem.*, **31**, 91
20. Guilbault, G. G., Sadar, S. H. and McQueen, R. (1969). *Anal. Chim. Acta*, **45**, 1
21. Guilbault, G. G., Smith, R. and Montalvo, J. (1969). *Anal. Chem.*, **41**, 600
22. Guilbault, G. G. and Montalvo, J. (1969). *J. Amer. Chem. Soc.*, **91**, 2164
23. Guilbault, G. G. and Montalvo, J. (1969). *Anal. Lett.*, **2**, 283
24. Guilbault, G. G. and Montalvo, J. (1969). *J. Amer. Chem. Soc.*, **92**, 2533
25. Guilbault, G. G. and Hrabankova, E. (1970). *Anal. Chim. Acta*, **52**, 287
26. Guilbault, G. G. and Hrabankova, E. (1970). *Anal. Lett.*, **3**, 53
27. Guilbault, G. G. and Hrabankova, E. (1970). *Anal. Chem.*, **42**, 1779
28. Guilbault, G. G. and Hrabankova, E. (1971). *Anal. Chim. Acta*, **56**, 285
29. Guilbault, G. G. and Shu, F. (1972). *Anal. Chim. Acta*, **56**, 333
30. Guilbault, G. G. and Sadar, S. H. Unpublished work
31. Faway, G. and Dahl, K. (1963). *Lebanese Med. J.*, **16**, 169
32. Mondzac, A., Ehrlich, G. and Seegmiller, J. (1965). *J. Lab. Clin. Med.*, **66**, 526
33. Reichelt, K. L., Kjamme, E. and Tviet, B. (1964). *Scand. J. Clin. Lab. Invest.*, **16**, 433
34. Kirsten, E., Gerez, C. and Kirsten, R. (1963). *Biochem. Z.* **337**, 312
35. Roch-Ramel, F. (1967). *Anal. Biochem.* **21**, 372
36. Rubin, M. and Knott, L. (1967). *Clin. Chim. Acta*, **18**, 409
37. Schulz, D. W., Passonneau, J. V. and Lowry, O. H. (1967). *Anal. Biochem.* **19**, 300
38. Faway, E., Roth, I. and Faway, G. (1966). *Biochem. Z.*, **344**, 212
39. Baum, P. and Czok, R. (1959). *Biochem. Z.*, **332**, 121
40. Adler, E., Gunther, G. and Plass, M. (1939). *Biochem. J.*, **33**, 1028
41. Blaedel, W. J. and Hicks, G. P. (1964). *Advances in Analytical Chemistry and Instrumentation*, Vol. 3, 105. (Reilley, C. N. editor). (New York: Interscience)
42. Warburg, O. and Christian, W. (1938). *Biochem. Z.*, **298**, 150
43. Straub, F. B. (1939). *Biochem. J.*, **33**, 787
44. Huennekens, F. M. and Felton, S. P. (1957). *Methods of Enzymology*, 950. (New York: Academic Press)
45. White, E. H., McCapra, F. and Field, G. F. (1963). *J. Amer. Chem. Soc.*, **85**, 337
46. Wahl, R. and Kozloff, L. (1962). *J. Biol. Chem.*, **237**, 1953
47. Chase, A. M. (1960). *Methods of Biochemical Analysis*, Vol. 8, (Glick, D. editor). (New York: Interscience)
48. Keller, H. (1965). *Naturwissenschaften*, **39**, 109
49. Kitz, R. J. (1964). *Acta Anaesthes. Scand.*, **8**, 197
50. Giang, P. A. and Hall, S. A. (1951). *Anal. Chem.*, **23**, 1830
51. Kramer, D. N. and Gamson, R. M. (1957). *Anal. Chem.*, **29**, (12) 21A
52. Guilbault, G. G., Tyson, B., Kramer, D. N. and Cannon, P. (1962). *Anal. Chem.*, **34**, 1437
53. Guilbault, G. G. and Sadar, M. H. (1969). *Anal. Chem.*, **41**, 366
54. Guilbault, G. G., Sadar, M. H. and Zimmer, M. (1969). *Anal. Chim. Acta*, **44**, 361
55. Cohen, J., Dosterbaan, R. and Berends, F. (1967). *Methods Enzymol*, **11**, 686
56. Delga, J. and Foulhoux, P. (1969). *Prod. Probl. Pharm.*, **24**, 57
57. Guilbault, G. G., Kuan, S. and Sadar, M. H. *J. Agr. Food Chem.*
58. Zhuravskaya, S. and Bobyreva, T. (1968). *Uzb. Biol. Zhur.*, **12**, 55
59. Soliman, S. (1966). *Ain Shams Sci. Bull.*, **9**, 127
60. Goszezynska, K. and Stycyznska, B. (1968). *Rocz. Panstiv. Zakl. Hig.*, **19**, 491
61. Voss, G. (1968). *Bull. Environ. Contam. Toxicol*, **3**, 339
62. Guilbault, G. G., Sadar, M. H. and Kuan, S. (1970). *Anal. Chim. Acta*, **51**, 83
63. Guilbault, G. G., Sadar, M. H. and Kuan, S. (1970). *Anal. Chim. Acta*, **52**, 75
64. Guilbault, G. G. and Sadar, M. H. (1971). *J. Agr. Food Chem.*, **19**, 357
65. Pardue, H. L. (1964). *Anal. Chem.*, **36**, 633
66. Pardue, H. L. (1964). *Anal. Chem.*, **36**, 1110

67. Pardue, H. L., Frings, C. S. and Delaney, C. J. (1965). *Anal. Chem.*, **37**, 1426
68. Mark, H., Weichsellbaum, T. and Plumpe, W. (1969). *Anal. Chem.*, **41**, 103A
69. Mark, H., Plumpe, T., Adams, R. and Hagerty, J. (1969). *Anal. Chem.*, **41**, 725
70. Pardue, H. and Deming, S. (1969). *Anal. Chem.*, **41**, 986
71. Stewart, J. and Lum, P. (1969). *Amer. Lab.*, p. 91. (November)
72. Chen, R., Schechter, A. and Berger, R. (1969). *Anal. Biochem.*, **29**, 68
73. Sajo, I. (1968). *Talanta*, **15**, 578
74. Roginskii, S., Aliev, R., Berman, A., Lokteva, N., Semensko, E. and Yanovskii, M. (1967). *Dokl. Akad. Nauk. SSSR*, **176**, 1114
75. Klatt, L. and Blaedel, W. (1968). *Anal. Chem.*, **40**, 512
76. Cordos, E., Crouch, S. and Malmstadt, H. (1968). *Anal. Chem.*, **40**, 1812
77. Bard, Y. and Lapidus, L. (1968). *Catal. Rev.*, **2**, 67
78. James, G. and Pardue, H. (1968). *Anal. Chem.*, **40**, 796
79. James, G. and Pardue, H. (1968). *Anal. Chem.*, **40**, 796
80. Parker, R., Pardue, H. and Willis, B. (1970). *Anal. Chem.*, **42**, 56
81. Toren, E. and Davis, J. (1968). *Anal. Letts.*, **1**, 289
82. Crouch, S. (1969). *Anal. Chem.*, **41**, 880
83. Dunn, R. and Frings, C. (1969). *Clin. Chem.*, **15**, 810
84. Keyser, A. (1969). *Clin. Chem.*, **15**, 810
85. Toren, E., Eggert, A., Sherry, A. and Hicks, G. (1969). *Clin. Chem.*, **15**, 811
86. Tanaka, M., Funahashi, S. and Shirai, K. (1967). *Anal. Chim. Acta*, **39**, 437
87. Margerum, D., Pausch, J., Nyssen, G. and Smith, G. (1969). *Anal. Chem.*, **41**, 233
88. Pausch, J. and Margerum, D. (1969). *Anal. Chem.*, **41**, 226
89. Lukawiewicz, R. and Fitzgerald, J. (1969). *Anal. Letts.*, **2**, 159
90. Guilbault, G. G. and McQueen, R. (1968). *Anal. Chim. Acta*, **40**, 251
91. Guilbault, G. G. and Kramer, D. (1966). *Anal. Chem.*, **38**, 834
92. Cordos, E., Crouch, S. and Malmstadt, H. (1968). *Anal. Chem.*, **40**, 1812
93. Crouch, S. (1969). *Anal. Chem.*, **41**, 880
94. Crouch, S. and Malmstadt, H. (1967). *Anal. Chem.*, **39**, 1084
95. Crouch, S. and Malmstadt, H. (1967). *Anal. Chem.*, **39**, 1064
96. Javier, A., Crouch, S. and Malmstadt, H. (1969). *Anal. Chem.*, **41**, 239
97. Lorentz, K. (1968). *Z. Anal. Chem.*, **237**, 32
98. Crouch, S. and Malmstadt, H. (1968). *Anal. Chem.*, **40**, 1901
99. Mottola, H., Haro, M. and Freiser, H. (1968). *Anal. Chem.*, **40**, 1263
100. Legradi, L. (1968). *Z. Anal. Chem.*, **237**, 426
101. Babkin, M. (1968). *Zh. Analit. Khim.*, **23**, 637
102. Rodziewicz, W., Kwiatkowska, I. and Kwiatkowski, E. (1969). *Chem. Analyt.* (Warsaw), **14**, 55
103. Tikhonova, V. (1968). *Zh. Analit. Khim.*, **23**, 1720
104. Blaedel, W. and Petitjean, D. (1958). *Anal. Chem.*, **30**, 1958
105. Burgess, A. and Lantham, J. (1966). *Analyst*, **91**, 343
106. Banting, S. C. (1954). *Arch. Biochem. Biophys.*, **52**, 272
107. Lee, T. S. and Kolthoff, I. M. (1951). *Ann. New York Acad. Sci.*, **53**, 1093
108. Garmon, R. (1961). *Ph. D. Thesis*, Univ. of North Carolina, Chapel Hill, N. C.
109. Garmon, R. and Reilley, C. N. (1962). *Anal. Chem.*, **34**, 600
110. Siggia, S. and Hanna, J. (1961). *Anal. Chem.*, **33**, 896
111. Munnely, T. (1968). *Anal. Chem.*, **40**, 1494
112. Lohman, F. and Mulligan, T. (1969). *Anal. Chem.*, **41**, 243
113. Benson, D. and Fletcher, N. (1966). *Talanta*, **13**, 1207
114. Toren, E. and Gnuse, M. (1968). *Anal. Letts.*, **1**, 295
115. Mark, H. and Greinke, R. (1970). *J. Chem. Educ.*, **47**
116. Greinke, R. and Mark H. (1966). *Anal. Chem.*, **38**, 1001
117. Greinke, R. and Mark, H. (1966). *Anal. Chem.*, **38**, 340
118. Mark, H. B., Greinke, R. and Papa, L. (1965). *Div. of Analyt. Chem., 150th Amer. Chem. Soc. Meeting, Atlantic City,* 1965
119. Bond, B. D., Cullion, H. S. and Conduit, C. P. (1965). *Anal. Chem.*, **37**, 147
120. Hanna, J. and Siggia, S. (1964). *Anal. Chem.*, **36**, 2022
121. Siggia, S., Hanna, J. and Serecha, N. (1964). *Anal. Chem.*, **36**, 638
122. Siggia, S., Hanna, J. and Serecha, N. (1964). *Anal. Chem.*, **36**, 227
123. Siggia, S. and Hanna, J. (1961). *Anal. Chem.*, **33**, 896

124. Lee, T. S. (1954). *Organic Analysis,* Vol. 2, 237. (New York: Interscience)
125. Lee, T. S. and Kolthoff, I. M. (1951). *Ann. New York Acad. Sci.,* **53,** 1093
126. Kreevoy, M. and Taft, R. (1955). *J. Amer. Chem. Soc.,* **77,** 5590
127. Skrabal, A. and Zlatewa, M. (1926). *Z. Physik. Chem.,* **119,** 305
128. Roberts, J. and Regan, C. (1952). *Anal. Chem.,* **24,** 360
129. Hammett, L. P. (1940). *Physical Organic Chemistry,* (New York: McGraw-Hill)
130. Krieble, V. K. and Holst, K. (1938). *J. Amer. Chem. Soc.,* **60,** 2976
131. Krieble, V. K. and Noll, C. (1939). *J. Amer. Chem. Soc.,* **61,** 560
132. Hanna, J. G. and Siggia, S. (1962). *Anal. Chem.,* **34,** 547
133. Greinke, R. and Mark, H. (1966). *Anal. Chem.,* **38,** 1001
134. Greinke, R. and Mark, H. (1967). *Anal. Chem.,* **39,** 1952
135. Zuman, P. (1951). *Coll. Czech. Chem. Commun.,* **15,** 839
136. Davies, W. C., Evans, E. and Hulbert, F. (1939). *J. Chem. Soc.,* 412
137. Kolthoff, I. M., Lee, T. and Mairs, M. (1947). *J. Polymer. Sci.,* **2,** 199, 206, 220
138. Saffer, A. and Johnson, B. (1948). *Ind. Eng. Chem.,* **40,** 538

6
Catalytic Methods of Analysis

K. B. YATSIMIRSKII
Academy of Sciences of the Ukrainian Soviet Socialist Republic, Kiev

6.1   INTRODUCTION

In the past two decades kinetic methods of analysis, i.e. quantitative analytical methods which are based on the measurement of the reaction rate and the use of the value obtained for the determination of concentration have been extensively developed[1]. All kinetic methods of analysis may be grouped into two main classes: (a) methods using catalytic reactions and (b) methods using non-catalytic reactions.

Those belonging to the first group are widely used and prove to be rather important for the solution of many analytical problems especially in trace-element analysis. Moreover, these methods are characterised by a number of specific features inherent in catalysis itself. Hence, it may be expedient at present to consider only those kinetic methods of analysis based on the application of catalytic reactions and call them *catalytic methods of analysis*. It is these *catalytic or catalimetric methods of analysis* that are considered in the present review.

The application of non-catalytic reactions has been covered in a mono-graph[2] and in some reviews[3]*.

To determine an element catalytically one must obviously use a reaction in which the element acts as a catalyst. Such a reaction is called an indicator reaction with reference to a given element.

6.2 BASIC REQUIREMENTS

In most cases an indicator reaction may be described by the general scheme:

$$A + B \rightarrow X + Y \qquad (6.1)$$

where A and B are the initial substances and X and Y the reaction products. For example, the oxidation of iodide by hydrogen peroxide in acidic solution

$$H_2O_2 + HI \rightarrow I_2 + H_2O \qquad (6.1a)$$

proceeds in the presence of homogeneous catalysts such as molybdenum(VI), tungsten(VI), iron(III), niobium(V) and other elements. It can, therefore, be used as an indicator reaction for all the above-mentioned elements.

The indicator reaction in a catalytic method of analysis must satisfy the following conditions, namely (a) the concentration of reaction products or reactants should be measured by a fast and simple means; (b) the reaction rate should fall within certain limits because too fast or too slow rates are ob-jectionable for analytical purposes. In the latter case, however, the use of new techniques[4] further extends the lower limit of time-dependent measure-ments (down to 10^{-12} s).

The reaction rate may be determined by changes both in the concentration of the reaction products (Δx) and reactants (Δa) for a given time interval (Δt). In the first case, one can use the simplest differential modification of the catalytic method in which the ratio $\Delta x/\Delta t$, rather close to the actual reaction rate, dx/dt, is taken. In the simplest case (when the induction period is absent and the time reading starts from the moment of mixing) to characterise the reaction rate one may use the ratio x/t (x is the concentration of the reaction product X, t is the time).

As was mentioned above, the main initial value to be measured in kinetic (including catalytic) methods of analysis is the reaction rate, V, which in a differential modification is characterised by a value:

$$V = \Delta x/\Delta t \approx dx/dt \qquad (6.2)$$

In many cases this value is obtained as the slope of the straight line con-structed on the coordinates x v. t. One can also obtain the slope of the straight line (which is part of the whole kinetic curve) constructed on the same co-ordinates, or the tangent to this curve. This method which is called *the method of tangents* is the most used at present.

---

*See Chapter 5.

The essence of the *fixed time method* is that the reaction is allowed to proceed for a strictly determined time interval Δt_o (Δt_o = const). In this case *a measure of the rate* is taken as the change in the concentration, Δx, provided that the straight part of the kinetic curve can be observed.

$$V = \Delta x/\Delta t_o \qquad (6.3)$$

Finally, one can measure the time interval during which a definite change in the concentration of the substance X occurs or the time interval required for the concentration of one of the products to reach a specified value, x, provided the time reading takes place from the moment of reactant mixing.

This modification of the differential kinetic method of analysis is called the *fixed concentration method*.

As in this case Δx_o is constant, a measure of the rate is the value of the reciprocal of the time interval required to reach this value

$$V = \Delta x_o/\Delta t_o \qquad (6.4)$$

Some reactions, complicated from the viewpoint of kinetics are characterised by the occurrence of an induction period during which *visible* signs of the reactions are not observed. The length of an induction period depends on many factors including the concentrations of the catalyst and the reactants. It actually characterises the length of one or more reaction steps. Thus, the fixed concentration method involves the determination of an induction period. Indeed, the end of an induction period is determined as a rigidly defined change in the concentration of the same substance (in the simplest case from zero to a measurable value).

The relationship between catalyst concentration and the length of an induction period may prove more complicated than is shown in equation (6.4)[1].

The differential modification of the kinetic method of analysis is an advantage for those cases where the concentration of one or more reaction *products* is measured. When the reaction rate is measured on the basis of changes in reactant concentrations, it is necessary to take into account that in the course of the reaction rate measurements the reactant concentration is not constant. This makes it impossible to use the rate law in a differential form.

Most reactions used in catalimetry (short for catalytic methods of analysis) are first order with respect to each of the reactants or, in any event, can be carried out under conditions which give a linear relationship of reaction rate v. concentration. The reaction is commonly carried out in such a way that one of the reactants is taken in a great excess compared to the other one. In this case, the simplest form of the rate law for reaction (6.1) is as follows

$$V = dx/dt = kC_Kab \qquad (6.5)$$

where k is the rate constant and C_K, a and b are the concentrations of the catalyst and reactants A and B respectively.

When the reactant B is present in excess, integration of equation (6.5) gives

$$\ln(a_o/a) = kC_Kb_ot \qquad (6.6)$$

where a_0 and b_0 are the initial concentrations of A and B reactants respectively; a is the current concentration of the reactant A. In this case it is reasonable to construct the rate curve on the coordinates $\ln a$ $v.$ t or $\ln \eta$ $v.$ t, where

$$\eta = a_0/a \tag{6.7}$$

is the extent of reaction; the more the extent of reaction, the greater the value of η.

In case of curves (or straight lines) of the integral modification of catalimetry all three methods of analysis may be used: (a) the method of tangents, (b) the fixed time method (c) the fixed concentration method.

It should be noted that in the integral modification of catalimetry the reaction rate cannot be used directly.

As can be seen in equation (6.5) changing the concentration of reactants (for example, A) leads to a significant change in the reaction rate. The ratio $(1n\ \eta)/t$ is a value to some extent analogous to the reaction rate.

$$(\ln \eta)/t = kC_K b_0 \tag{6.8}$$

but comparing equation (6.5) with equation (6.8) one can see that it is the reaction rate value divided by the concentration of one of the reactants (in our case it is A)

$$W = (\ln \eta)/t = V/a \tag{6.9}$$

All that has been said about the differential modification may be applicable to the integral modification provided a value W is taken into account. This value can most conveniently be expressed as the ratio of a change in the logarithm of reactant concentration $(\ln a)$ to a definite time interval (t)

$$W = \Delta \ln a/\Delta t \tag{6.10}$$

Equation (6.10) is analogous to equation (6.4).

A comparison of equations (6.2), (6.5), (6.6) and (6.10) shows that the catalyst concentration can be obtained on the basis of the two expressions:

$$C_K = \Delta x/(\Delta t k a_0 b_0) \quad \text{(differential modification)} \tag{6.11}$$

$$C_K = \Delta \ln a/(\Delta t k b_0) \quad \text{(integral modification)} \tag{6.12}$$

The catalyst concentration may be calculated using equations (6.11) or (6.12) or from calibration curves constructed on the coordinates:

(a) $\Delta x/\Delta t$ $v.$ C_K by the method of tangents (differential modification) or $\Delta \ln a/\Delta t$ $v.$ C_K (integral modification).

(b) x $v.$ C_K by the fixed time method, $t_0 = $ const, (differential modification) or $\Delta \ln a - C_K$ (integral modification).

(c) $1/t$ $v.$ C_K by the fixed concentration method, $\Delta x_0 = $ const. (differential and integral modifications).

The difference between the two modifications of the fixed concentration method lies in the conditions: in the first case, $\Delta x_0 = $ const, and in the second one, $\Delta \ln a_t = $ const.

While constructing the calibration curve the values a_0 and b_0 are known (all tests are carried out at the same initial reactant concentrations) and the values of the constant k are given indirectly (they may easily be obtained from the slope of calibration curves).

In catalimetric determinations *two* parameters must be measured: *the concentration* of reaction product or reactants and the time interval during which a change in the concentration is observed.

It follows from the equations given above that the sensitivity of catalimetric methods is extremely high.

It is possible to determine the concentration of the catalyst down to 10^{-16} mol l^{-1} (when the molecular weight is of the order of 100, it is $c.$ 10^{-17} g cm^{-3}). If the catalytic coefficient is equal to 10^8 (time, in minutes), an observation time interval of the order of 10 min and reactant concentration of the order of 1 mol l^{-1} are chosen and changes in reactant concentration are $c.$ 10^{-7} mol l^{-1}.

This means that 10^{-19} mole or 60 000 molecules of the catalyst being estimated could be found in 1 cm^3 of solution.

The possibility of detecting individual molecules with the help of catalytic reactions becomes apparent if we take into account the use of ultramicroanalytical techniques such as spectrophotometric measurements of very small volumes.

It would seem that even the above-mentioned value (10^{-17} g cm^{-3}) is not the limit of sensitivity of catalimetric methods. Even more sensitive methods of measuring reactant concentration, or reaction products, can be proposed and a time interval of more than 10 min can be chosen for an observation. In addition, reactant concentrations can exceed molar and catalytic reaction rate constants can be greater than 10^8.

Nevertheless, the above limit has not been reached as yet. The most sensitive catalytic reaction allows concentrations of 10^{-12} g cm^{-3} to be measured[5], i.e. the sensitivity reached at present is some orders of magnitude less than the theoretical value.

Catalytic methods of analysis commonly used are equal to activation and fluorometric analysis methods in sensitivity. They differ from the first by virtue of their greater simplicity of measurement and from the second by their greater universality. They have been proposed at present as a means of determining more than 40 elements and they are almost universally applicable.

There are several reasons for the sensitivity obtained in practice being significantly less than the theoretically attainable value. The chief of these lies in that an indicator reaction catalysed by a substance to be determined can also be catalysed by other substances present in the solution. Alternatively the reaction may proceed via a non-catalytic pathway. Thus, a background is formed and it varies due to insufficient purity of the reagents or solvents and cleanliness of reaction vessels.

The total reaction rate, therefore, is the sum of the two rates:

$$V_{total} = V_{cat} + V_{background} \tag{6.13}$$

The rate of the catalytic reaction is found as the difference between the total reaction rate determined experimentally and the rate of any non-catalytic reaction. The operator should minimise background variations or, at any rate, try to maintain as constant a background as possible.

At very small catalyst concentrations, the reproducibility becomes poor owing to many factors, some of which are difficult to avoid, namely, the presence of adsorbed substances on the walls of reaction vessels, the effect of solution microheterogeneity, the effect of suspended particles, etc.

The reproducibility of catalimetric measurements is essentially controlled by the precision of concentration measurements of reaction products or reactants and time-interval measurements.

The reproducibility of measurements of time intervals of the order of several minutes (namely such time intervals as are commonly used in catalimetry) is, as a rule, very high even if simple measuring instruments are used such as a stop-watch; the error is usually less than 1 %. On the other hand, errors in the determination of concentration by optical techniques are frequently of the order of several per cent.

If one takes into account the inevitable variations in temperature that lead to rather appreciable variations in reaction rate constants and probable inaccuracies in the determination of the initial concentrations of reactants as well as background differences in the construction of calibration curves and in conducting analysis, the reproducibility of the order of 10–15 % commonly attainable in catalimetric determinations will be easily understood.

A variety of ways exist to increase the reproducibility of results. It may be markedly improved by using strictly thermostatically controlled solutions, extremely pure reactants and solvents and special reaction vessels yielding minimal quantities of impurities.

The reproducibility may also be improved by using the method of standard additions[6]. The essence of this method is that the rate of the indicator reaction is determined twice, (a) in the solution being analysed and (b) in a similar solution to which a known quantity of the element being analysed has been added. Naturally, the background remains practically constant.

Fairly good reproducibility of results can be attained by a catalimetric titration in which the solution of the catalyst is titrated with a solution of inhibitor (or vice versa) and the end-point is determined with the help of a catalytic indicator reaction. In catalimetric titration, catalimetry is used only to determine the end-point while the values being measured are the volumes of reactants.

Apart from the accuracy in measurements of reactant volumes (which is rather high), the error of determination depends on the steepness of the titration curve at the end-point which is known to be dependent on the ratio.

$$\overline{C}_{cat}/C_{cat}^{o} = \gamma \qquad (6.14)$$

where $\overline{C}_{cat}$ and C_{cat}^{o} are catalyst concentrations at the end-point and in the initial solution, respectively.

It can be seen from this equation that when a titration accuracy of the order of 0.01–0.001 % (a necessary condition for titration) is obtainable the *sensitivity* of concentration measurements of catalimetric titration proves to be *c.* 2–3 orders of magnitude less than in 'conventional' catalimetric methods. However, the error (in catalimetric titration) can be reduced to several tenths or hundredths of a per cent.

The specificity of catalimetric methods of analysis varies considerably. There are reactions that are catalysed by only one element and, hence, extremely selective. However, most reactions are catalysed by many elements and cannot be selective. Yet even in the latter case the specificity may be increased by varying the reaction conditions (for example, solution acidity) or by using substances that mask one group of catalysts without affecting the other.

It can be seen, therefore, that the correct choice of an indicator reaction is of great importance. Indicator reactions are continually being sought for in investigations carried on in the field of catalimetry.

6.2.1 Classification of catalytic reactions

It is expedient to group known indicator reactions into three classes:

(a) Oxidation–reduction reactions (including electrochemical processes).

(b) Heterolytic reactions (exchange reactions proceeding without changing the oxidation states of atoms of reactants).

(c) Enzymatic reactions.

6.3 OXIDATION–REDUCTION REACTIONS

Redox reactions form the largest group of catalytic reactions.

In recent years, the kinetics and mechanism of redox reactions have been studied in considerable detail and several monographs have been published[8, 9] on this problem.

In accordance with our classification[10] oxidants and reductants may be grouped into:

(a) s,p-oxidants and reductants which react, as a rule, slowly even in the presence of catalysts,

(b) d-oxidants and reductants which usually react quickly and are able to act as catalysts themselves.

The total number of indicator redox reactions known at present exceeds 100. They may be classified as follows:

(i) Reactions of various reducing agents with *oxidants* that belong to group (a) H_2O_2, O_2, $S_2O_8^{2-}$, ClO_3^-, BrO_3^-, IO_3^-, ClO_4^-, IO_4^-, FeO_4^{2-}, NO_3^-, NO_2^-, organic substances.

(ii) Reactions between various oxidants and *reducing agents* that belong to group (a) Hg_2^{2+}, Sn^{2+}, $H_2PO_2^-$, AsO_2^-, $S_2O_3^{2-}$, I_2, N_2H_4, organic substances (amines, oxalates, phenols, etc.).

(iii) Heterogeneous redox reactions (involving MnO_2, silver halogenide, etc.).

Reactions of oxidants with reducing agents of Group (ii) may also proceed slowly. In the latter case, the reaction rate may be low owing to like electric charges of the reacting species. This makes their approach and contact difficult.

6.3.1 Reactions of hydrogen peroxide

One of the most important reactions used in catalimetry is considered to be the oxidation of various reducing agents by hydrogen peroxide. These reactions are rather easily carried out and pure reactants can usually be obtained. Usually the reaction rate can easily be controlled if suitable reducing agents are used.

Numerous reduction reactions of hydrogen peroxide in an acidic medium are catalysed by the same catalysts. As a rule, they are ions and molecules which form complexes with H_2O_2. In such complex compounds, a hydrogen peroxide molecule is activated, its geometry and the distribution of electric charges are changed, and even heterolytic decomposition to OH^+ and OH^- ions becomes possible[7]. It is interesting to note that the catalytic activity of hydroxy complexes appears to be higher than that of simple ions[1]. This finding may be used to increase reaction specificity considerably and to determine one element-catalyst in the presence of another (for example, zirconium and hafnium present in solution simultaneously)[1].

A list of catalysts and their rather close sensitivity of indicator reactions (see Table 6.1) points to the similarity of catalyst action in the reactions of hydrogen peroxide with different reducing agents.

The oxidation of iodide with hydrogen peroxide was one of the first reactions proposed for quantitative catalimetric determinations[1]. However, this reaction is still under investigation[11–15]. New element-catalysts have been

Table 6.1   Sensitivity of pC ($-\log C$)* catalytic reduction of hydrogen peroxide with different reducing agents.

Reducing agents	Catalysts										
	Ti	Zr	Hf	Th	V	Nb	Ta	Cr	Mo	W	Fe^{III}
I$_-$	7	7	7	6.7	—	6	6.7	6.6	7.7	8	6.3
$S_2O_3^{2-}$	8	7	7	—	8.3	6.7	8	—	9	7.7	—
e$^-$	†—	—	5	—	9	—	—	—	8	6	5

*C is the concentration given in g/cm^3. The sensitivity is expressed as $pC = -\log C$.
†Polarographic reduction at a mercury electrode.

found[11, 15], the possibility of the analysis of technical materials by catalimetric methods have been discussed[12], and modifications of known methods have been proposed[13, 14].

The same holds for the reduction of H_2O_2 by thiosulphate[16–19]. Investigations of catalytic waves originating in the cathodic polarographic reduction of H_2O_2 are now being undertaken[20–22].

The oxidation of colourless substances to give colourless products presents considerable difficulties. Therefore, in the overwhelming majority of indicator reactions involving hydrogen peroxide, coloured substances or others that give coloured oxidation products are used as reducing agents. Such substances frequently include simple and substituted aromatic amines and phenols. New techniques for the measurement of reaction rate have been proposed[23].

The mechanism of the catalytic action may be expected to change considerably in neutral or weakly basic media. Under such conditions the mechanism of catalytic action is significantly changed and alternative redox mechanisms may occur. Elements exhibiting variable valency and relatively small redox potentials, may act as catalysts, namely, copper[24, 25], manganese[26], iron[27], cobalt[28, 29], ruthenium[30, 31] and chromium[32, 33].

Analogous mechanisms of catalyst action may be observed in the oxidation of more complex dyes of different kinds, such as Methyl Orange, Alizarin S, Eriochrome Black, Pyrocatechol Violet, etc. Methods for the determination of chromium[34, 35], manganese[36-39], cobalt[40] and iron[41-44] have been proposed on the basis of these indicator reactions. No doubt, the composition and structure of a substrate (reducing agent) have an important bearing upon the sensitivity of catalytic reactions but this question needs further consideration.

Several new catalytic chemiluminescence oxidation reactions of hydrogen peroxide have been proposed[49].

The kinetics and mechanism of hydrogen peroxide decomposition to give water and oxygen are under investigation. This reaction has previously

Table 6.2   Principal catalytic indicator reactions of hydrogen peroxide

Reactions	*Catalysts*
$H_2O_2 + I^-$	Ti, Zr, Hf, Th, V, Nb, Ta, Cr, Mo, W, Fe[III]
$H_2O_2 + S_2O_3^{2-}$	Ti, Zr, Hf, V, Nb, Ta, Mo, W
$H_2O_2 + e^-$	Hf, V, Mo, W, Fe[III]
$H_2O_2 + H_2C_2O_2N_2H_2$	Mo, W
$H_2O_2 + PhOH, PhNH_2$	Cr, Mn, Fe, Co, Cu, Ru, Os
$H_2O + dyes$	Cr, Mn, Fe, Co, Ni, Cu
$H_2O \rightarrow H_2O + O_2$	V, Mn, Fe, Cu, Pd, Os
$H_2O + C_2O_4^{2-}$	Co

been used to determine copper, manganese, iron and palladium[1]. Recently it was suggested for the determination of element-catalysts such as copper (in complexes)[45, 46], osmium[47] and vanadium[48].

The most important catalytic indicator reactions involving hydrogen peroxide are given in Table 6.2.

An examination of the experimental data presented in Table 6.2 shows that catalytic reactions of hydrogen peroxide can be useful for the determination of 18 element-catalysts, i.e. about half of all the elements determined by catalimetric methods.

However, it is evident that the possibilities of reactions of this type have not been used to full advantage as yet. It is probable that the known reactions will be used for the determination of other elements and that reactions involving new substrate-reducing agents will be proposed. So far, those reactions, the rate of which may be followed by photometry, i.e. reactions involving changes in light absorption have been used mainly. New methods of measuring reaction rates will undoubtedly allow use of the oxidation of colourless substrates to yield colourless products.

Aerial oxidation is similar to the type of reaction under consideration (i.e. oxidations by hydrogen peroxide). Formerly, it was used only to determine copper[1, 50] but now the aerial oxidation of several substrates has been proposed for the determination of iron[51, 52]. The photochemical oxidation of Methyl Orange catalysed by iron ($pC = 7.6$) deserves special mention[53].

Finally, in this category there exists one more reagent involving a linkage between two atoms of oxygen, namely the persulphate ion. The catalytic reduction of persulphate with different organic compounds has been used to determine silver, iron and vanadium[1]. The sensitivity of these reactions was increased by activators[54-57]. New reactions involving the oxidation of different inorganic and organic substrate reducing agents with persulphate catalysed by silver and iron have been studied[58, 59].

Catalimetric titration has proved very useful for the determination of iodide and bromide using the oxidation of Cadion with persulphate as an indicator reaction[60]. The oxidation of a series of organic dyes with persulphate may be catalysed by micro-concentrations of lead[61, 62] ($pC_{Pb} = 8$).

6.3.2   Reactions of oxyanions

The reduction reactions of anions of oxygen acids form a large group of catalytic redox reactions.

This type of reaction will be discussed starting with the reduction of anions of the halogen acids in their highest oxidation state (VII).

In catalimetric determination, the perchlorate ion can only be reduced electrochemically[1]. Periodate is reduced by different aromatic anions in the presence of catalysts. Such reactions are useful in the determination of vanadium and manganese which act as catalysts[1].

In recent years, some investigations have been carried out to increase the sensitivity of the manganese-catalysed periodate reduction by organic amines[63, 64]. The reduction of periodate by benzidine and its derivatives has been proposed for the determination of ruthenium with high sensitivity[65, 66] ($pC = 11$).

The reductions of the halogenates (ClO_3^-, BrO_3^-, IO_3^-) by various reducing agents are commonly used in catalimetry. The reduction of chlorates and bromates has much in common, but the reduction of iodates is somewhat different.

The following substances are used as reducing agents for chlorates: iodide ion (in the determination of vanadium, rhenium, ruthenium and osmium), aromatic amines and phenols (in the determination of vanadium and osmium), dyes of complex composition (in the determination of osmium). Bromates are also reduced by iodide (the determination of vanadium), aromatic amines and phenols (in the determination of vanadium and molybdenum)[1].

In recent years the reduction of halogenates by iodide and aromatic amines has been shown to be catalysed by osmium compounds as well[67]. New rather sensitive reactions for the determination of osmium ($pC_{Os} = 9$–10.4) have been worked out on this basis. New amines have been proposed as reducing agents for reactions catalysed by vanadium[70, 71]. However, this has not led to the development of reactions with better sensitivities.

The oxidation of thiocyanate by bromate appears to be catalysed by vanadium compounds[72]. These also catalyse the reduction of bromate by ascorbic acid[73].

The reduction of chlorate by phenylhydrazine p-sulphonic acid is catalysed by vanadium[74] and selenium(IV)[75]. This reaction may be useful for the determination of selenium since it has a sensitivity of the order of $pC_{Se} = 6$.

A number of dyes (Indigo Carmine[76], Bordeaux B[77] and H-acid[78]) have been proposed for the reduction of bromate, all these reactions being catalysed by vanadium compounds. The reduction of meturin is catalysed by chromate[79] and this may be used for the determination of chromium ($pC_{Cr} = 8.5$). The decomposition of bromate in acidic media is catalysed by low concentrations of bromide and iodide[80] and this reaction may serve as an indicator reaction for their determination ($pC_{Br} = 8.5, pC_I = 8$). Numerous investigations are dedicated to the kinetics and mechanism of the reduction of halogenates[81–90]. In a series of papers[91–94] several carboxylic acids and oxyquinoline derivatives have been proposed to increase the sensitivity of the catalytic reduction of chlorate and bromate by amines and phenols. The reaction of iodate with arsenite has been studied by polarographic methods[95]. The well known copper-catalysed oxidation of manganese(II) to permanganate by hypobromite has now been proposed[96] as a most sensitive reaction for rhodium ($pC_{Rh} = 10$).

Tellurate (TeO_4^{2-}) and selenate (SeO_4^{2-}) are the only other anions of oxygen acids formed by elements of the sixth group of the periodic table (persulphate has already been mentioned and with peroxides also play a role in catalytic reactions). The reduction of tellurate by tin(II) compounds is catalysed by rhenium and molybdenum[1]. Analogously, selenate is reduced by tin(II) and catalysed by extremely small quantities of molybdenum. This catalysis ($pC_{Mo} = 9$), may be used for the determination of molybdenum[97].

Several catalytic reduction reactions of nitrate and nitrite have been proposed for the determination of the platinum metals and iodide[1]. Changes in concentrations of the reaction between nitrite and thiocyanate results in a sharp increase in the sensitivity of the catalytic reaction for iodide ($pC_I = 8.7$)[98].

Numerous anions of oxygen acids of elements with sp-orbitals possess too low a potential to act as oxidants, while elements with vacant d-orbitals form anions of oxygen acids that react strongly (MnO_4^-, MnO_4^{2-}, CrO_4^{2-}, VO_4^{3-}, ReO_4^{2-}, TeO_4^{2-}, etc.). However, in this case, the anions of this group form iso-polyions and the oxidation can only proceed slowly as it is preceded by fairly slow decomposition of the polyacids to smaller highly active structural units[99, 100].

Several reactions involving reduction of molybdate to 'molybdenum blue' by iodide[99, 100], stannous chloride[101], ascorbic acid[102, 103] have been proposed. These reactions are catalysed by phosphate[99, 101–105] and germanium[100]. The sensitivities of these reactions correspond to $pC_P = 8$ and $pC_{Ge} = 7.3$.

The mechanism of the reduction of molybdate by iodide has been studied in the presence of phosphorus and germanium[104] and investigations of the reduction of phosphoric heteropoly acids with thiourea catalysed by copper compounds have also been reported[105, 106].

Next we shall consider those redox reactions that proceed with a rate that is characterised by the nature of the reducing agent. By analogy with reactions of the peroxide compounds mentioned above we can consider them as catalytic reactions involving bond breaking between the same two atoms. The simplest reactions of this type are catalytic oxidation of the diatomic mercury(I) ion, Hg_2^{2+}. The oxidation of Hg_2^{2+} with cerium(IV) is catalysed by iridium compounds[107, 108] ($pC_{Ir} = 8$). This reaction is also catalysed by compounds of gold[109] ($pC_{Au} = 8$) manganese(II)[110] and silver[111].

The oxidation of hydrazine with cerium(IV) is catalysed by ruthenium compounds[112]. The oxidation of water proceeding with the formation of oxygen is related to the aerial oxidation of many organic substances and also occurs slowly. The oxidation of water by cerium(IV), for example, proceeds slowly with the evolution of molecular oxygen. It is catalysed by iridium compounds and used for the determination of iridium ($pC_{Ir} = 8$)[113, 114]. The sensitivity of the well known oxidation of chloride with cerium(IV) catalysed by silver salts has been considerably increased[115]. The catalytic indicator reactions of arsenite oxidation with iodate by means of which micro-concentrations of iodide can be determined has already been mentioned[95]. The oxidation of arsenite with permanganate is catalysed by micro-concentrations of osmium compounds[1].

The oxidation of arsenite with cerium(IV) catalysed by osmium and iodide has been studied in considerable detail[1]. Automatic and semi-automatic units have been proposed for assessing the reaction rate by means of spectrophotometry and potentiometry[116–118]. The utilisation of these processes minimises the time required for analysis and increases the sensitivity and accuracy of the methods[119]. The kinetics and mechanism of this indicator reaction have been investigated[119, 120] and the effects of different acids on the rate of the oxidation have been established particularly with HNO_3. It proved to be possible to determine osmium and iodine present in solution together. After the sum of the two elements had been determined the iodide was inactivated with mercury[122]. The oxidation of arsenite by cerium(IV) may be useful as an indicator reaction for iridium ($pC_{Ir} = 8.7$) and ruthenium[123] ($pC_{Ru} = 8.3$). Ruthenium and osmium present in solution together can also be determined by using differences in their catalytic behaviour[124] and it is possible to determine palladium ($pC_{Pd} = 7$) by catalimetric titration with iodide[125].

Investigations of the well-known oxidation of hypophosphite and tin(II) have been continued and new application of these reactions have been proposed. The former was used to determine ruthenium, osmium, palladium, platinum and tellurium[1]. More recently, the oxidation of hypophosphite by 1,4,6,11-tetraazonaphthacene was proposed as an indicator reaction for the determination of selenium[126] ($pC_{Se} = 7$). The oxidation of tin(II) by iron(III) salts formerly used as an indicator reaction only for molybdenum and copper(II) is now proposed for ruthenium[127] ($pC_{Ru} = 7.5$), and osmium[128] ($pC_{Os} = 8.5$). Tin(II) chloride slowly reduces arsenious acid to metallic arsenic but the reaction may be catalysed by palladium compounds to yield a useful procedure for palladium[129] ($pC_{Pd} = 6.7$). So far we have considered only reactions involving *slowly* reacting substances which had already been used in redox processes. Now let us pass to the discussion of a fairly small number of reactions with new slowly reacting substances. The reduction of plutonium(IV) by uranium(IV) is catalysed by micro-concentrations of chromium(III) and orthophosphate[130]. The oxidation of formate by permanganate in basic solution (permanganate is reduced to manganate) is catalysed by compounds of nickel ($pC_{Ni} = 6.7$), cobalt ($pC_{Co} = 6.7$), silver ($pC_{Ag} = 8$) and some of the platinum metals[131].

This reaction is an example of a redox reaction in which a 'limiting substance' is rather a simple organic substance, the formate ion. There are several such cases. For instance, some organic dyes ($C_{23}H_{25}N_2Cl$, Malachite Green, $C_{33}H_{32}Cl$, Victoria Blue B) are reduced by $TiCl_3$ slowly, but the reaction is catalysed by compounds of vanadium, molybdenum, tungsten, uranium and osmium. The indicator reaction for the reduction of Methylene Blue with sulphide catalysed by selenium is also known. The reduction of Basic Blue with titanium(III) compounds is catalysed by compounds of tungsten[132] ($pC_W = 8.7$). There exist, as mentioned above, reactions in which organic compounds behave as substrate-reducing agents. Such reactions include the oxidation of diphenylamine with cerium(IV), catalysed by iridium compounds[133] ($pC_{Ir} = 8$), and the oxidation of ethylenediaminetetra-acetate with dichromate catalysed by manganese(II) compounds[134] ($pC_{Mn} = 6$).

Indicator reactions have also been proposed in which both the oxidant and the reductant are organic substances. The oxidation of phenol with chloramine T catalysed by trace quantities of copper ($pC_{Cu} = 7.3$) is an example of this[135]. The interaction of p-nitrobenzaldehyde with o-dinitrobenzene catalysed by extremely low concentrations of cyanides, ($pC_{CN} = 9$) is another[136, 137]. This reaction, as indeed many others, may be used to determine inhibitors, i.e. ions which form stable complexes with the catalyst (Ag^+, Hg^{2+}, Cu^{2+}, Ni^{2+}, Zn^{2+}, Cd^{2+}). The sensitivity is dependent on the stability of cyanide complexes formed. Its value ranges from 8.5 to 5 (in pC_M units).

A photochemical reaction for the determination of the trace quantities of iron has already been mentioned[53]. The photochemical decomposition of oxalic acid was proposed as an indicator reaction for the determination of uranium(VI) which catalyses this reaction[138] ($pC_U = 5.5$).

The catalytic reduction of pre-irradiated silver halide emulsions by different reducing agents has proved rather promising[1]. In the presence of micro-concentrations of sulphide, rhodium or palladium chlorides the trace quantities of metallic silver, silver sulphide and other compounds formed in the treatment of the irradiated AgBr emulsion, provide sensitive techniques for the determination of sulphide sulphur[139, 140] ($pC_S = 10$), rhodium and palladium[141] ($pC_{Rh} = 7.3$, $pC_{Pd} = 6$).

6.3.3 Inhibition of catalysis

As has already been mentioned, a great variety of substances can be determined by using their inhibiting action[1, 60, 125]. The number of such reactions is great. A catalytic hydrogen wave may be detected in 5-sulpho-8-mercaptoquinoline solution in the presence of cobalt. Antimony(III) compounds display a marked inhibiting action on this catalysis which may be used to determine micro-concentrations of antimony[142] ($pC_{Sb} = 8.3$). Numerous ligands which form complexes with titanium (sulphosalicylate, phenols and others) may be determined by their inhibiting action on the catalytic wave of titanium(IV)[143]. The aerial oxidation of ascorbic acid catalysed by copper compounds may be inhibited by extremely small quantities of cyanide[144] ($pC_{CN} = 7.5$). A similar method of analysis may also be applied to the determination of cystine, salicylate, ethylendiaminetetra-acetate,o-phenanthroline and ethylenediamine[145]. The catalytic oxidation of Malachite Green with periodate in the presence of manganese(II) used as an indicator reaction in catalimetric titration may be used to determine ethylendiaminetetra-acetate by its inhibiting action[146]. Finally, cystine in concentrations as low as 10^{-9} mol l^{-1} may be determined by measuring its inhibition of the Co-catalysed hydrogen wave[147].

6.4   HETEROLYTIC REACTIONS

Heterolytic catalytic reactions which do not involve a change in oxidation state of the elements of the reactants are used far less than catalytic redox processes. As a rule, they are characterised by lower sensitivity than the first group of indicator reactions. However, the use of this group of catalytic reactions allows considerable extension of the list of element-catalysts and includes a number of elements of the main groups of the periodic table. The number of heterolytic reactions is relatively small and, can be easily enumerated up to 1966, namely, the isotope exchange between Ce^{4+} and Ce^{3+} catalysed by fluoride ions ($pC_F = 5.7$)[1], partial or complete substitution of cyanide ions in the complex ion $Fe(CN)_6^{4-}$, catalysed by compounds of mercury, silver and gold[1] and the alkaline hydrolysis of ethyl cysteine ester catalysed by lead ions[1].

It has been reported that the substitution of cyanide ions in $Fe(CN)_6^{4-}$ can be catalysed not only by ions of mercury but also by compounds of platinum, silver and gold[148].

A more detailed investigation of the alkaline hydrolysis of the ethyl ester of cysteine has shown that this reaction is catalysed only by lead ions but that the ions of zinc, cadmium and mercury accelerate the reaction by formation of a labile complex which is easily hydrolysed. Zinc, cadmium and mercury compounds are parts of the reaction products[149–151].

It has been shown that the decarboxylation of acetoacetic acid may be catalysed by copper, zinc, cadmium, tin or aluminium ions. The sensitivity of this reaction is fairly low, however, being only of the order of 0.1–several micrograms[152–154].

An interesting type of heterolytic catalytic reaction was revealed in 1963 in which an exchange reaction of polydentate ligands takes place between two metal complexes, for example

$$Cu(EDTA)^{2-} + Ni(TRIEN)^{2+} \rightarrow Cu(TRIEN)^{2+} + Ni(EDTA)^{2-} \quad (6.15)$$

where EDTA^{4-} denotes ethylenediaminetetra-acetate and TRIEN – triethylenetetramine[155].

Such reactions were shown to proceed via a chain non-radical mechanism, i.e. the chains were initiated and propagated by different *ligand species* (EDTA and TRIEN), which catalyse these heterolytic exchange reactions. Methods for the determination of micro-concentrations of many ions and substances that act as inhibitors for this reaction have been proposed[156, 157].

A great number of anions catalyse complex formation between zirconyl ions and metallochromic indicators such as Xylenol Orange and Methyl Thymol Blue. Anion-catalysts include fluoride, phosphate, arsenate, sulphate and anions of organic polybasic acids. The sensitivity is of the order of

10^{-7} g for catalyst ions such as fluoride and phosphate. The kinetochromic technique for the determination of fluoride with the help of this type of reaction has been developed in detail[158-160].

Although heterolytic reactions have been studied much less than redox processes and their sensitivity is, as a rule, not very high, as has already been remarked, they considerably increase the number of elements that can be analysed by catalimetric methods. Moreover, there appears to be some similarity between heterolytic reactions and the majority of enzymatic processes.

6.5 ENZYME-CATALYSED REACTIONS

An unusual and very interesting group of catalytic reactions are those involving enzymes. Several authors separate this group of reactions into a special section — enzymatic methods of analysis. They are of particular interest in biochemistry because the concentration of enzymes, co-enzymes, substrates and inhibitors may best be determined by their influence on reaction rates. However, it should be noted that many metal ions can act as inhibitors or activators of enzymes. Fairly sensitive methods for determining a number of metals have been suggested on this basis. Copper ions for example inhibit the hydrolytic splitting of starch by the enzyme diastase[1].

Magnesium, manganese and cobalt activate the enzyme dehydrogenase and sharply accelerate the conversion of isocitric acid into α-ketoglutarate[161]. The activity of invertase on the hydrolysis of saccharose is much affected by silver ions ($pC_{Ag} = 8$), thiourea ($pC_{CSN_2H_4} = 8.5$), cyanide, sulphide and iodide ions[162, 163]. Beryllium ($pC_{Be} = 7.5$), and zinc ($pC_{Zn} = 6$) inhibit the phosphatase-enzyme which catalyses hydrolysis of phosphate esters[164] and zinc and calcium act as activators in analogous reactions involving similar enzymes[165]. Finally, mercury and silver in very small concentrations inhibit the dehydrogenase catalysed oxidation of alcohols[166]. It is evident that the use of the activating or inhibiting action of ions on enzymic catalysis should permit the determination of ions of many different groups of metals particularly in the transition and main groups of the periodic table.

The total number of elements that can be determined by catalimetric methods is great. The types of reactions used for the determination of 45 given elements and the sensitivity obtained for each are presented in Table 6.3 below. Only the most sensitive indicator reactions are given.

Table 6.3 shows that the sensitivity of the overwhelming majority of catalimetric determinations falls within the nanogram range, i.e. the sensitivity is of the order of 10^{-9} g cm^{-3} or $pC = 9$. Such sensitivity considerably exceeds that of normal solution spectrophotometric techniques and proves to be comparable with the sensitivity of activation analysis.

In the past 5 years the number of elements to be determined by catalimetric methods has increased from 39 to 45, and during this time more sensitive catalimetric methods have been proposed for twelve elements of the previous group of 39.

The overwhelming majority of indicator reactions belong to group A (redox reactions). Such reactions are, as a rule, applicable to transition

Table 6.3 Sensitivity of catalimetric determination of elements*

H ABC																	
Li —	Be C7											B —	C —	N —	O —		F B7
Na —	Mg C											Al B6	Si A5	P A8	S A10		Cl —
K —	Ca C6	Sc —	Ti A7	V A10	Cr A9	Mn A10	Fe A9	Co A12	Ni A9	Cu A9	Zn B6C7	Ga	Ge A7	As A5	Se A7		Br A8
Rb —	Sr —	Y —	Zr A7	Nb A7	Mo A10	Tc —	Ru A11	Rh A10	Pd A9	Ag A10	Cd B6	In —	Sn —	Sb A8	Te A10		I A10
Cs —	Ba —	La-Lu A7	Hf A7	Ta A7	W A9	Re A9	Os A10	Ir A8	Pt A7	Au A8	Hg B9	Tl —	Pb A8B7	Bi —	Po —		At —
Fr —	Ra —	Ac A7	Th A7	Pa —	U A9												

*Table lists the negative logarithms of sensitivity, g cm^{-3}.
A – redox, B – heterolytic, C – enzymatic catalytic reactions.

elements, although the use of inhibiting action and complex formation makes it possible to utilise them for phosphorus, germanium, antimony, etc.

6.6 FUTURE DEVELOPMENT OF CATALIMETRY

The theory of homogeneous catalytic redox reaction is fairly well developed at present. In many cases, the mechanism of these reactions is complex and each reaction involves several stages. These are best established and the total rate law determined by the topological graph method[179].

In a number of cases the mechanisms of a catalytic redox reaction may be classified into one of two main types:

(a) Activation of complexing

$$A + K \rightleftharpoons KA \tag{6.16}$$

$$KA + B \rightarrow X + Y + K \tag{6.17}$$

where, K is Mo^{VI}, W^{VI}, Fe^{III}, etc; A is H_2O_2 and B is a reducing agent. For example, oxidations by hydrogen peroxide in acidic solution proceed via this scheme.

(b) Alternate oxidation–reduction of the catalyst:

$$A + K \rightarrow X + K^* \tag{6.18}$$

$$K^* + B \rightarrow K + Y \tag{6.19}$$

where K and K* are the oxidised and reduced forms of the catalyst, respectively. Numerous catalytic oxidation reactions of the halogenates proceed in this way.

It can be seen from these schemes that some definitive *thermodynamic* requirements exist in searching for new catalytic indicator reactions.

In the first type it is necessary that a thermodynamically stable complex of the type KA should be formed. Indeed, oxidations with hydrogen peroxide in acidic solution are catalysed by substances that form complexes with it, namely, molybdenum(VI), tungsten(VI), iron(III), vanadium(V), titanium(IV), etc.

In the second type, it is necessary that the redox potential value of the catalyst K ($K \rightleftharpoons K^* + e$) should be intermediate *between* those of the oxidant A, and the reductant, B. It will be easily understood from this point of view that when the differences in potentials of systems involving A and B are high a large number of substances may catalyse the reactions. But, if these differences are small, the number of catalysts will be smaller and the reactions will appear to be more specific.

It should be noted that in the second type the electron transfer from catalyst, K, to substrate, A, or vice versa, can be preceded by the formation of a complex KA in which the electron transfer occurs. It has been shown that in many cases these complexes behave as charge transfer complexes and their nature has been studied in some detail.

6.6.1 Inhibition and activation

A great variety of substances act as inhibitors (giving, for example inactive complexes with the catalyst) or as activators, and their potential mechanisms of activation or inhibition have been discussed in considerable detail[168]. Reactions in which the same substance may act as an inhibitor for one catalyst and as an activator for another are of particular interest. For example, in the oxidation of p-phenetidine by bromate, sulphosalicylic acid activates catalysis by vanadium compounds and completely masks that of iron compounds[89]. The catalytic indicator reaction of the oxidation of p-phenetidine by bromate may be used for the determination of osmium. On the addition of ethylenediaminetetra-acetate to the solution, most interfering elements are bound in catalytically inactive complexes[89].

In some instances a combination of catalimetric and separation methods proves rather effective for the determination of some elements. An example of this is the determination of a number of platinum metals after their separation by partition chromatography[108], electrophoresis[31] or ion exchange[169].

The analysis of traces of different elements in technical materials by catalimetry following the application of general schemes of analysis and separation of elements has become fairly common. The application of kinetic methods of analysis is, however, often limited by the absence of convenient techniques for measuring changes in concentrations. Most modern applications are based on spectrophotometric measurements, and, hence, are only suitable for indicator reactions involving changes in light absorption. There are, however, many catalytic indicator reactions in which there is no change in light absorption.

Therefore, investigations in which techniques of measuring concentration other than spectrophotometry are used, deserve special attention, e.g. chemical methods of analysis (titration after reaction being abruptly arrested), gas volumetric methods, turbidimetry, refractometry, potentiometry, polarography, amperometry, luminescent analysis[1].

Predictably in recent years optical methods of measuring concentrations in solution in catalimetry have again predominated but investigations involving the application of other techniques such as titrimetry[138], potentiometry[117, 150, 151], manometry[152, 153], polarography and amperometry[21, 22, 95, 142] have also been reported. Automatic and semi-automatic devices have also been proposed for measuring reaction rates by different physical methods[13, 14, 116, 118, 170].

Calorimetric methods of measuring reaction rates using either liberation or absorption of the heat in the course of the reaction[171, 172] have proved useful and in this case high sensitivity was obtained by employing thermistors[2].

Catalimetric methods may also be applicable to the analysis of extremely small quantities of solutions. A procedure of spot catalimetric analysis similar to chromatography has been described[173].

6.6.2   Catalimetric titration

In recent years, the catalimetric titration procedure proposed in 1962 has been developed further[174]. A solution of an inhibitor, L, is titrated with a solution of the catalyst. For the simplest case the reaction proceeds according to the equation

$$L + K \rightleftharpoons KL \qquad (6.20)$$

where KL is a weakly dissociated compound or has a low solubility. When the end-point is reached, an excess of the catalyst in solution is revealed by a suitable indicator reaction. An example, is the titration of iodide by silver to form insoluble silver iodide. Beyond the end-point the excess of silver

Table 6.4   Catalimetric titration procedures

Catalyst	Inhibitor	Indicator reaction	Ref.
I^-	Ag^+	$Ce^{IV} + As^{III}$	174
I^-	Pd^{2+}	$Ce^{IV} + As^{III}$	125
I^-	Hg^{2+}	$Ce^{IV} + As^{III}$	176
Ag^+	I^-, Br^-	$S_2O_8^{2-} + cation$	60
Cu^{2+}	CN^-	Ascorbic acid $+ O_2$	144
Cu^{2+}	Cystine, salicylic acid $EDTA^{4-}$, En, Phen	Ascorbic acid $+ O_2$	145
Mn^{2+}	$EDTA^{4-}$	Malachite Green $+ IO_4^-$	146
Co^{2+}	$EDTA^{4-}$	Tiron $+ Perborate$	175

ions is determined by means of their catalytic indicator reaction involving the oxidation of arsenite by cerium(IV). The equilibrium constant of the reaction (6.20) has an important bearing on the accuracy of the catalimetric titration (see equation 6.14).

A theoretical discussion of the possibilities of catalimetric titration and some new examples have been given in two fairly recent publications[177, 178]. Some examples of catalimetric titration are given in Table 6.4.

6.6.3   Assessment of status and prospects

In recent years, many catalytic methods have been developed and are now generally accepted. At present they are applicable to at least 45 elements and this number is likely to increase in the near future. Redox indicator reactions are most commonly used in catalimetry and since their theory of action is well developed it is possible to plan the search for new indicator reactions.

Heterolytic reactions and reactions involving enzymes are now also in general use and are finding new applications. The sensitivity of most catalimetric methods ranges from nanogram to microgram amounts of elements per cubic centimetre though in some cases it is possible to determine elements in quantities much less than a nanogram.

It can reasonably be assumed that in the future the availibility of very pure substances, activators and precise measuring techniques will permit the determination of substances in the picogram range. The combination of inhibition and activation as well as the use of selective separation methods should permit the attainment of even higher specificity in catalimetric determinations.

References

1. Yatsimirskii, K. B. (1967). *Kinetic Methods of Analysis*, 2nd ed. (Moscow: Khimia)
2. Mark, H. B., Jr, Rechnitz, G. A. (1968). *Kinetics in Analytical Chemistry*. (New York: Interscience)
3. Rechnitz, G. A. (1966). *Anal. Chem.*, **38**, 513R; **40**, 455R
4. Caldin, E. F. (1968). *Fast Reactions in Solution*. (Oxford: Blackwell)
5. Bognar, J. and Jellinek, O. (1961). *Mag. Kem. foly.*, **67**, 147
6. Poluektov, N. S. (1941). *J. Appl. Chem. (USSR)*, **14**, 695
7. Yatsimirskii, K. B. (1952). *J. Anal. Chem. (USSR)*, **7**, 206
8. Turney, T. A. (1965). *Oxidation Mechanisms*. (London: Butterworths)
9. Taube, H. (1970). *Electron Transfer Reactions of Complex Ions in Solution*. (New York: Academic Press)
10. Yatsimirskii, K. B. (1965). *Theoret. Exp. Chem. (USSR)*, **1**, 343; (1971). *Pure Appl. Chem.*, **27**, 251
11. Hadjüoannou, T. (1965). *Talanta*, **15**, 535
12. Babko, A. K., Lisetskaya, G. S. and Tsarenko, G. F. (1968). *J. Anal. Chem. (USSR)*, **23**, 1342
13. Hadjüoannou, T. (1966). *Anal. Chim. Acta*, **35**, 360
14. Wilson, A. M. (1966). *Anal. Chem.*, **38**, 1784
15. Litvinenko, V. A. (1966). *Ukr. Chem. J. (USSR)*, **32**, 1160
16. Litvinenko, V. A. (1970). *News of the Higher Institutes of Chemistry and Chemical Technology (USSR)*, **13**, N2, 139
17. Babko, A. K. and Litvinenko, V. A. (1966). *J. Anal. Chem. (USSR)*, **21**, 301
18. Litvinenko, V. A. (1969). *Ukr. Chem. J. (USSR)*, **35**, 311
19. Litvinenko, V. A. (1968). *J. Anal. Chem. (USSR)*, **23**, 12, 1807
20. Toropova, V. F. and Zabarova, R. S. (1969). *News of the Higher Institutes of Chemistry and Chemical Technology (USSR)*, **12**, (No. 11), 1487
21. Toropova, V. F., Chovnyk, N. G., Vekslina, V. A. and Vashchenko, V. V. (1970). *J. Analyt. Chem. (USSR)*, **25**, 2464
22. Sharipov, R. and Songina, O. (1966). *J. Anal. Chem. (USSR)*, **21**, 800
23. Sajo, J. (1968). *Talanta*, **15**, 578
24. Gregorowicz, Z. (1967). *Chem. Analit. (Warsaw)*, **12**, N4, 911
25. Gregorowicz, Z. and Suwinska, T. (1970). *Chem. Analit. (Warsaw)*, **15**, (2), 295
26. Bartkus, P. and Yasinskene, E. J. (1970). *Proceedings of the Higher Institutes of Chemistry and Chemical Technology of the Lithuanian SSR*, **23**, 19GD
27. Kriss, E. E., Savichenko, Ya. S., Yatsimirskii, K. B. (1969). *J. Anal. Chem. (USSR)*, **24**, 875
28. Kucharkowski, R. (1970). *Z. Analyt. Chem.*, **249**, (1), 22
29. Kucharkowski, R. and Doege, H. (1968). *Z. Anal. Chem.*, **238**, 241
30. Morozova, R. P. and Yatsimirskii, K. B. (1969). *J. Anal. Chem. (USSR)*, **24**, 1183
31. Varshal, G. M., Kashsheeva, I. Ya., Morozova, R. P. and Konopleva, O. N. (1971). *J. Anal. Chem. (USSR)*, **26**, N5, 939
32. Dolmanova, I. F., Zolotova, G. A. and Peshkova, V. M. (1969). *J. Anal. Chem. (USSR)*, **24**, 1035
33. Dolmanova, I. F., Zolotova, G. A., Shekhovtsova, T. N. and Peshkova, V. M. (1970). *J. Anal. Chem. (USSR)*, **25**, N11, 2136
34. Yasinskene, E. I. and Bilidene, E. B. (1968). *J. Anal. Chem. (USSR)*, **23**, 143
35. Yasinskene, E. I. and Bilidene, E. B. (1967). *J. Anal. Chem. (USSR)*, **22**, 741
36. Janjic, T., Milovanovic, C. and Celap, M. (1970). *Anal. Chem.*, **42**, 27
37. Blank, A. B. and Voronova, A. Ya. (1965). *Zavod Lab. (USSR)*, **31**, 1299
38. Bartkus, P. I. and Yasinskene, E. I. (1968). *J. Anal. Chem. (USSR)*, **23**, 1622
39. Sychev, A. Ya. and Tiginyanu, Ya. D. (1969). *J. Anal. Chem. (USSR)*, **24**, 1842
40. Vershinin, V. I., Chuiko, V. T. and Reznik, B. E. (1971). *J. Anal. Chem. (USSR)*, **26**, N9, 1710
41. Yasinskene, E. I. and Birmantas, I. I. (1963). *News of the Higher Institutes of Chemistry and Technology (USSR)*, **6**, 918
42. Kreingold, S. U., Bozhevolnov, E. A. and Drapkina, D. A. (1967). *J. Anal. Chem. (USSR)*, **22**, 218
43. Kreingold, S. U. and Sosenkova, S. I. (1971). *J. Analyt. Chem. (USSR)*, **26**, N2, 332

44. Kreingold, S. U., Bozhevolnov, E. A. and Antonov, V. A. (1968). *Zavod. Lab. (USSR)*, **34**, 260
45. Sychev, A. Ya. (1969). *News of the Higher Institutes of Chemistry and Chemical Technology (USSR)*, **12**, N12, 1666
46. Sharma, O. S., Schubert, J., Brooks, H. B. and Sicilio, F. (1970). *J. Amer. Chem. Soc.*, **92**, 822
47. Domka, F. and Marciniec, B. (1969). *Chem. Analit. (Warsaw)*, **14**, 145
48. Yurchenko, G. K., Frolova, L. A., Vorobeva, N. A. and Bogdanov, G. A. (1969). *News of the Higher Institutes of Chemistry and Chemical Technology (USSR)*, **12**, N11, 1499
49. Babko, A. K., Dubovenko, L. I. and Terletskaya, A. (1966). *Ukr. Chem. J. (USSR)*, **32**, 1326
50. Ilicheva, L. P. and Yatsimirskii, K. B. (1968). *News of the Higher Institutes of Chemistry and Chemical Technology (USSR)*, **11**, N5, 520
51. Joshino, J., Takeuchi, T., Kinoshita, H. and Uchida, S. (1968). *Bull. Chem. Soc. Jap.*, **41**, 765
52. Taylor, J. E., Jan, J. F. and Wang, J. (1966). *J. Amer. Chem. Soc.*, **88**, 1663
53. Kharlamov, I. P., Dodin, E. I. and Mantsevich, A. D. (1967). *J. Anal. Chem. (USSR)*, **22**, 371
54. Yasinskene, E. I. and Yankauskene, E. K. (1966). *J. Anal. Chem. (USSR)*, **21**, 940
55. Yasinskene, E. I. and Rasevichute, N. I. (1970). *J. Anal. Chem. (USSR)*, **25**, N3, 458
56. Bontschev, P. R., Alexiev, A. and Dimitrova, B. (1969). *Talanta*, **16**, 597
57. Bonchev, P. R., Alexiev, A. and Dimitrova, B. *Mikrochim. Acta*, **1970**, 1104-1108
58. Alexiev, A. A. and Bonchev, P. R. *Mikrochim. Acta*, **1970**, (1), 13-19
59. Budanov, V. V. (1965). *News of the Higher Institutes of Chemistry and Chemical Technology (USSR)*, **8**, N1, 23
60. Tamarchenko, L. M. (1970). *J. Anal. Chem. (USSR)*, **25**, N3, 567
61. Yasinskene, E. I. and Kalesnikaite, S. Z. (1968). *J. Anal. Chem. (USSR)*, **23**, 1169
62. Yasinskene, E. I. and Kalesnikaite, S. Z. (1970). *J. Anal. Chem. (USSR)*, **25**, N1, 87
63. Dolmanova, I. F., Poddubenko, V. P. and Peshkova, V. M. (1970). *J. Anal. Chem. (USSR)*, **25**, N11, 2146
64. Alekseeva, I. I. and Danylova, Z. P. (1971). *J. Anal. Chem. (USSR)*, **26**, N9, 1786
65. Kalinina, V. E. Yatsimirskii, K. B. and Zimina, T. S. (1969). *J. Anal. Chem. (USSR)*, **24**, 1178
66. Kalinina, V. E. (1971). *Kinetics and Catalysis (USSR)*, **12**, 100
67. Alekseeva, I. I., Smirnova, I. B. and Yatsimirskii, K. B. (1970). *J. Anal. Chem. (USSR)*, **25**, N3, 539
68. Filippov, A. P., Zyatkovsky, V. M. and Yatsimirskii, K. B. (1970). *J. Anal. Chem. (USSR)*, **25**, 1769
69. Bognar, J. and Sarosi, S., (1962). *Magyar kem. foly.*, 53
70. Lazarev, A. E. and Tronina, E. M. (1965). *Zavod. Lab. (USSR)*, **31**, 270
71. Lazarev, A. I. and Lazareva, V. I. (1969). *J. Anal. Chem. (USSR)*, **24**, 395
72. Jendrzejevsky, V. I. and Yatsimirskii, K. B. (1966). *J. Anal. Chem. (USSR)*, **21**, 314
73. Bognar, J. and Jellinek, O. *Mikrochim. Acta*, **1970**, 1017-1021
74. Tanaka, M. and Awata, N. (1967). *Anal. Chim. Acta*, **39**, 485
75. Kawashima, T., Nakano, S. and Tanaka, M. (1970). *Anal. Chim. Acta*, **49**, (3), 443-447
76. Budanov, V. V. and Panfilova, E. I. (1965). *News of the Higher Institutes of Chemistry and Chemical Technology (USSR)*, **8**, N2, 208
77. Fuller, C. W. and Ottaway, J. M. (1970). *Analyst*, **95**, 41-46
78. Kreingold, S. U. and Bozhevolnov, E. A. (1965). *Zavod. Lab. (USSR)*, **31**, N7, 784
79. Kreingold, S. U., Bozhevolnov, E. A., Supin, G., Antonov, V. and Panteleimonova, A. (1969). *J. Anal. Chem. (USSR)*, **24**, 853
80. Toropova, V. F. and Tamarchenko, L. M. (1967). *J. Anal. Chem. (USSR)*. **22**, 576
81. Yatsimirskii, K. B. and Nikolov, G. St. (1970). *J. Phys. Chem. (USSR)*, **44**, N5, 1129
82. Yatsimirskii, K. B. and Nikolov, G. St. (1970). *J. Phys. Chem. (USSR)*, **44**, N6, 1400
83. Nikolov, G. St. and Yatsimirskii, K. B. (1963). *Theoret. Exp. Chem. (USSR)*, **5**, 773
84. Bontschev, P. R. and Jeliazkowa, B. (1967). *Mikrochim. Acta*, **1**, 116
85. Fuller, C. W. and Ottaway, J. M. (1970). *Analyst*, **95**, 28
86. Fuller, C. W. and Ottaway, J. M. (1970). *Analyst*, **95**, 34
87. Filippov, A. P., Zyatkovsky, V. M., Yatsimirskii, K. B. (1969). *Ukr. Chem. J. (USSR)*, **35**, 451

88. Zhelyazkova, B. G., Mityaeva, M. J. and Bonchev, P. R. (1970). *Ukr. Chem. J. (USSR)*, **34**, 40
89. Yatsimirskii, K. B., Filippov, A. P. and Zyatkovsky, V. M. (1969). *Ukr. Chem. J. (USSR)*, **35**, 233
90. Yatsimirskii, K. B. and Kalinina, V. E. (1965). *News of the Higher Institutes of Chemistry and Chemical Technology (USSR)*, **8**, 378
91. Bonchev, P. R. and Yatsimirskii, K. B. (1965). *J. Phys. Chem. (USSR)*, **39**, 1995
92. Yatsimirskii, K. B. and Kalinina, V. E. (1969). *J. Anal. Chem. (USSR)*, **24**, 390
93. Bonchev, P. R. and Zhelyazkova, B. G. (1967). *Coll. Catalytic Reactions in Liauid Phases*. "Nauka", Alma-Ata, 473
94. Bontschev, P. R. and Evtimova, B (1968). *Mikrochim. Acta*, **3**, 492, 498
95. Toporova, V. F. and Tamarchenko, L. M. (1967). *J. Anal. Chem. (USSR)*, **22**, 23, 4
96. Morozova, R. P., Yatsimirskii, K. B. and Egorova, I. T. (1970). *J. Anal. Chem. (USSR)*, **25**, N10, 1954
97. Lazarev, A. I. (1967). *J. Anal. Chem. (USSR)*, **22**, 1836
98. Proskuryakova, G. F. (1967). *J. Anal. Chem. (USSR)*, **22**, 802
99. Alekseeva, I. I., Nemzer, I. I. and Tolstych, E. A. (1969). *Zavod. Lab. (USSR)*, **34**, N11, 1305
100. Alekseeva, I. I. and Nemzer, I. I. (1969). *J. Anal. Chem. (USSR)*, **24**, 1393
101. Kriss, E. E., Rudenko, V. K., Yatsimirskii. K. B. and Vershinin, V. I. (1970). *J. Anal Chem. (USSR)*, **25**, N8, 1303
102. Yatsimirskii, K. B., Kriss, E. E. and Rossolowski, S. (1970). *J. Anal. Chem. (USSR)*, **23**, 325
103. Rosolovski, S. (1970). *Chemia Analit.*, **15**, (1), 157
104. Alekseeva, I. I., Nemzer, I. I. and Rysev, A. G. (1970). *News of the Higher Institutes of Chemistry and Chemical Technology (USSR)*, **13**, N10, 1423
105. Tsyganok, L. P., Chuiko, E. and Reznik, V. (1967). *Zavod Lab. (USSR)*, **33**, 5
106. Reznik, B. E. and Tsyganok, L. P. (1965). *News of the Higher Institutes of Chemistry and Chemical Technology (USSR)*, **8**, N3, 392
107. Tikhonova, L. P. and Yatsimirskii, K. B. (1968). *J. Anal. Chem. (USSR)*, **23**, 1413
108. Varshal, G. M., Tikhonova, L. P., Sychkova, V. A. and Shulik, L. S. (1970). *J. Anal. Chem. (USSR)*, **25**, 12, 2427
109. Frumina, N., Mustafik, J. and Kykleva, L. (1966). *News of the Higher Institutes of Chemistry and Chemical Technology (USSR)*, **9**, 554
110. McCurdy, M. H. Jr. and Guilbault, G. G. (1960). *Anal. Chem.*, **32**, 647
111. Guilbault, G. G. and McCurdy, M. H., Jr. (1966). *J. Phys. Chem.*, **70**, 656
112. Romanov, V. F. (1971). *J. Inorg. Chem. (USSR)*, **16**, 1119
113. Ginzburg, S. I. and Yuzko, M. J. (1966), *J. Anal. Chem. (USSR)*, **21**, 79
114. Ginzburg, S. I. and Yuzko, M. J. (1965), *J. Inorg. Chem. (USSR)*, **10**, 823
115. Bontchev, P. R. and Alexiev, A. A. (1968). *Microchim. Acta*, 875
116. Pardue, H. and Habig, R. (1966). *Anal. Chim. Acta*, **35**, 383
117. Pardue, H. and Habig, R. L. (1966). *Anal. Chim. Acta*, **35**, N3, 383-90
118. Knapp, G. and Spittzy, H. (1969). *Talanta*, **16**, 1361
119. Habig, R., Pardue, H. and Worthington, J. B. (1967). *J. Anal. Chem.*, **39**, 600
120. Rodriguez, P. and Pardue, H. (1969). *Anal. Chem.*, **41**, 1369
121. Knapp, G. and Spitzy, H. (1969). *Talanta*, **16**, 1353
122. Rodziguez, P. and Pardue, H. (1969). *Anal. Chem.*, **41**, 1376
123. Gillet, A. C. Jr., *Mikrochim. Acta*, (1970). 855-863
124. Worthington, J. B. and Pardue, H. L. (1970). *Anal. Chem.*, **42**, 1157
125. Fedorova, T. I. and Yatsimirskii, K. B. (1967). *J. Anal. Chem. (USSR)*, **22**, 283
126. Kawashima, T. and Tanaka, M. (1968). *Anal. Chim. Acta*, **40**, 137
127. Reznik, B. E. and Bednyak, N. A. (1968). *J. Anal. Chem. (USSR)*, **23**, 1502
128. Alekseeva, I. I. and Zhir-Lebed, L. N. (1970). *News of the Higher Institutes of Chemistry and Chemical Technology (USSR)*, **13**, 1260
129. Fedorova, T. I., Shvedova, I. V. and Yatsimirskii, K. B. (1970). *J. Anal. Chem. (USSR)*, **25**, N2, 307
130. Biddlle, P., Miles, J. H. and Waterman, M. J. (1966). *J. Inorg. Nucl. Chem.*, **28**, 1736
131. Mealor, D. and Townshend, A. (1967). *Anal. Chim. Acta*, **39**, 235
132. Omarova, E. S., Speranskaya, E. F. and Kozlovsky, M. T. (1968). *J. Anal. Chem. (USSR)*, **23**, 12, 1826

133. Tikhonova, L. P., Yatsımirskii, K. B. and Svarkovskaya, I. P. (1970). *J. Anal. Chem. (USSR)*, **25**, N9, 1766
134. Sanchez-Pedreno, C., Sierra Jimenes, F. and Martinez Lozano, C. (1970). *Intei on Quim. Analit. pura apl. Ind.*, **24**, 39
135. Yachiyo, K. M., Namiki, and Goto, H. (1966). *Talanta*, **13**, 1561
136. Guilbault, G. and Kramer, D. (1966). *Anal. Chem.*, **38**, 834
137. Guilbault, G. and Mequeen, R. J. (1968). *Anal. Chim. Acta*, **40**, 251
138. Babko, A. K. and Ginzburg, L. M. (1966). *J. Anal. Chem. (USSR)*, **21**, 1070
139. Babko, A. K. and Maksimenko, T. S. (1967). *J. Anal. Chem. (USSR)*, **22**, 570
140. Babko, A. K., Markova, L. V. and Kaplan, M. I. (1963). *Zavod. Lab. (USSR)*, **34**, 1053
141. Kaplan, M. I., Pilipenko, A. T. and Markova, A. B. (1970). *J. Anal. Chem. (USSR)*, **25**, N12, 244
142. Toropova, V. F., Anisimova, L. A., Pavlichenko, L. A. and Bankovsky, Yu. A. (1964). *J. Anal. Chem. (USSR)*, **24**, 7, 1031
143. Saksin, E. V. and Turyan, Yu. I. (1970). *J. Anal. Chem. (USSR)*, **25**, 12, 2362
144. Mottola, H. and Freiser, H. (1966). *Anal. Chem.*, **38**, 1266
145. Mottola, H., Haro, M. and Freiser, H. (1968). *Anal. Chem.*, **40**, 1263
146. Mottola, H. A. and Freiser, H. (1967). *Anal. Chem.*, **39**, 1294
147. Gilbert, D. (1969). *Anal. Chem.*, **41**, 1567
148. Hadjüonannou, F. P. (1966). *Anal. Chim. Acta*, **35**, 351
149. Yatsimirskii, K. B. and Tikhonova, L. P. (1965). *J. Inorg. Chem. (USSR)*, **10**, 2070
150. Tikhonova, L. P. and Yatsimirskii, K. B. (1966). *J. Inorg. Chem. (USSR)*, **11**, 2259
151. Yatsimirskii, K. B. and Tikhonova, L. P. (1967). *J. Inorg. Chem. (USSR)*, **12**, 417
152. Bontchev, P. R. and Michaylova, V. (1967). *J. Inorg. Nucl. Chem.*, **29**, 2945
153. Michaylova, V, Evtimowa, B. and Bontchev, P. R. (1968). *Mikrochim. Acta*, 922
154. Michaylova, V. and Bontchev, P. R. (1970). *Mikrochim. Acta*, 344
155. Olson, D. C. and Margerum, D. W. (1963). *J. Amer. Chem. Soc.*, **85**, 297
156. Margerum, D. W. and Steinhous, R. R. (1965). *Anal. Chem.*, **37**, 222
157. Stehl, R. H., Margerum, D. W. and Lattevell, J. J. (1967). *Anal. Chem.*, **39**, 1346
158. Cabello-Thomas, M. and West, T. S. (1969). *Talanta*, **16**, 781
159. Hems, R. V., Kirkbright, G. F. and West, T. S. (1969). *Talanta*, **16**, 789
160. Hems, R. V., Kirkbright, G. F. and West, T. S. (1970). *Talanta*, **17**, 433
161. Kratochvil, B., Boyer, S. L. and Hicks, S. (1967). *Anal. Chem.*, **39**, 45
162. Mealor, D. and Townshend, A. (1968). *Talanta*, **15**, 1477
163. Mealor, D. and Townshend, A. (1968). *Talanta*, **15**, 1371
164. Townshend, A. and Vaughan, A. (1969). *Talanta*, **16**, 929
165. Townshend, A. and Vaughan, A. (1970). *Talanta*, **17**, 289
166. Townshend, A. and Vaughan, A. (1970). *Talanta*, **17**, 299
167. Yatsimirskii, K. B. (1965). *Theoret. Exp. Chem. (USSR)*, **1**, 343
168. Bonchev, P. R. and Yatsimirskii, K. B. (1965). *Theoret. Exp. Chem. (USSR)*, **1**, 179
169. Hesselbarth, A. (1969). *Z. Anal. Chem.*, **248**, 289
170. Weisz, H., Klockow, D. and Ludwig, H. (1969). *Talanta*, **16**, 921
171. Vaugham, G. A. and Swithenbank, J. J. (1965). *Analyst*, **90**, 594
172. Vajgand, V. J. and Gaab, F. F. (1967). *Talanta*, **14**, 345
173. Pavlova, V. K. and Yatsimirskii, K. B. (1969). *J. Inorg. Chem. (USSR)*, **24**, 1347
174. Yatsimirskii, K. B. and Fedorova, T. I. (1962). *Proc. USSR Acad. Sci*, **143**, 143
175. Weisz, H. and Muschelknautz, U. (1966). *Z. Anal. Chem.*, **215**, 17
176. Weisz, H. and Klockow, D. (1967). *Z. Anal. Chem.*, **232**, 321
177. Weisz, H. and Janjic, T. (1967). *Z. Anal. Chem.*, **227**, 1
178. Mottola, N. (1969). *Talanta*, **16**, 1267
179. Yatsimirskii, K. B. (1971). *The Use of Graphical Methods in Chemistry*, (Kiev: Naukova Dumka)

7

Molecular Fluorescence Spectroscopy

G. G. GUILBAULT
Louisiana State University in New Orleans

7.1 INTRODUCTION TO LUMINESCENCE

7.1.1 Theory of luminescence

When a quanta of light impinges on a molecule it is absorbed in c. 10^{-15} s and an electronic transition can take place to a higher electronic state (Figure 7.1). This absorption of radiation is highly specific, and radiation of a particular energy is absorbed only by a characteristic structure.

In the ground state of most molecules each orbital electron in the lower energy levels is paired with another electron whose spin is opposite to its own spin. Such a state is called a singlet state, S. When the molecule absorbs radiation, the electron is raised to an upper excited singlet state, S_1, S_2, and so forth. These singlet transitions are responsible for the visible and ultra-violet absorption spectra observed for molecules. The absorption transitions usually originate in the lowest vibrational level of the ground electronic state.

During the time the molecule can spend in the excited state, 10^{-4}–10^{-8} s, some energy in excess of the lowest vibrational energy level is rapidly dissipated. The lowest vibrational level ($v = 0$) of the excited singlet state S is attained. If all the excess energy is not further dissipated by collisions with other molecules, the electron returns to the ground electronic state with the emission of energy. This phenomenon is called fluorescence. Because some energy is lost in the brief period before emission can occur, the emitted energy (fluorescence) is of longer wavelength than the energy that was absorbed.

The phenomenon of phosphorescence involves an intersystem crossing or transition from the singlet to the triplet state. A triplet state results when the spin of one electron changes so that the spins are the same or unpaired. The transition from the lowest triplet state, T, to the singlet ground state is quantum mechanically 'not allowed', consequently, transition times of 10^{-4}–10 s are observed. Hence, a characteristic feature of phosphorescence is an afterglow, i.e. emission which continues after the exciting source is removed. Because of the relatively long lifetime of the triplet state, molecules in this state are much more susceptible to radiationless deactivation processes, and generally only those substances dissolved in a rigid medium phosphoresce.

Some of the better general references in luminescence spectroscopy are listed at the end of this article. Books by Guilbault[1], Hercules[2], Passwater[3], Phillips and Elevitch[4] and Udenfriend[5] are worth obtaining. Excellent

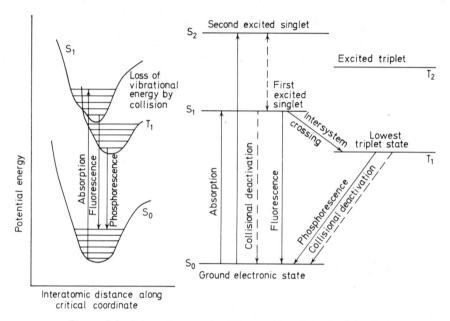

Figure 7.1   Schematic energy-level diagram for a diatomic molecule

chapters by Weissler and White on fluorescence appear in the *Handbook of Analytical Chemistry*[6] and in Scott's *Standard Methods of Chemical Analysis*[7]. Workers in fluorescence should also consult the reviews by White in *Analytical Chemistry* every two years[8] and receive the free pamphlets published monthly by American Instrument Company[9] and by Turner Instrument Company[10].

7.1.1.1   Excitation spectrum

Any fluorescent molecule has two characteristic spectra: the excitation spectrum (the relative efficiency of different wavelengths of exciting radiation to cause fluorescence) and the emission spectrum (the relative intensity of radiation emitted at various wavelengths).

The shape of the excitation spectrum should be identical to that of the absorption spectrum of the molecule and independent of the wavelength at which fluorescence is measured. This is seldom the case, however, the differences being due to instrumental artifacts. Examination of the excitation spectrum indicates the positions of the absorption spectrum that give rise to fluorescence emission. The excitation spectrum of the Al chelate of acid

Alizarin Garnet R (Figure 7.2), for example, indicates peaks at 350, 430 and 470 nm. The absorption spectrum exhibits peaks at 270, 350 and 480 nm. To obtain the true or 'corrected' spectra of this compound the apparent curve would have to be corrected for changes with frequency of (a) the

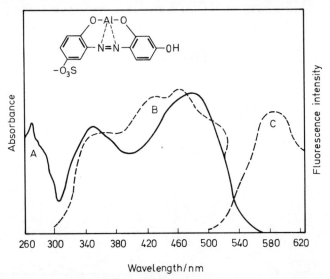

Figure 7.2 Aluminium complex with acid Alizarin Garnet R (0.008 %). (A) the absorption spectrum; (B) the fluorescence excitation spectrum; and (C) the fluorescence emission spectrum

photomultiplier, (b) the band-width of the monochromator and (c) the changing transmission of the monochromator.

A general rule of thumb is that the longest wavelength peak in the excitation spectra is chosen for excitation of the sample. This minimises possible decomposition caused by the higher energy radiation at the shorter wavelengths.

7.1.1.2 Emission or fluorescence spectra

The emission or fluorescence spectrum of a compound results from the re-emission of radiation absorbed by that molecule. The quantum efficiency and the shape of the emission spectrum are independent of the wavelength of exciting radiation. If the exciting radiation is at a wavelength different from the wavelength of the absorption peak, less radiant energy will be absorbed and hence less will be emitted. The emission spectrum of the Al- acid Alizarin Garnet R complex indicates a fluorescence peak at 580 nm (Curve C, Figure 7.2).

Each absorption band to the first electronic state will have a corresponding emission or fluorescence band. These two bands or spectra will be approximately mirror images of each other. In fact, this mirror image principle is useful in distinguishing whether an absorption band is another vibrational band in the first excited state or a higher electronic level. Additional fluorescence peaks other than the mirror image of the absorption spectrum indicate scatter or the presence of impurities. Rayleigh and Tyndall scatter could be observed in the emission spectrum at the same wavelength as the excitation wavelength, and also at twice this value (second-order grating effect). With very dilute solutions one may also observe Raman scatter. The wider the fluorescence band is, the more complex and less symmetrical the compound.

Figure 7.3 shows the absorption and emission spectra of anthracene and quinine. Four major absorption peaks are observed in the anthracene spectrum—all correspond to transitions from S_0 to S_1^*, but denote transitions to different vibrational levels. Four major emission peaks, each a mirror image of the peaks in the absorption spectrum, are likewise observed. For quinine two excitation peaks are observed, one at 250 nm corresponding to a $S_0 \rightarrow S_2^*$ transition, and a second at 350 nm corresponding to a $S_0 \rightarrow S_1^*$ transition. Only one emission peak, corresponding to the $S_1^* \rightarrow S_0$ transition, is observed.

The fact that some compounds possess several excitation and/or emission peaks is of analytical utility. If two compounds have overlapping excitation bands, as in the case of anthracene and quinine, both could be excited together, then differentiated by their emission spectrum. Quinine could be measured at a λ_{em} of 450 nm, whereas anthracene could be monitored at a λ_{em} of 400 nm. Similarly, if two compounds emit radiation at the same

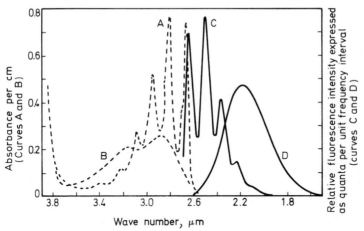

Figure 7.3 Absorption and fluorescence spectra of anthracene (in ethanol) and quinine (in 0.05 mol l^{-1} sulphuric acid). (A) anthracene absorption; (B) quinine absorption; (C) anthracene fluorescence; (D) quinine fluorescence

wavelength, they could still be measured together in the same solution if they had different, non-overlapping, excitation peaks. This, in fact, is one of the major advantages of fluorescence over absorption spectroscopy.

Any portion of the spectrum where absorption occurs may produce fluorescence since in solution, emission almost always takes place from the lowest vibrational level of the first excited singlet state regardless of which vibrational level or to which state the molecule is originally excited. The fluorescence peak will be at the same wavelength regardless of the excitation wavelength; however, the intensity of the fluorescence will vary with the relative strength of the absorption (or the sum total of all the absorptions).

7.1.1.3 Fluorescence quantum efficiency and lifetime of the excited state

Every molecule possesses a characteristic property which is described by a number called the quantum efficiency, Φ. This is the ratio of total energy emitted per quanta of energy absorbed:

$$\Phi = \frac{\text{number of quanta emitted}}{\text{number of quanta absorbed}} = \text{quantum yield}$$

The higher the value of Φ, the greater the fluorescence of a compound. A non-fluorescent molecule is one whose quantum efficiency is zero or so close to zero that the fluorescence is not measurable. All energy absorbed by such a molecule is rapidly lost by collisional deactivation. Some typical quantum yields of various substances are given in Table 7.1.

Table 7.1   Quantum yields of various substances

Compound	Solvent	Φ
Fluorescein	Water, pH 7	0.65
	0.1 M NaOH	0.92
Rhodamine B	Ethanol	0.97
Riboflavin	Water, ;pH 7	0.26
Anthracene	Ethanol	0.30
Naphthalene	Alcohol	0.12
Phenol	Water	0.22
Chlorophyll a	Ethanol	0.23
Chlorophyll b	Ethanol	0.10

The value of Φ can be determined by measuring the fluorescence of a dilute solution (F_1) of a substance whose quantum efficiency is known, Φ_1, such as quinine sulphate. The fluorescence and absorbance of a solution of the substance whose Φ is to be determined is then measured and the quantum efficiency is calculated as follows:

$$\frac{F_2}{F_1} = \frac{\Phi_2}{\Phi_1} \cdot \frac{\text{absorbance of 2}}{\text{absorbance of 1}}$$

Use $\Phi_1 = 0.55$ for quinine in 0.05 mol l^{-1} H_2SO_4.

The fluorescence lifetime of most organic molecules is in the nanosecond region. The fluorescence lifetime, τ, refers to the mean lifetime of the excited state; the probability of finding a given molecule that has been excited still in the excited state after time t is $e^{-t/\tau}$. The general equation relating the fluorescence intensity, I, and the lifetime, τ, is,

$$I = I_o e^{-t/\tau}$$

where I = fluorescence intensity at time t, I_o = maximum fluorescence intensity during excitation, t = time after removing source of excitation, τ = average lifetime of excited state.

The average lifetimes of the excited state of some typical compounds is given in Table 7.2.

Table 7.2 Average lifetime of the excited state (τ) of some compounds

Compound	τ/s
DPNH	5×10^{-10}
Fluorescein anion	5.1×10^{-9}
Quinine	4×10^{-8}
Chlorophyll	3×10^{-8}
Anthracene	2.5×10^{-7}
Naphthalene	3.3×10^{-5}

7.1.2 Relation between fluorescence intensity and concentration

The basic equation defining the relationship of fluorescence to concentration is,

$$F = \Phi I_0 (1 - e^{-\varepsilon bc})$$

where, Φ is the quantum efficiency, I_0 is the incident radiant power, ε is the molar absorptivity, b is the path length of the cell, and c is the molar concentration.

The basic fluorescence intensity-concentration equation indicates that there are three major factors other than concentration that affect the

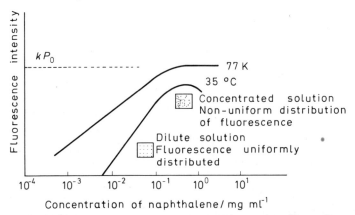

Figure 7.4 Dependence of fluorescence upon concentration of fluorophor and temperature

fluorescence intensity: (a) the quantum efficiency, Φ. The greater Φ, the greater will be the fluorescence. (This was discussed above.) (b) The intensity of incident radiation, I_0. Hence, theoretically, the most intense source will yield the greatest fluorescence. In actual practice, a very intense source can cause photodecomposition of the sample. Hence, one compromises on a source of moderate intensity, i.e. a Hg or Xe arc lamp is used. (c) The molar absorptivity of the compound, ε. For a molecule to emit radiation it must first absorb radiation. Hence, the higher the molar absorptivity, the better will be the fluorescence intensity of the compound. It is for this reason that saturated non-aromatic compounds are non-fluorescent.

For very dilute solutions, the equation reduces to one comparable to Beer's law in spectrophotometry,

$$F = K\Phi I_o \varepsilon bc.$$

Thus a plot of fluorescence v. concentration should be linear at low concentrations, then reach a maximum at higher concentrations (Figure 7.4). At high concentrations, quenching becomes so great that the fluorescence intensity decreases (inner cell effect). The linearity of fluorescence as a function of concentration holds over a very wide range of concentration (Figure 7.4). Measurements down to $0.000\,01\ \mu g\ ml^{-1}$ are feasible and linearity commonly extends up to $100\ \mu g\ ml^{-1}$ or higher.

7.1.3   Introduction to experimentation

The fundamental principles of fluorescence measurement are illustrated by the following simplified schematic representation of a filter fluorimeter (Figure 7.5). The desired narrow band of wavelengths of exciting radiation are selected by a filter (called the primary filter) placed between the radiation source and the sample. The wavelength of fluorescence radiant energy to be

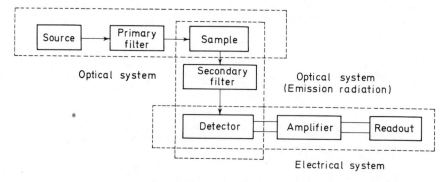

Figure 7.5 Schematic diagram of the optical components of a typical fluorimeter

measured is selected by a second optical filter (called the secondary filter) placed between the sample and a photodetector located at a 90 degree angle to the incident optical path. The output of the photodetector, a current which is proportional to the intensity of the fluorescent energy, is amplified to give a reading on a meter or a recorder. In a spectrofluorimeter, the filters are replaced by prism or grating monochromators and an x–y recorder is used to display the excitation and emission spectra.

Further details on fluorescence apparatus will be discussed below.

7.1.4   Use of luminescence

Fluorescence was noted in solutions and minerals in the early 1800s. Sir David Brewster observed the red emission from chlorophyll in 1833 and in 1852 Sir G. G. Stokes suggested the name fluorescence for the emission process, naming the phenomenon after the mineral fluospar which produces

Table 7.3   Applications of luminescence

Clinical pathology
Electrolytes: Calcium, magnesium, inorganic sulphate, inorganic phosphate
Steroids: Corticosteroids, oestrogens, progesterone, androgens, testosterone, bile acids
Lipids: Lipoproteins, phospholipids, cholesterol, triglycerides
Proteins: Serum albumin, protein electrophoresis
Amino acids and metabolites: Tryptophan, serotonin, phenylalanine, tyrosine, catecholamines, 3-*o*-methylcatecholamines, homovanillic acid, DOPA, tyramine and 3-methoxytyramine, histidine and histamine, creatine, kynurenic acid, xanthurenic acid
Immunology: Fluorescent antibodies, fluorescent antigens, blood typing
Enzymes: Dehydrogenases, transaminases, phosphatases, proteases, lipases, creatine kinase, LDH-isoenzymes, peroxidases
Drugs: Barbiturates, salicylates, quinidine, LSD, tetracyclines
Metabolites: Blood glucose, porphyrins, carboxylic acids and ketones
Other: BUN, ammonia, hippuric acid, hematin iron

Inorganic
Metals: Anions – Cyanide, fluoride, sulphate, silicate, iodide
             Cations – Aluminium, arsenic, beryllium, boron, cadmium, cerium, calcium, gallium, iron, lithium, magnesium, rare earths, selenium, silicon, tin, tungsten, uranium, zinc, zirconium

Agricultural chemistry
Inorganic: As noted above, especially selenium, magnesium, boron, fluorides, aluminium, tin
Tracing techniques: Insecticide and pesticide spray coverage studies, residue evaluations
Natural products: Gibberellic acid, chlorophylls, pigments
Vitamins: A, B_1, B_2, B_6, C, D, and E
Proteins: Protein in milk

Public health
Pollution control: Insecticide aerial drift studies, water pollution studies, spent sulphide liquor
Bacteriology: Identification and counting of bacteria
Metal poisoning: Beryllium, boron, lead, uranium, cadmium
Immunology: Fluorescent antibody control
Screening programmes: P.K.U., histidemia

a blue–white fluorescence. The emission from minerals such as barite, was observed early in the fifteenth century and was named phosphorescence from the Greek, 'light bearing'.

Fluorescence, phosphorescence and chemiluminescence provide some of the most sensitive and selective methods of analysis for many compounds.

Some typical examples of analysis in clinical pathology, inorganic analysis, agricultural chemistry and public health are listed in Table 7.3.

7.1.5   Practical considerations

7.1.5.1   *Advantages of fluorescence*

Molecular emission (fluorescence and phosphorescence) is a particularly important analytical technique because of its extreme sensitivity and good specificity. Fluorimetric methods can detect concentrations of substances as low as one part in ten billion (10^{10}), a sensitivity 1000 times greater than that of most absorption spectrophotometric methods. The main reason for this increased sensitivity is that in fluorescence the emitted radiation is measured *directly*, and can be increased or decreased by altering the intensity of the exciting radiant energy. An increase in signal over a zero background signal is measured by fluorimetric methods. In absorption spectrophotometric methods the analogous quantity, absorbed radiation, is measured indirectly as the difference between the incident and the transmitted beams. This small decrease in intensity of a very large signal is necessarily measured in spectrophotometry with a correspondingly large loss in sensitivity.

The specificity of fluorescence is the result of two main factors. One, there are fewer fluorescent compounds than absorbing ones because all fluorescent compounds must necessarily absorb radiation, but not all those compounds that absorb radiation emit. Second, there are two wavelengths used in fluorimetry compared to one used in spectrophotometry. Two compounds that absorb radiation at the same wavelength will probably not emit at the same wavelength. The difference between the excitation and emission peaks range from 10 to 280 nm.

Materials that possess native fluorescence, those that can be converted to fluorescent compounds (fluorophors) and those that extinguish the fluorescence of other compounds can all be determined quantitatively by fluorimetry.

7.1.5.2   *Limitations of fluorescence*

The principal disadvantage of fluorescence as an analytical tool is its serious dependence on the environment (temperature, pH, ionic strength, etc.). The ultraviolet light used for excitation may cause photochemical changes or destruction of the fluorescent compound giving a gradual decrease in the intensity reading.

Fluorescence is not usually suited for the determination of the major constituents of a sample because for larger amounts, the accuracy is considerably less than that attainable by gravimetric or titrimetric methods.

Quenching, the reduction of fluorescence by a competing deactivating process resulting from the specific interaction between a fluorophor and another substance present in the system, is also frequently a problem.

The general mechanism for the quenching process can be denoted as:

$$M + h\nu \longrightarrow M^* \text{ (light absorption)}$$
$$M^* \longrightarrow M + h\nu \text{ (fluorescence emission)}$$
$$M^* + Q \longrightarrow Q^* + M \text{ (quenching)}$$
$$Q^* \longrightarrow Q + \text{Energy}$$

One of the most notorious quenchers is dissolved O_2 which causes a reduction of fluorescence intensity and a complete destruction of phosphorescent intensity. Small amounts of iodide and nitrogen oxides are very effective quenchers and interfere.

Small amounts of highly absorbing substances, such as dichromate interfere by robbing the fluorescing species of the light available for excitation. For this reason many workers prefer not to wash their cuvettes with dichromate cleaning solution.

7.1.5.3 Effect of temperature

As the temperature is increased, fluorescence decreases. This is illustrated in Figure 7.6, which shows the effect of temperature on the fluorescence of three common substances. The degree of temperature dependence varies from compound to compound. Rhodamine B, tryptophan, 9-methylanthracene and indoleacetic acid are compounds that vary greatly with temperature. It

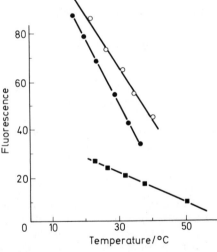

Figure 7.6 Variations in fluorescence intensity of several compounds as a function of temperature. All compounds were dissolved in 0.1 mol l^{-1} phosphate buffer, pH 7.0, except quinine. $\bigcirc$ $-\bigcirc$, tryptophan or indoleactic acid; $\bullet$ $-\bullet$, indoleactic acid in buffer saturated with benzene; $\blacksquare$ $-\blacksquare$, quinine in 0.05 mol l^{-1} sulphuric acid

is likely that increasing temperature increases molecular motion and collisions which, in turn, robs energy from the molecule of interest. For these reasons some control of temperature is important in fluorescence measurements.

7.1.6 Types of luminescence

The various types of luminescence can be classified according to the means by which energy is supplied to excite the luminescent molecule.

When molecules are excited by interaction with photons of electromagnetic radiation the form of luminescence is called *photoluminescence*. If the release of electromagnetic energy is immediate or from the singlet state, the process is called *fluorescence*, whereas *phosphorescence* is a delayed release of energy from the triplet state. Some molecules exhibit a *delayed fluorescence* which might incorrectly be assumed to be phosphorescence. This results from two intersystem crossings, first from the singlet to the triplet, then from the triplet to the singlet, i.e.

$$S_2^* \leadsto T^* \leadsto S_1^* \to S_o$$

If the excitation energy is obtained from the chemical energy of reaction, the process is *chemiluminescence*. In *bioluminescence* the electromagnetic energy is released by organisms.

Triboluminescence (Greek *tribo*, to rub) is produced as a release of energy when certain crystals, such as sugar, are broken. The energy stored on crystal formation is released in this process.

Other types of luminescence — *cathodoluminescence*, resulting from a release of energy produced by exposure to cathode rays or *thermal luminescence*, which occurs when a material existing in high vibrational energy levels emits energy at a temperature below red heat, after being exposed to small amounts of thermal energy — are much less commonly encountered.

7.2 INSTRUMENTATION

7.2.1 Component parts

The basic components of any instrument designed to measure luminescence are: light source, wavelength selectors, sample compartment and detector system. The instrumentation is the same as that used in spectrophotometry, with two exceptions: (a) the detector is rotated 90 degrees to the incident light path and (b) a second wavelength selector is placed in front of the detector. Thus any good spectrophotometer can be adopted to fluorescence work at a small additional cost.

In fluorescence, the detector is placed at 90 degrees to the path of incident light so that little of the incident radiation will strike the detector. Only light emitted from the sample reaches the detector, so that this device will register zero signal when no luminescence occurs. Then, an increase in signal indicates an emission from the sample. This is the major reason for the greater sensitivity of luminescence over absorption spectrophotometric methods.

The second wavelength selector is placed before the detector to remove all radiation except that emitted by the sample. This provides another degree of specificity to the analysis.

7.2.1.1 Sources

The most common sources used in luminescence instrumentation are the mercury and high-pressure xenon arc lamps.

The mercury vapour lamp is widely used in filter fluorimeters because of its intense line emission and stability. An elaborate power supply is not needed, hence the lamp is economical. The most commonly used wavelength in the Hg vapour lamp is the 365 nm resonance line.

A low-pressure mercury vapour lamp can be modified to provide energy at wavelengths other than the resonance lines. The inner surface of the lamp is painted with a thin-layer of crystalline phosphors which absorb the mercury vapour's resonance radiation and generate a broad band at longer wavelength. The Aminco and Turner filter fluorimeters, for example, use a 4 W low-pressure mercury phosphor lamp with a broad emission band having its peak at 360–365 nm. A Green Lamp is available with a 520–560 nm continuum and a Blue Lamp with 405 and 436 nm peaks is sold.

Most grating instruments, such as the Aminco SPF use a high-pressure xenon arc lamp (commonly 150 W). The xenon lamp has a good continuum. but does not have the intensity that the mercury lamp has at the resonance lines of mercury (at those lines the mercury lamp has twice the intensity of the xenon lamp). Moreover, it is necessary to use an expensive d.c. converter and stabiliser with this lamp. A mercury lamp cannot be used with a scanning spectrofluorimeter, however, because the excitation spectrum obtained would simply be that of the mercury resonance lines superimposed on the sample's excitation spectrum. A lamp with a smooth continuum is needed in the ultraviolet region also.

The life expectancy of ionised-gas lamps ranges from 200 to 900 h depending on various conditions of power and use. Any lamp should be allowed to stabilise for 15–30 min before any readings are taken.

7.2.1.2 Monochromators

The purpose of the monochromator is to isolate narrow bands of electromagnetic radiation from the source. It is a wavelength selector. There are two types of monochromator used in luminescence equipment — a filter or a grating. In fact, instruments are classified as filter fluorimeters (non-scanning) or grating spectrofluorimeters (scanning). Prisms are not used in instruments because they give their greatest dispersion in the ultraviolet not the visible where most measurements are made and to obtain adequate sensitivity a large prism would be needed. This is expensive.

There are three types of filters:

(a) Monochromatic filters — these transmit one limited band or window of radiation.

(b) Cut-off filters — this type of filter is used to cut off stray or unwanted radiation since it produces a sharp cut-off in the spectrum with complete transmission on one side of the cut-off and little or no transmission on the other.

(c) Neutral density — these give nearly constant transmission over a wide range.

There are several typical filters which are useful in conjunction with the mercury vapour lamp (Table 7.4).

Many other filters are available for isolating almost any spectral region, and the reader is advised to consult Kodak, Turner or Aminco catalogues for the wide selection available.

Grating monochromators select the desired wavelength by dispersing the light into a spectrum and mechanically selecting the wavelength desired in

Table 7.4 Wratten filters useful with mercury vapour lamps

Filter No.	Principal lines transmitted/nm
18A	365
22	577 and 579
50	436
74	546

the spectrum. Unlike the prism, the grating provides a linear dispersion throughout the visible and u.v. However, a grating produces several orders of spectra – the first-order spectrum is the primary image that is reflected or transmitted by a grating, the second-order spectrum is the second image, etc. To observe only one order spectrum, filters must be used. This is the only major disadvantage of gratings. Further advantages of gratings are that: (a) large gratings may be manufactured less expensively than prisms, (b) 75–80% of incident radiation is transmitted in the first order and (c) all wavelengths can be dispersed.

There are two important terms to remember in selecting a grating: ruling and blaze. The number of lines of ruling on a grating determines the dispersing power of the grating. The larger the ruled area, the greater the speed, the sensitivity and the resolving power of the instrument. A typical good ruling as used in the Aminco SPF is 50 mm per side which gives a resolution of 30 000 in the first order (0.02 nm at 577 nm).

The blaze is the wavelength at which the maximum output of the grating is concentrated. If a grating is blazed at 500 nm, its maximum output is 500 nm. The Aminco SPF, for example, has its excitation monochromator blazed at 300 nm and its emission monochromator blazed at 500 nm.

7.2.1.3 Cell compartment

Pyrex cells are useful for measurements above 320 nm (which compose 95% of all common analyses). Only below 320 nm are quartz or fused silica cells required. Hence, for all practical purposes, the large additional cost of quartz or fused silica (Supersil, etc.) is not justified.

A few words on cell configuration are desirable at this time. One could use any cell configuration from 0 to 180 degrees in fluorescence. A 30 to 45 degree configuration works as well as a 90 degree one. However, the lowest backgrounds from incident radiation are obtained at 90 degrees, and, for this reason, this configuration is most commonly used in all instruments.

7.2.1.4 Detector

The light emitted by a sample in the u.v. and visible region of the spectrum is best measured with a photomultiplier tube. This detector multiplies the light signal up to a million times and presents a signal output. Photomultiplier tubes can be made to respond to different wavelengths by varying the materials used on the photosensitive surface.

7.2.2 Basic instruments

As we have previously mentioned, all instruments available for the measurement of luminescence fall into two classes: filter instruments and grating instruments.

7.2.2.1 Filter instruments

Filter fluorimeters are inexpensive, very sensitive and simple in design. The instrument uses a mercury lamp, a primary filter to pass only certain wavelengths to excite the sample, a secondary filter to pass only the fluorescence emission, and not the incident energy, strong radiation and light scatter, to the detector which is usually a photomultiplier. The filter fluorimeter is generally more sensitive than the grating instrument, and can do almost anything the latter can do, except scan. In Tables 7.5 and 7.6 the characteristics of some commonly available filter fluorimeters are compared.

Table 7.5 Characteristics of some typical fluorimeters

Instrument	Optics	Dispersing device	Photodetector	Circuit
Coleman model 12C	Glass	Glass filters	Phototube	Single-beam direct reading
Photovolt model 540	Quartz	Filters	1P21 Photomultiplier	Single-beam, direct reading
Klett	Glass	Glass filters	Barrier-layer cells	Double-beam, potentiometric balance
Lumetron model 402-EF	Glass	Glass filters	Barrier-layer cells	Double-beam, bridge circuit
Farrand	Quartz	Interference filters	1P21 Photomultiplier	Single-beam
Turner model 110	Quartz	Glass filters	1P21 Photomultiplier	Double-beam, optical balance
Hilger Spekker	Glass	Glass filters	Photomultiplier	Double-beam, optical balance
Hitachi type FPL-2	Glass	Interference filters	Photomultiplier	—
Beckman ratio	Vycor	Glass filters	Photomultiplier	Double-beam, ratio recording

To use a filter instrument one must first determine the wavelengths of excitation and emission in order to choose filters to use for maximum sensitivity. To do this, one first runs the absorption spectrum of the compound on any spectrophotometer. Since the excitation spectrum should match the absorption spectrum, one need simply pick a primary filter to reach a peak around the λ_{max} of the compound. Then one places a dilute solution (10^{-5} mol l^{-1}) of the compound to be assayed into the fluorimeter

Table 7.6 Comparison of common filter fluorimeters

Instrument	Lamp	Lowest detectable conc. ($S/N = 1$)	Cost	Monochromator	Remarks
Aminco fluoromicro photometer	Hg (85W)	0.0002 μg ml^{-1} QS*	$1150 $1625 (for solid state)	Glass filters	Photomultiplier detector; single beam temperature control; 7 scales; quartz optics
Beckman ratio fluorimeter	Hg (Phosphor coated sleeve)	0.0010 ug ml^{-1} QS	$1075	Glass filters	Photomultiplier detector; double beam, ratio recording; Vycor optics
Coleman 12C	Hg Arc	0.003 ug ml^{-1}	$480	Glass filters	Phototube detector; single beam; one scale; glass optics
Technicon	Hg (85W)	0.001 μg ml^{-1} QS	$2755	Glass filters	Photomultiplier detector; double beam
Turner 110, 111	Hg (4W)	0.003 μg ml^{-1} QS	$1195 (110) $1685 (111)	Glass filters	Photomultiplier detector; double beam; optical balance; temperature control; quartz optics; 4 scales; 110 = Null balance; 111 = Recording

*QS = Quinine sulphate

Table 7.7 Comparison of spectrofluorimeters

Instrument	Lamp	Monochromator (slits)	Lowest detectable conc. (at $S/N = 1$)	Cost	Resolution	Remarks
Aminco SPF	Xenon (150 W)	Grating (1–30 μm)	0.0002 ug ml^{-1} QS	$4850 ($8900 for corrected spectra)	1.6 nm (0.5 nm optional)	1P21 photomultiplier; 7 scales; temperature control
Baird Atomic Fluorispec	Xenon (150 W)	Dual grating (2–32 nm)	0.0001 μg ml^{-1} QS	$5775	1.6 nm	1P28 photomultiplier; 4 scales at 1 and 0.1 s time constants
Farrand MK-1	Xenon (150 W)	Grating (0.5–20 nm)	0.0001 μg ml^{-1} QS	$5300	0.5 nm	1P28 photomultiplier; 10 scanning speeds
Perkin-Elmer MPF-2A	Xenon (150 W)	Grating (1–40 nm)	$0.000\,05$ μg ml^{-1} QS at 10 nm bandpass	$7450	1 nm	R-106 photomultiplier; ratio recording; temperature control
Perkin-Elmer 203	Xenon arc (150 W)	Grating (10 nm)	$0.000\,05$ μg ml^{-1} QS at 10 nm bandpass	$3750	10 nm	R-212 photomultiplier; meter read-out with 36 scales
Turner 210	Xenon (75 W)	Grating (0.5–25 nm)	0.002 μg ml^{-1} QS at 15 nm bandpass	$18 750	0.5 nm	Corrected spectra; double beam; temperature control; can be used as spectrophotometer

and tries several secondary filters (realising that the emission peak is usually separated from the excitation peak by 20–150 nm) until a maximum signal is obtained.

7.2.2.2   Grating instruments

A grating instrument generally uses a xenon lamp as the source of radiation, two gratings to disperse and select the desired excitation and fluorescence energy and a photomultiplier as the detector. Various slits in the focal planes of the gratings determine the band pass and the intensity of the energy striking the sample or the detector.

The grating spectrofluorimeter is more versatile than the filter instrument and can be used for luminescence research projects. A comparison of some spectrofluorimeters is presented in Table 7.7.

7.3   STRUCTURAL AND ENVIRONMENTAL EFFECTS

In order to utilise luminescence effectively as an analytical tool, it is necessary that every researcher knows the basis of the effects of structure and the environment on the emission process. Under structural effects we shall briefly consider what types of compounds fluoresce, and how we might increase the total emission by changes in structure. Under environmental effects we shall study how pH, the solvent, other ions and other factors change the luminescence.

7.3.1   Structural effects

Fluorescence phenomena are not generally sensitive to the finer details of molecular structure. Among the huge number of known organic compounds, only a small fraction exhibits intense luminescence. Therefore, the mere fact that a molecule fluoresces can in itself provide significant information regarding its structural features. Aromatic hydrocarbons possessing large conjugated systems often exhibit fluorescence. In these systems π electrons, which are less strongly held than σ electrons, can be promoted to π^* anti-bonding orbitals by absorption of electromagnetic radiation of fairly low energy without extensive disruption of bonding (Figure 7.7). The $\pi \rightarrow \pi^*$ (K band) has a high transition probability and hence is likely to occur.

Characteristically, fluorescence is the light emission from a π, π^* singlet. Fluorescence is less likely from an n, π^* state. The extinction coefficient for the n $\rightarrow \pi^*$ transition is low, hence less likely to occur. Secondly, the characteristics of the resulting n, π^* state are such that other energy-dissipating processes compete successfully against the emission of light. Other processes ($\sigma \rightarrow \sigma^*$, $\sigma \rightarrow \pi^*$) never lead to fluorescence since they are of such high energy that bond breaking occurs before emission.

Only a relatively few aliphatic and saturated cyclic compounds fluoresce or phosphoresce. This is because the electrons in aliphatic compounds are normally tightly bound and participate in sigma bonding. When aliphatic and saturated cyclic molecules absorb u.v. energy, photodecomposition usually results. Aliphatic aldehydes and ketones, where the carbonyl oxygen can be excited to an antibonding π^* orbital, do fluoresce, though weakly.

Because they possess a large number of π electrons, aromatic compounds

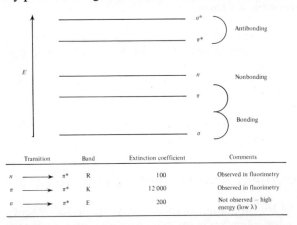

Figure 7.7 Types of transition involved in the absorption process

possess a high fluorescence probability. As the degree of conjugation increases, the intensity of fluorescence increases and a bathochromic shift (shift to longer wavelengths) is observed (Table 7.8).

Certain ring structures normally produce intense fluorescence, such as quinone, pyrrol, fluorescein, furan, etc. Heterocyclic compounds usually do not fluoresce in non-polar solvents, but many do in polar solvents or acid solutions. Some typical heterocyclics possessing native fluorescence are the indoles, quinolines, coumarins and resorufin derivatives.

Table 7.8   Effect of conjugation on fluorescence

Compound	Excitation maximum/nm	Fluorescence maximum/nm
Benzene	204	278
Naphthalene	286	321
Anthracene	375	400
Naphthacene	390	480
Pentacene	580	640

The nature of substituent groups (especially chromophoric groups) plays an important role in the nature and extent of fluorescence by a molecule. Fluorescence yields (intensities) and energies of aromatic and heterocyclic hydrocarbons are usually altered by ring substitution. Unfortunately, we cannot often make many broad generalisations. Substituent effects upon the chemical and physical properties of organic molecules in their ground electronic states constitute a lively area of investigation at present. Furthermore, only little is known about the influence of substituents on the behaviour of excited states. Both effects must be understood before generalisations concerning the effect of various substituent groups can be made.

In general, the substitution of certain electron-donating functional groups for hydrogen, such as F, OH, OMe, NH_2, NHMe or NMe_2, usually produces a more intensely fluorescing compound, while electron-withdrawing groups such as halides (except F), NO_2 or CO_2H often yield a compound of decreased fluorescence intensity (Table 7.9).

Table 7.9 Fluorescence of monosubstituted benzenes

Compound	Substituent	λ_{ex}/nm	λ_{em}/nm	Relative intensity
Ortho-para directing				
Benzene	H	269	291	1
Aniline	NH_2	290	345	46
Dimethylaniline	NMe_2	297	363	114
Fluorobenzene	F	269	285	13
Chlorobenzene	Cl	281	294	0.02
Bromobenzene	Br	—	None	0
Iodobenzene	F	—	None	0
Phenol	OH	279	302	112
Meta-directing				
Benzoic acid	CO_2H	—	—	0
Nitrobenzene	NO_2	—	—	0
Benzaldehyde	CHO	—	—	0
Benzonitrile	CN	287	294	45

The only exception to the general rule of substitution, is cyanide, which is *meta*-directing, yet which always causes an increase in intensity. In all cases, except that of phenol, the ionised hydroxy group is more fluorescent than the OH group. Certainly the electronic configuration of a molecule changes when a proton is extracted from a functional group. The changes, however, may not necessarily be of the same order of magnitude in both the ground and excited electronic states. In many cases, such as the fluoresceins, the presence of a carboxylic acid functional group can act to increase or decrease fluorescence depending on whether or not the group is ionised.

In many cases two compounds with almost identical structures will exhibit vastly different fluorescent properties. Consider, for example, phenolphthalein (1) which is non-fluorescent and fluorescein (2) which is highly fluorescent. This indicates that ring closure is conducive to fluorescence.

(1)   (2)

Structural rigidity reduces vibrational amplitudes that promote radiationless losses, and is a prime factor in fluorescence capability.

The introduction of heavy atoms (e.g. halogens) decreases the fluorescence in favour of phosphorescence. Thus fluoro- or chloro-fluorescein is fluorescent, but iodofluorescein is phosphorescent but not fluorescent. (See discussion of phosphorescence in Section 7.4 below.)

7.3.2 Environmental effects

Environmental factors can strongly influence the fluorescence of polyatomic molecules. Molecular environment constitutes an important parameter which can be used by the analyst to increase the sensitivity and selectivity of fluorimetry.

A large number of environmental effects are of importance. We will only discuss a few of these: pH, heavy atoms, metal ions, oxygen and temperature. It cannot be too strongly emphasised that pure solvents must be used in fluorimetry to obtain good results. It is not generally sufficient to demonstrate that the solvent does not itself fluoresce since non-luminescent impurities can act as quenchers.

7.3.2.1 Solvent effects

An electronic transition must occur while the photon is in the vicinity of the absorbing molecule in about 10^{-15} s. This is 10^3–10^4 times faster than the rate of bond stretching, so that an electronic transition occurs before any change in interatomic distance will occur. This observation is known as the Franck–Condon principle, and is pictured in Figure 7.8. The molecule, unchanged except for the electronic transition, finds itself in a metastable, excited state, called the Franck–Condon excited state. The molecule then readjusts to its new environment by a solvent reorientation in c. 10^{-12} s and reaches the equilibrium excited state. Emission then occurs from the equilibrium excited state to the Franck–Condon ground state in c. 10^{-8} s. The molecule then re-orientates itself to its environment and reaches the equilibrium ground state.

Thus, in solutions, the molecule enters into one excited state and leaves from another excited state, each state having different energy levels. Therefore, the wavelengths of absorption and emission are different. The fact that the molecule is excited to the Franck–Condon excited state first, is the principal reason that the fluorescence spectrum is subject to different solvent effects than the absorption spectrum.

In most polar molecules the excited state is more polar than the ground state. Hence an increase in the polarity of the solvent produces a greater stabilisation of the excited rather than the ground state. Consequently, a shift in both the absorption and fluorescence spectra to lower energy or longer wavelength is observed (Table 7.10). This type of behaviour is observed for almost all cases, even when the solute and solvent are not polar, because of the induced dipole in the excited state. The magnitude of the change, however, is not as great.

Concerning intensity, electrostatic solvent–solute interactions do not produce significant variations in fluorescent yields if both solute and solvent

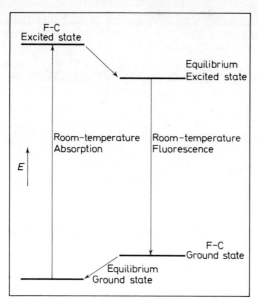

Figure 7.8 Schematic representation of equilibrium and Franck–Condon electronic states

Table 7.10   Electrostatic solvent effects on anthracene fluorescence

Solvent	ν_F/cm^{-1}	λ_F/nm	$\lambda_F - \lambda_A/nm$
(vapour)	27380	365	180
Hexane	26517	377	156
Methanol	26461	378	177
Dioxane	26220	381	280
Toluene	26170	382	228
Benzene	26116	383	262
Chlorobenzene	26064	384	279
Acetonitrile	25972	385	285
Formamide	25508	392	347
N-Methylformamide	25112	398	368

Table 7.11   External heavy-atom effect on phosphorescence intensity

Compound	Concentration/μg ml^{-1}	$(P/P_0)^*$
Naphthalene	13	0.1
Anthracene	18	0.9
Triphenylene	2.3	0.05
Naphthacene	6.0	4.6
1,2-Benzanthracene	23	1.4
3,4-Benzpyrene	25	3.5
1,2-Benzfluorene	22	13
2,3-Benzfluorene	22	25

*Ratio of phosphorescence intensity in ethanol–ethyl iodide (5/1 v./v.) to that in ethanol

are non-polar. Some solvents may appear to be better than others because of solvent quenching:

$$^1A^* + S \rightleftharpoons [A^-S^+] \rightarrow {}^3A^* + S$$

radiationless

A

where, $^1A^*$, excited singlet state; S, solvent molecule; $^3A^*$, triplet excited state; A, ground state.

For polar solutes or solvents, electrostatic intensity perturbations are relative to those produced by hydrogen bonding or other specific interactions.

7.3.2.2   Effect of external heavy atoms and paramagnetic ions

The effect of an external heavy atom (iodide as solvent, for example) or of a paramagnetic ion (Cu^{2+}, Fe^{3+} in solution) is to increase the rate of inter-system crossing ($S_1^* \rightarrow T_1^*$) and also the rate of $T_1^* \rightarrow S_0$ transition. The net result is an increase in the phosphorescence/fluorescence ratio (Φ_P/Φ_F). The effect is believed due to the formation of weak charge-transfer complexes of the substance A and the heavy atom X:

$$A^* + X \rightarrow [A \rightarrow : X] \rightarrow {}^3A^* \xrightarrow{hv} A$$

A shift in the spectra to longer wavelengths is also frequently observed. This effect can be used to distinguish n,π^* and π,π^* triplets in carbonyls. Phosphorescence from the n, π^* state is much less susceptible to heavy atom enhancement than the π, π^* state.

The effect of a heavy atom (ethyl iodide) on the phosphorescence intensity of various solutes is given in Table 7.11.

7.3.2.3   Effect of pH

Most molecules that contain an ionisable hydroxy group will exhibit an increase in fluorescence intensity as the pH is increased — one of the exceptions to this is phenol, as was previously mentioned. Most hydroxy compounds also possess a lower pK_a in the excited state than they do in the ground state due to a stabilisation of the $O—O^-*$ species.

The acid–base properties of 2-naphthol serve as an excellent example of excited state dissociation. The pK_a for ground-state 2-naphthol is $c.$ 9.5. The 2-naphthol molecule exhibits a single fluorescence peak at 359 nm, whereas, the 2-naphtholate anion exhibits a fluorescence peak at 429 nm. The large energy separation of these two fluorescence peaks makes it easy to measure the fluorescence of either molecular 2-naphthol or ionic 2-naphthol-ate, without interference from the other species. In the pH region where molecular 2-naphthol predominates the overall process may be represented as:

$$C_{10}H_7OH + h\nu_M \qquad\qquad C_{10}H_7O^- + h\nu_I$$

$$\uparrow (ii) \qquad\qquad\qquad\qquad \uparrow (v)$$

$$C_{10}H_7OH + h\nu \xrightarrow{\;(i)\;} C_{10}H_7OH^* \underset{\;(iv)\;}{\rightleftharpoons} C_{10}H_7O^{-*} + H^+$$

$$\downarrow (iii) \qquad\qquad\qquad\qquad \downarrow (vi)$$

$$C_{10}H_7OH \qquad\qquad\qquad\qquad C_{10}H_7O^-$$

(i) Absorption of radiant energy to produce the excited *molecule*. Remember the pH is less than 9.5 and so 2-naphthol predominates.

(ii) Deactivation of the excited molecule by molecular fluorescence (359 nm).

(iii) Radiationless deactivation of the excited molecule.

(iv) Dissociation of the *excited* molecule producing a proton and an excited anion.

(v) Deactivation of the excited anion by ionic fluorescence (429 nm).

(vi) Radiationless deactivation of the excited anion.

In the pH range below 9.5, molecular 2-naphthol predominates in the ground state and so it may be possible to observe an excited state dissociation reaction by measuring the fluorescence of 2-naphtholate ion (429 nm) while exciting 2-naphthol (solution of pH less than 9.5). The excited state acidity of 2-naphthol is found to be 3.1, over six orders of magnitude different from the ground state. This means that if the fluorescence of 2-naphthol is measured at pH values >4, then a large increase in sensitivity results.

In a practical sense, analytical schemes for mixtures of aromatic compounds without separations can be designed based on a pH control.

7.3.2.4 Effect of temperature

As we previously mentioned, a decrease in temperature usually produces an increase in fluorescence intensity. The reason for this observation is that competing radiationless processes cannot compete as effectively at low temperatures as they can at higher temperatures. In fact, at very low temperatures one finds the vibrational structure of the molecule fully resolved. The use of low-temperature glasses, however, offer little advantage for fluorescence; they do for phosphorescence, however, due to a bimolecular quenching of the triplet state by oxygen.

A blue shift (shift to shorter wavelengths) is also observed at low temperatures because the Franck–Condon excited state is frozen in and is the same as the equilibrium excited state. The result is a larger ΔE and hence a longer wavelength.

7.3.2.5 Effect of oxygen

Oxygen quenches the excited singlet and triplet states of many molecules; this is a paramagnetic effect since other paramagnetic gases such as NO are also excited state quenchers. The effect of oxygen is devastating. As a result, no phosphorescence is observed at room temperature, only in rigid glass (frozen) media.

The effect of oxygen on fluorescence varies from compound to compound. Many substituted aromatics and some heterocyclics are almost insensitive to oxygen; the fluorescence from unsubstituted aromatic and aliphatic aldehydes and ketones is very sensitive to oxygen. In most cases only a 20% decrease in intensity is observed, so it is not worth the trouble to eliminate oxygen from the solution.

7.4 PHOSPHORESCENCE

7.4.1 Theoretical considerations

Each energy level can be occupied by two electrons which must have opposite spins, designated as plus and minus. If all the electrons are 'paired' in this way, the system is in the singlet state. However, if the atom or molecule has two unpaired electrons, both having the same spin, it is then in a triplet state. The lowest energy level available, the 'ground state' is a singlet state. If the absorption of energy causes one of the electrons to be raised to a higher vacant level, without change of spin, the result is an excited singlet state. If a change in spin occurs, the result is an excited triplet state.

There are a variety of possible electron-energy transitions for a molecule, accompanied by an absorption or emission of light. If a pair of π electrons are excited to a higher π level, an antibonding state designated as π^*, the resulting state is a π,π^* singlet if no change in spin has occurred, but a π,π^* triplet if the spin has flipped over to the opposite sign.

The light emission from a π,π^* singlet is fluorescence. If the excited state is a π,π^* triplet, the higher improbability of a spin-flipping transition back to the ground state $(T_1^* \rightarrow S_0)$ causes the light emission to be greatly delayed and the result is phosphorescence.

Also, because more energy is lost in the process, the wavelength of phosphorescence is shifted to longer wavelengths than for fluorescence.

7.4.2 Instrumentation

The instrumentation used to study phosphorescence is very similar to that used for fluorescence. To see phosphorescence in the presence of fluorescence we must take advantage of the slight time difference involved between the absorption and emission of radiant energy. This is accomplished with a mechanical device called a phosphoroscope which modulates the radiation from the light source incident on the sample and simultaneously modulates the luminescence radiation from the sample which is incident on the photodetector. The modulation is periodic and out-of-phase so that no incident exciting or luminescent radiation reaches the photodetector during one phase, whereas, only long decaying luminescence (phosphorescence) radiation reaches the photodetector during the other phase. Therefore, the main function of the phosphoroscope is to allow measurement of phosphorescence in the presence of fluorescence and scattered radiation.

At room temperature the energy of the triplet state is readily lost by a collisional deactivation process involving the solvent, and phosphorescence is not observed. At reduced temperatures a solidified sample does not lose energy readily and phosphorescence can be easily observed. The usual procedure in phosphorescence studies is to place the samples in small quartz tubes which are then placed in liquid nitrogen (77 K) and held in a quartz

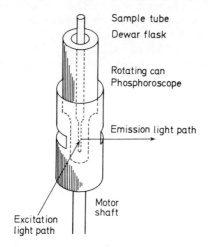

Figure 7.9 Schematic diagram of a rotating-can phosphoroscope

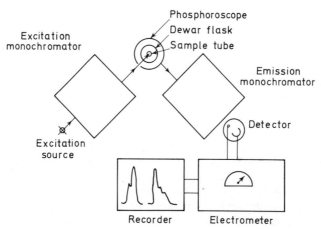

Figure 7.10 Schematic diagram of a spectrophosphorimeter

Dewar flask. A typical quartz Dewar flask is shown in Figure 7.9. The incident radiation energy passes through the unsilvered part of the Dewar flask, and luminescence is observed through the same part of the flask at right angles to the incident beam. The sample cell is immersed directly into the coolant, which is usually liquid nitrogen.

The solvents used in phosphorimetry at liquid nitrogen temperatures must·form clear, rigid glasses, must have good solubility characteristics for the compounds to be studied, must be readily available and inexpensive, and must neither absorb strongly nor luminesce greatly in the spectral regions of interest. The most commonly mixed solvent is EPA (5:5:2 (volume) diethylether–isopentane–ethanol).

Phosphorescence attachments which can be attached to a commercially available fluorimeter are available for the Aminco Bowman SPF, the Baird Atomic Fluorispec, the Farrand MK1 spectrofluorimeter, and the Aminco filter fluorimeter. The last (Accessory D2-63019, Figure 7.9) is the cheapest and most readily available. Source light passes through rotating shutter to excite the sample to phosphoresce: emitted light alternately passes to the detector.

7.4.3   Structural effects

The effect of halogen substitution on the luminescence of aromatic hydro-carbons is of considerable importance to chemists. One generally observes that as the substituent series F, Cl, Br, I is traversed, phosphorescence is increasingly favoured relative to fluorescence. This effect is illustrated in Table 7.12 which shows the ratio of the quantum efficiency of phosphorescence to fluorescence, Φ_p/Φ_f, for a number of substituted naphthalenes and fluoresceins. The ratio Φ_p/Φ_f increases across the series F → I. The trends for yields

Table 7.12   Substituent effects upon luminescence of naphthalenes and fluoresceins in EPA

Compound	Φ_p/Φ_f	Φ_f	λ_{Fluor}/nm	λ_{Phos}/nm
Naphthalene	0.093			
1-Chloronaphthalene	5.2			
1-Bromonaphthalene	6.4			
1-Iodonaphthalene	>1,000			
Fluorescein	0	0.83	527	
2',7'-Dichlorofluorescein	0	0.79	538	
4',5'-Dibromofluorescein	0.21	0.29	540	650
4',5'-Di-iodofluorescein	1.05	0.054	544	667

of luminescence and excited-state lifetimes in halogenoaromatics can be rationalised only if one postulates that heavy-halogen substitution increases the rate of intersystem crossing from a singlet to a triplet state (radiationless process) and subsequent triplet to singlet transition (a radiative process). Halogen substitution must, therefore, increase the extent of spin–orbit coupling in aromatic systems, the increase being larger for heavier halogens. This perturbation is commonly termed the 'heavy-atom effect'. In the case of the substituted naphthalenes the increase in the ratio of phosphorescence to fluorescence is primarily due to an increase of probability of singlet to triplet transitions. However, in the fluorescein series the primary effect is the result of an increased non-radiative transition from the excited state back to the ground state.

Since fluorescein and dichlorofluorescein are highly fluorescent but do not phosphoresce, whereas the corresponding bromo- and iodo-substituted derivatives are weakly fluorescent but do phosphoresce, a study of the luminescent properties of the fluorescein family provides an interesting introduction to some of the structural factors affecting luminescence.

Paramagnetic cations also effect a transition from the singlet to the triplet with a resulting increase in phosphorescence intensity. Metal chelates of such ions are generally phosphorescent.

7.4.4 Applications

Phosphorimetry has been used for the analysis of agriculturals, human blood and urine, amino acid and proteins, nucleic acids, nucleotides, nucleosides, purines and pyrimidines, enzymes, pharmaceuticals, petroleum products and air pollutants. Typical examples of analyses are reported below.

Moye and Winefordner[11] reported the phosphorescence characteristics of 52 pesticides. Of the 32 pesticides that phosphoresce, the authors listed the spectral characteristics, lifetimes, limits of detection and linear region. Phosphorescence was much more sensitive than fluorescence for a number of these compounds, especially the carbamates.

Hollifield and Winefordner[12] described a method for the analysis of sulphur drugs in blood. The procedure avoids the necessity of prior separation steps.

Steele and Szent-Gyorgyi[13] reported the phosphorescence characteristics of a number of purines and pyrimidines: adenine, adenosine, AMP, ADP and ATP. The latter was very phosphorescent.

Winefordner and co-workers have described procedures for the assay of a number of pharmaceuticals: aspirin[14], procaine, phenobarbital, cocaine and chloropromazine[15], sulphanamides[16] and anticoagulants[17].

Drushel and Sommers[18] presented an extensive survey of the luminescence characteristics of more than 100 aromatic compounds containing sulphur, oxygen and nitrogen. Compound identification was easily accomplished with the aid of phosphorimetry.

Sawicki et al.[19-21] have reported a number of thorough examinations of air pollutants using phosphorimetry as an analytical tool. The complementary nature of fluorescence and phosphorescence in the assay of air pollutants was pointed out by these workers.

Phosphorescence excitation and emission spectra and phosphorescence decay times and quantum efficiencies provide valuable information concerning the structure, chemical behaviour and environment of organic molecules in real, complex systems. Phosphorimetry is similar to fluorimetry in methodology and areas of application, and offers a powerful complementary tool to fluorimetry.

7.5 DETERMINATION OF INORGANIC IONS

7.5.1 General considerations

The combination of an inorganic ion with a non-fluorescent organic ligand to form a highly fluorescent metal chelate can provide a very sensitive and highly specific method for the determination of a metal ion. Procedures for the analysis of nearly thirty different metals have been devised using this basic approach. In addition, elements and ions such as cyanide, fluoride, iron, copper, and oxygen may be determined indirectly by measuring the amount of quenching of the fluorescence of a chelate or by causing the release of a ligand which can then react to form a fluorescent product as in the determination of cyanide:

$$CN^- + Pd(\text{quinolin-8-ol})\text{-5-sulphonate} \rightarrow \text{quinolin-8-ol-5-sulphonate} +$$

(Non-fluorescent chelate) $\qquad\qquad\qquad Pd(CN)_6^{4-}$

$$\downarrow Mg^{2+}$$

(Fluorescent chelate)

The organic ligand should be an aromatic molecule containing oxygen or nitrogen, which is itself non-fluorescent. The presence of non-bonding n electrons makes it probable that the excited state will be n, π^*, which is non-fluorescent or weakly fluorescent. Upon complex formation with a metal ion, the n electrons are utilised in bonding with the metal, and thus become less accessible for excitation. The π, π^* state then results upon excitation, and strong fluorescence is observed.

The organic ligands that have most commonly been used to react with metal ions to form fluorescent chelates are the 2,2'-dihydroxyazo dyes; quinolin-8-ol and its derivatives, flavanols, salicycylidenes and 2,2'-dihydroxymethines, salicylic acid, benzoin, Rhodamine B, G and S, salicylaldehydes, β-diketones, hydroxynaphthoic acid and hydroxyanthraquinones. Other ligands have found specific applications and new compounds are continually being added to the list.

In the course of a study of azo compounds to determine the minimum structural requirements for combination of an azo compound with calcium and magnesium, for example, Diehl et al.[22] found that o,o'-dihydroxyazobenzene was unique in reacting with magnesium and not calcium. At pH 10, this ligand reacts with magnesium to produce a stable, orange fluorescent chelate, but forms no coloured or fluorescent complex with calcium. The chelate is fluorescent in water solution at pH values greater than 11, but the

Non-fluorescent

Fluorescent
(λ_{ex} = 470 nm;
λ_{em} = 580 nm)

fluorescence intensity is increased in ethanol–water solutions at pH values of 10–11.4. At more acidic pH values, protons compete more effectively with the Mg^{2+} for the hydroxy oxygens and consequently the chelate is not formed quantitatively. At pH values greater than c. 11.5, magnesium hydroxide may form in appreciable amounts.

7.5.2   Assay of cations and anions

Some typical fluorimetric methods for the determination of inorganic substances are listed in Table 7.13. Most of these methods are highly selective and sensitive and represent the best analytical methods for these substances.

Table 7.13   Fluorescent methods for the assay of inorganics

Ion	Reagent	Excitation/nm	Emission/nm	Sensitivity/p.p.m.	Ref.
Al	Alizarin Garnet R	470	580	0.007	23
	SOAP	—	Green	0.0003	24
Au	Rhodamine B	550	575	0.01	25
B	Benzoin	370	480	0.040	26
Be	Morin	470	570	0.004	27
CN^-	Quinone	440	500	0.10	28
Eu	Hexafluoro-β-diketone	312	544	0.10	29
F^-	Ternary complex with Zr + Calcein Blue	350	410	0.010	30
Ga	Rhodamine B	550	575	0.010	31
Li	Quinolin-8-nol	370	580	0.20	32
Mg	N,N'-Bis-salicylidene–ethylenediamine	355	440	0.0002	33
NH_4^+	Enzyme–glutamate dehydrogenase + NAD	320	410	0.010	34
PO_4^{3-}	Enzyme–glycogen + phosphorphase	320	410	0.01	35
Se	2,3-Diaminonaphthalene	365	525	0.0005	36, 37
Zn	Dibenzothiazolylmethane	365	415	0.05	38
Zr	Flavonol	400	465	0.10	39

In some cases, many different reagents have been described for the analysis of an ion. For example, aluminium could be assayed by chelating with Alizarin Garnet R, 3-hydroxy-2-naphthoic acid, Morin, Pontachrome Blue Black R, quinolin-8-ol and salicylidene-o-amino-phenol (SOAP). Of these, the best are Alizarin Garnet R and salicylidene-o-amino-phenol.

7.6   DETERMINATION OF ORGANIC COMPOUNDS

Very sensitive and highly selective analytical procedures have been developed for the assay of hundreds of organic compounds, including amino acids, vitamins, steroids and drugs.

Some organic compounds are naturally fluorescent, possessing structures that are rigid, co-planar and possess labile π electrons.

Fluorescein (3), resorufin (4), indoxyl (5) and umbelliferone (6), for example, all fluoresce in the sub-nanogram region.

Amino acids, such as phenylalanine, tyrosine and tryptophan, likewise possess a native fluorescence that can be used for their assay.

The fluorescence of these amino acids demonstrates the effect of structures on the luminescence. Phenylalanine, with only a benzene ring and a CH_2 side chain is weakly fluorescent. Add a hydroxy group as in tyrosine and the fluorescence goes up by 20; add the indole ring as in tryptophan and the relative fluorescence is 200 times better.

Fluorescence has been a valuable aid in identifying the mechanism of photosynthesis, since the different chlorophylls, a, b, c, d, etc., pheophytin, protochlorophyll, bacteriopheophytin and other photosynthesis precursors

Fluorescein
(λ_{ex} = 525 nm; λ_{em} = 575 nm)

(3)

Resorufin
(λ_{ex} = 560 nm; λ_{em} = 580 nm)

(4)

Indoxyl
(λ_{ex} = 495 nm; λ_{em} = 570 nm)

(5)

Umbelliferone
(7-Hydroxy coumarin)
(λ_{ex} = 325 nm; λ_{em} = 440 nm)

(6)

Phenylalanine
(λ_{ex} = 260 nm; λ_{em} = 282 nm)

Rel Fl = 0.5

(7)

Tyrosine
(λ_{ex} = 275 nm; λ_{em} = 303 nm)

Rel Fl = 9

(8)

Tryptophan
(λ_{ex} = 287 nm; λ_{em} = 348 nm)

Rel Fl = 100

(9)

and products are fluorescent at different wavelengths. An excellent chapter on this subject can be found in Guilbault's book[1].

Other organic compounds, themselves non-fluorescent or weakly fluorescent, can be converted to good fluorophores by a simple chemical reaction.

Luminol is measured by its intense luminescence, produced via the following reaction:

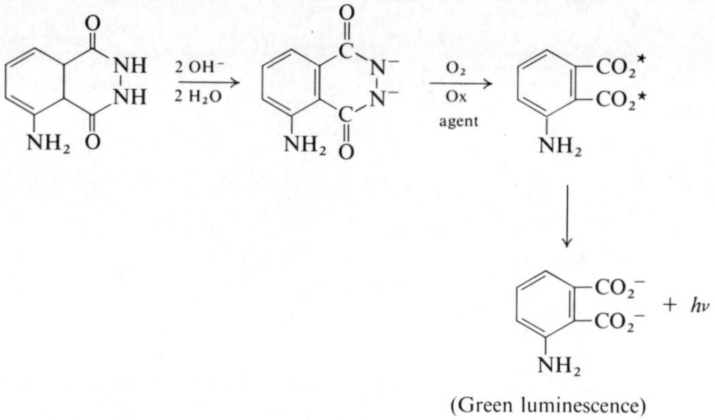

(Green luminescence)

The luminol reaction has been used for the determination of oxidising agents, such as peroxide, and for metal ions such as Cu or Co, which catalyse the reaction. As little as 2 p.p.b. Co or 30 p.p.b. Cu can be determined[40].

Acetol can be determined by a condensation with o-aminobenzaldehyde to produce the fluorophore 3-hydroxyquinaldine[41]:

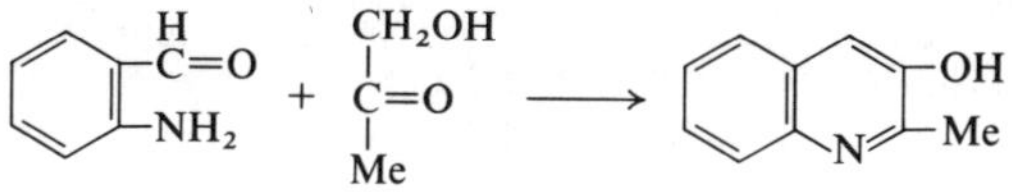

$\lambda_{ex} = 365$ nm; $\lambda_{em} = 440$ nm

Organic acids, like malic acid, can be assayed by a condensation with resorcinol to yield umbelliferone derivatives[42].

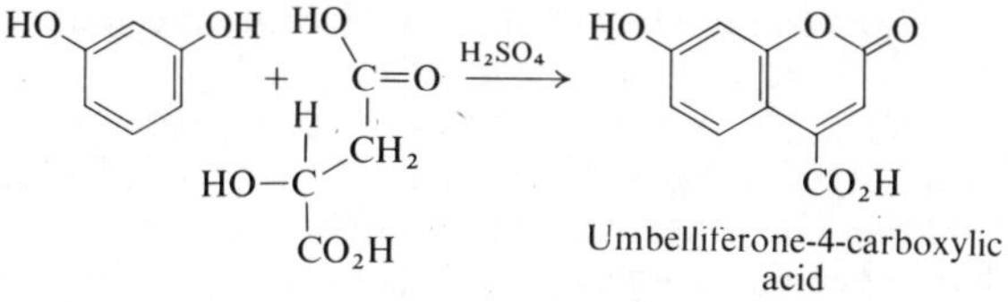

Umbelliferone-4-carboxylic acid

Some of the organic acids assayable and the relative fluorescences are given in Table 7.14.

Table 7.14

Acid	Colour of fluorescence	Relative fluorescence
Malic	Blue–Violet	22
Fumaric	Blue–Violet	24
Succinic	Yellow–Green	20
Isocitric	Light Blue	58
Citric	Sky Blue	89

(From Frohman and Orten[42], by courtesy of the American Society of Biological Chemists)

Adrenaline and dopamine[43] are similarly assayed via fluorophore formation to highly fluorescent indoxyl derivatives:

Adrenaline

$\lambda_{ex} = 420$ nm; $\lambda_{em} = 520$ nm

$\lambda_{ex} = 345$ nm; $\lambda_{em} = 410$ nm

Some vitamins, as Vitamin A, possess a native fluorescence, and can be measured directly[44] ($\lambda_{ex} = 327$ nm; $\lambda_{em} = 510$ nm for Vitamin A in the 0–10 p.p.m. range). Others, like Thiamine and Riboflavin, are best converted to fluorophores by simple dehydration reactions:

$$\text{Thiamine} \xrightarrow{-H} \text{Thiochrome}$$
$$(\lambda_{ex} = 365 \text{ nm}; \lambda_{em} = 435 \text{ nm})$$

$$\text{Riboflavin} \xrightarrow[\text{Light}]{OH^-} \text{Lumiflavin}$$
$$(\lambda_{ex} = 440 \text{ nm}; \lambda_{em} = 550 \text{ nm})$$

Vitamins D_2 and D_3 are treated with trichloroacetic acid to give fluorophores measured at 480 nm[45].

Cholesterol is commonly assayed fluorimetrically by treatment with H_2SO_4 to yield a red–orange fluorophore ($\lambda_{ex} = 546$ nm; $\lambda_{em} = 590$ nm). From 0.1–2 µg of cholesterol are assayable[46]. Similarly, other steroids can be assayed by fluorophore formation with H_2SO_4, and the reader is referred to the chapter on Steroids in Udenfriend's book[47].

Fluorescent methods similar to the representative areas quoted above have been described for the assay of many other organic compounds. The readers are referred to other references, such as the books by White[48] and Udenfriend[5] for specific details.

References

1. Guilbault, G. G., editor (1967). *Fluorescence. Theory, Instrumentation and Practice* (New York: Marcel Dekker)
2. Hercules, D. D., editor (1966). *Fluorescence and Phosphorescence Analysis* (New York: Interscience)
3. Passwater, R. A. (1967). *Guide to Fluorescence Literature* (New York: Plenum)
4. Phillips, R. E. and Elevitch, F. R. (1966). *Fluorometric Techniques in Clinical Pathology in Progress in Clinical Pathology*, Chapter 4 (New York: Grune and Stratton)
5. Udenfriend, S. (1966, Vol. I; 1970, Vol. II). *Fluorescent Assay in Biology and Medicine* (New York: Academic Press)
6. Weissler, A. and White, C. E. (1963). Fluorescence Analysis, Chapter 6, *Handbook of Analytical Chemistry* (L. Meites, editor) (New York: McGraw-Hill)

7. Weissler, A. and White, C. E. (1966). 'Fluorometric Analysis', Chapter 5 in *Standard Methods of Clinical Analysis*, Vol. 3A (F. W. Welcher, editor) (Princeton: Van Nostrand)
8. White, C. E., *Fluorometric Analysis, Fundamental Reviews. Analyt. Chem.*, (1949) **21**, 104; (1950) **22**, 69; (1952) **24**, 85; (1954) **26**, 129; (1956) **28**, 621; (1958) **30**, 729; (1960) **32**, 47R; (1962) **34**, 82R; (1964) **36**, 116R; (1966) **38**, 115R; (1968) **40**, 114R; (1970) **42**, 57R
9. *Fluorescence News Monthly* (Silver Spring, Md: American Instrument Co.)
10. *Traces*, Monthly (Palo Alto, Calif.: G. K. Turner Co.)
11. Moye, H. A. and Winefordner, J. D. (1965). *J. Agric. Food Chem.*, **13**, 516
12. Hollifield, H. C. and Winefordner, J. D. (1967). *Talanta*, **14**, 103
13. Steele, R. H. and Szent,-Gyorgyi, A. (1957). *Proc. Nat. Acad. Sci.*, U.S.A., **43**, 477
14. Winefordner, J. D. and Latz, H. W. (1963). *Anal. Chem.*, **35**, 1517
15. Winefordner, J. D. and Tin, M. (1964). *Anal. Chim. Acta*, **31**, 239
16. Hollifield, H. C. and Winefordner, J. D. (1966). *Anal. Chim. Acta*, **36**, 352
17. Hollifield, H. C. and Winefordner, J. D. (1967). *Talanta*, **14**, 103
18. Drushel, H. V. and Sommers, A. L. (1966). *Anal. Chem.*, **38**, 10
19. Sawicki, E., Stanley, T. W., Pfaff, J. D. and Elbert, W. L. (1964). *Anal. Chim. Acta*, **31**, 359
20. Sawicki, E. and Pfaff, J. D. (1965). *Anal. Chim. Acta*, **32**, 521
21. Pfaff, J. D. and Sawicki, E. (1965). *Chemist-Analyst*, **54**, 30
22. Diehl, H., Olsen, R., Spielholtz, G. and Jensen, R. (1963). *Anal. Chem.*, **35**, 1144
23. White, C. E. and Argauer, R. J. (1970). *Fluorescence Analysis, A Practical Approach* (New York: Marcel Dekker)
24. Dagnall, R. M., Smith, R. and West, T. S. (1966). *Talanta*, **13**, 609
25. Marienko, J. and May, I. (1968). *Anal. Chem.*, **40**, 1137
26. White, C. E., Weissler, A. and Busker, D. (1947). *Anal. Chem.*, **19**, 802; White, C. E. and Hoffman, D. E. (1957). *Anal. Chem.*, **29**, 1105
27. Sill, C. W. and Willis, C. P. (1959). *Anal. Chem.*, **31**, 598
28. Guilbault, G. G. and Kramer, D. N. (1965). *Anal. Chem.*, **37**, 918
29. Williams, D. E. and Guyon, J. C. (1971). *Anal. Chem.*, **43**, 139
30. Har, T. L. and West, T. S. (1971). *Anal. Chem.*, **43**, 136
31. Onishi, H. (1955). *Anal. Chem.*, **27**, 832
32. White, C. E., Fletcher, M. H. and Parks, J. (1951). *Anal. Chem.*, **23**, 478
33. White, C. E. and Cuttitta, F. (1959). *Anal. Chem.*, **31**, 2083
34. Rubin, M. and Knott, L. (1967). *Clin. Chim. Acta*, **18**, 409
35. Schulz, D. W., Passonneau, J. V. and Lowry, O. H. (1967). *Analyt. Biochem.*, **19**, 300
36. Parker, C. A. and Harvey, L. G. (1962). *Analyst*, **87**, 558
37. Watkinson, J. H. (1966). *Anal. Chem.*, **38**, 92
38. Trenholm, R. R. and Ryan, D. E. (1965). *Anal. Chim. Acta*, **32**, 317
39. Alford, W. C., Shapiro, L. and White, C. E. (1951). *Anal. Chem.*, **23**, 1149
40. Guilbault, G. G. Ref. I, p. 346–348
41. Bandisch, O. and Deuel, H. J. (1922). *J. Amer. Chem. Soc.*, **44**, 1586
42. Frohman, C. E. and Orten, J. M. (1953). *J. Biol. Chem.*, **205**, 717
43. Crout, R. J. (1969). *Standard Methods of Clinical Chemistry*, Vol. 3, 62 (D. Seligson, editor) (New York: Academic Press)
44. De, N. K. (1955). *Indian J. Med. Res.*, **43**, 3
45. Jones, S. W., Wilkie, J. B., Morris, W. W. and Friedman, L. (1960). *138th Meeting of the Amer. Chem. Soc.* (New York), 60C
46. Albers, R. W. and Lowry, O. H. (1955). *Anal. Chem.*, **27**, 1829
47. Udenfriend, S. Ref. 5, Vol. I, pp. 249–371
48. White, C. E. and Argauer, R. J. (1970). *Fluorescence Analysis, A Practical Approach* (New York: Marcel Dekker)

8
Microwave Spectroscopy

J. SHERIDAN
University College of North Wales, Bangor

8.1 INTRODUCTION

Spectroscopic analyses, in their proper sense of methods depending on the interaction of electromagnetic radiation with matter, depend for their feasibility on the same primary factors, whatever the frequency range. Absorptions or emissions at frequencies characteristic of the sample are resolved, or partly resolved, and measured. Procedures for characterisation and quantitative estimation vary for different ranges, being influenced by secondary factors affecting particular parts of the electromagnetic spectrum. A 'type of spectroscopy' normally implies a frequency range. This determines the techniques and, in conjunction with the sample conditions, decides the phenomena associated with spectral transitions. The potential of a type of spectroscopy for analytical purposes is a function of both primary and secondary factors, and the applicability or popularity of a given type of spectroscopy clearly depends on how favourable the various factors may be overall, set against cost, and any experimental inconvenience or slowness. These obvious generalisations are mentioned at the outset, because some of the special factors affecting microwave spectroscopy have, until recently, caused insufficient attention to be paid to it in works on spectroscopic analysis. Ways in which microwave spectroscopy notably fulfils primary analytical requirements, however, have been known for over 25 years[1–7, 48], while the rapid development of the techniques, and widespread use of them to reveal great detail of molecular structure, is now described in standard works[8–12]. Most of the reluctance to consider microwave spectroscopy as a method of analysis has sprung from the novelty of the techniques and the special factors associated with them. The rapid development of microwave electronics to provide familiar routine techniques in the last few years has greatly increased the number of potential applications of microwave spectroscopy to analytical problems.

8.2 THE MICROWAVE RANGE AND MICROWAVE TECHNIQUES

The microwave range strictly covers wavelengths from c. 0.3 mm to 1 m, but, for analytical purposes, the easily swept part of this range, with wavelengths between c. 7 mm and 40 mm, is almost invariably used. In terms of frequency, this is the range from 8 GHz to upwards of 40 GHz. The techniques are radio-techniques. The radiation sources are monochromatic, at any instant, but tuneable to desired frequencies. Measurement of frequency and detection of radiation are by radio methods. The basic spectrometer assembly, therefore, consists of a tuneable source oscillator, an absorption cell and a detector, which are recognisable as the main elements in the typical arrangement shown in Figure 8.1.

There is no dispersion element, nor any optical system. The radiation travels in waveguides, i.e. rectangular metal tubes with inner cross-sectional dimensions comparable with the wavelength, these dimensions conforming conventionally with one or more of the four radar bands covering the range stated. These are: 8–12.4 GHz (X-band), 12.4–18 GHz (J- or K_u-band), 18–26.5 GHz (K-band), and 26.5–40 GHz (Q- or R-band). Most analytical systems use only one such band, and Q-band (R-band in the U.S.A.), offering the largest frequency range coupled with other high performance features, is the most popular. The source is now usually a backward wave oscillator, commonly built into a standard 'sweeper', and its instantaneously mono-chromatic frequency can be read to a few MHz from the calibration scale. Since this measurement, however, is not accurate enough for the powers of characterisation permitted by the resolution, the frequency is also measured to within about 0.1 MHz or less by standard heterodyne methods (Figure 8.1). Frequencies of this accuracy can be read directly on counters, and marked as a frequency scale on the output signal trace.

Increasing use will undoubtedly be made of long-lived solid state source oscillators; their rapid development offers important economies in price

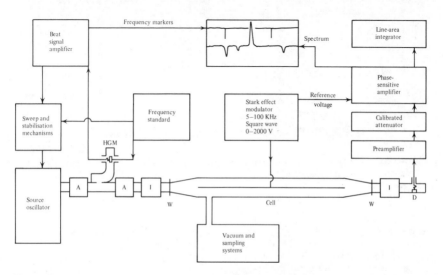

Figure 8.1   Basic components of a Stark-modulated microwave spectrometer. The line area measurement, and on occasion the source stabilisation, may be omitted in qualitative analysis. A = attenuator, I = isolator, D = detector, W = window, HGM = harmonic generator-mixer

and maintenance costs of microwave spectrometers. Absorption cells are discussed in Sections 8.3 and 8.4.

The detectors are crystal diodes, which, like all waveguide units employed, have essentially uniform performance over the band in use.

Development of such wide-banded components, and of sweepers, has greatly facilitated the operation of systems as laboratory routines. Since the radiation is coherent, it is necessary to minimise problems due to standing waves, and hence all components must be carefully designed and constructed to avoid mismatch. It is advantageous to include isolators at appropriate points to suppress standing wave effects.

In recording the spectrum, the frequency is swept over a chosen range, which may be up to the total width of the band, at a predetermined rate, while the detected signals are amplified and recorded against the frequency scale. On account of the monochromatic source, this scale can be magnified

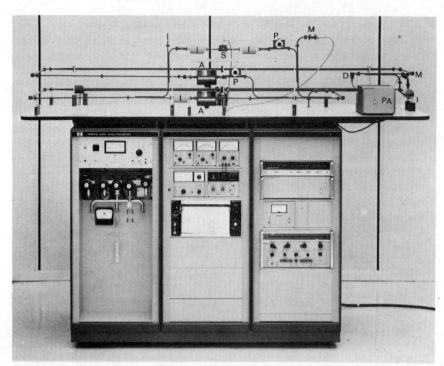

Figure 8.2 The Hewlett Packard 8460A spectrometer. The left-hand unit contains the sample admission and cell-pumping systems. The right-hand unit contains the frequency synthesiser and programmer to set the desired source frequency and sweep range. The central unit contains the control and measurement of the Stark modulation voltage, and the phase sensitive amplifier for the line signals, that are presented on the recorder, aligned with a frequency scale from the source control. It also contains instruments to measure the power level, and other conditions in the spectrometer. The source radiation emerges through the right-hand side of the table top, and passes from right to left through the devices shown in Figure 8.3, before traversing from left to right the absorption cell (larger, lower waveguide), and bridge balancing arm (upper waveguide), to the detector and signal pre-amplifier (A = attenuator, P = phase shifter, I = isolator, M = power monitor, S = signal calibrator, D = detector, PA = pre-amplifier). By exchanging waveguide units (apart from the absorption cell and balancing cell), and the source plug-in for each band, the four bands from 8 to 40 GHz are covered

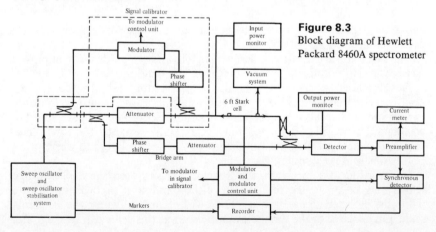

Figure 8.3
Block diagram of Hewlett Packard 8460A spectrometer

to whatever extent may be justified by the line-width, and there is virtually no limit to the electronically permitted resolving power. Since standing waves can never be entirely eliminated, there is always some purely instrumental variation of power reaching the detector as the frequency is swept. It is necessary to ensure that the power absorbed by the sample is distinguished from such instrumental effects, and is also a 'sharp' absorption, occurring predominantly in a frequency range much narrower than that over which the detected power varies on account of instrumental mismatch.

Figures 8.2 and 8.3 show the appearance and block diagram of a commercial microwave spectrometer, of somewhat more complex design than that shown in Figure 8.1. The use of some of the more refined features of the instruments in Figures 8.1 and 8.3 are indicated at appropriate points in the text.

8.3   CELL CONDITIONS FOR HIGH RESOLUTION

Although there are no limitations on resolution analogous to the effects of slit-widths and incomplete monochromation in optical spectroscopy, the need for sharp absorptions, both instrumentally (Section 8.2) and in order to take advantage of the analytical potential of the resolving power, places stringent limits on the condition of the sample in the cell.

8.3.1   Need for gaseous samples and large cells

The sample must be a gas at a pressure well below one atmosphere. Only in such gases are molecules sufficiently isolated for their energy levels to be sharp enough in comparison with a microwave quantum to give sharp absorptions. The original energy level of a molecule, E'', and the higher level, E', to which it is raised by absorption of a quantum, must, in accordance with the Bohr relationship, be separated by the quantum size, hv, where h is Planck's constant, and v the frequency absorbed. For microwave frequencies, the quanta are in the range of a few joules per mole. The levels, E'' and E', however, are only occupied by a molecule, on average, for the mean time between collisions of that molecule, and this is $c.$ 10^{-10} s for a molecule in a gas at one atmosphere. Since the uncertainty principle requires that the maximum 'sharpness' with which a level can be defined is given by Planck's constant divided by the average level lifetime, a simple calculation shows that, in a gas at one atmosphere, the molecular energy levels cannot be defined to better than a few joules per mole. Transitions between such levels accordingly correspond to absorptions which are blurred and spread over a large part of the microwave range. If we wish to make the absorptions occur predominantly over a few MHz or less, E'' and E' must be defined to within $c.$ 10^{-4} J mol^{-1}, which requires the mean inter-collision time of the molecules to be raised to 10^{-6} s or more, corresponding to a reduction in gas pressure to about 10^{-1} Torr or less.

Under such conditions, only a very small fraction of the incident power is absorbed by the sample per centimetre of track (Section 8.5), and, for a single passage of radiation through the sample, a cell from one to several metres in length is usual. It is a suitable type of waveguide, frequently of standard X-band cross-section, which, if straight and fitted with any necessary tapers to the source and detector systems, is effective over the whole 8–40 GHz range. Cells normally incorporate facilities for applying Stark effect modulation (Section 8.5).

8.3.2   Line widths and shapes

Consideration of these is necessary, not only to define the effective resolution, but also because an understanding of the factors involved is essential in evaluation of the special factors affecting analytical microwave spectroscopy.

There are several small fundamental contributions to line widths. For a sample of randomly moving gaseous molecules, to each of which the

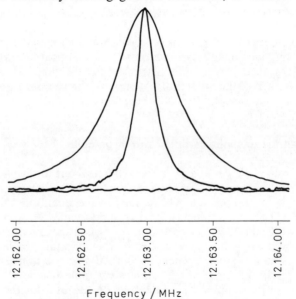

Frequency / MHz

Figure 8.4   Effect of sample pressure on the shape of a line due to carbonyl sulphide. The base line shows the trace with the cell evacuated, while the narrow and broader curves are for sample pressures of 8 and 39 m Torr respectively. The half-width due to gas-phase collision is proportional to pressure, as is the line area; the peak intensity is independent of pressure at 8 m Torr and higher

(Reproduced with permission by courtesy of the Hewlett Packard Company)

frequency of radiation is effectively slightly different, according to the molecular velocity component in the direction of radiation, the Doppler effect gives rise in an obvious way to a line width contribution. Line widths are described by Δv, the half-width at half peak intensity, and a typical Doppler contribution to Δv is some 20 kHz. In addition, in most cells, collisions of sample molecules with the walls of the cell contribute to the uncertainty of the energy levels, and commonly add a few kHz to the line width. The Doppler and wall collision contributions are both independent of the gas pressure. Modulation (Section 8.5) also adds to the line width, its contribution being typically between 20 and 50 kHz.

The above effects are important if extreme resolution is sought, and they indicate a minimum easily attained Δv of near 0.1 MHz. This is still so small that, unless the pressure is as low as $c.$ 10^{-3} Torr, there is a considerable, and often dominant, contribution to Δv from gas-phase collisions. At sample pressures near 10^{-2} Torr, the blurring of energy levels by such collisions is responsible for most of the line width, which is then typically $c.$ 1 MHz. This pressure region is approximately that usually employed, although somewhat lower pressures are used if the resolution required demands it. At successively higher sample pressures, the resolution is progressively lost, and the broadened lines ultimately become difficult to detect against the instrumental signal variations (Section 8.2). The peak intensity of the lines, moreover, does not increase as the pressure is increased above $c.$ 10^{-2} Torr. Since the collisional blurring of energy levels is inversely proportional to the time between collisions, and, therefore, proportional to the gas pressure, the line width is also proportional to this pressure. Since the integrated absorption, or area under the line, is also proportional to the pressure of the absorbing species, the peak intensity remains constant as the pressure rises. This effect is well shown in Figure 8.4.

Since higher sample pressures merely add to the width, and not to the peak intensities of absorptions, there is little point in the use of pressures much above 10^{-2} Torr, particularly since resolution is also lost. It should be mentioned, however, that pressure-broadened lines are less subject to the phenomenon of power saturation (Section 8.5) than the narrow lines obtained at low pressures, where the molecular relaxation processes are inefficient.

Besides power saturation, other factors which affect the widths and shapes of lines include poorly resolved fine structures (Section 8.4), and distortions due to incomplete resolution of Stark effects (Section 8.5).

8.3.3   Introduction of samples

The techniques associated with the introduction, with precision, of known amounts of substance into the absorption cell are essentially those of normal vacuum practice. For samples which are gaseous at room temperature, known amounts may be introduced by volume sharing between calibrated sections of the system and the cell itself. Liquid samples, provided they are completely vaporised at the temperature and pressure prevailing in the cell, may be introduced as precisely measured volumes of liquid. In either case, it may be necessary to correct for adsorption effects on the large inner surfaces of the cell. Good facilities for monitoring the actual pressure within the cell are necessary.

When the sample is not volatile enough to be vaporised at a pressure of $c.$ 10^{-2} Torr and at laboratory temperatures, it is necessary to heat the absorption cell, and to adjust the conditions for sample introduction accordingly. For analytical purposes, it is still necessary that the whole of the sample should be vaporised in the cell if possible, and known amounts of liquid samples may be injected in the manner already indicated. The complete vaporisation of known amounts of solid samples may require the introduction of these into a side-arm of the cell, which is maintained at a temperature slightly greater than that of the cell itself.

Modern commercial spectrometers contain sophisticated systems for sample introduction, which can be further adapted for special purposes. Good facilities for out-gassing the cell between samples, preferably accelerated by raising the cell temperature, are essential.

8.4 SPECIFICITY: SENSITIVITY OF SPECTRA TO MOLECULAR CONSTITUTION

8.4.1 Rotational spectra and molecular moments of inertia

Microwave spectra will result from any transition allowed in gaseous molecules, for which the energy levels are separated by a microwave quantum. In practice, apart from a few spectra in which vibrational effects contribute to the energy jump (notably the inversion spectrum of ammonia and the spectra of a few substances with internally rotating polar groups, such as alcohols and mercaptans), microwave spectra are predominantly pure rotation spectra of dipolar molecules. They are occasionally called molecular rotational resonances (MRR)[13]. The allowed rotational energies of free molecules are very sensitive to the molecular moments of inertia about the three mutually perpendicular axes in the molecular framework which are known as the principal axes. The directions of these axes are simply related to any symmetry in the molecule, and they are the axes about which the summed mass × (distance)^2 terms are at either maximum or minimum values. By convention, the axes of least, intermediate and greatest moments of inertia are called the A-, B- and C-axes, and the corresponding moments I_A, I_B and I_C. Well developed theory allows the energy levels of any molecule to be expressed in terms of the time-averaged reciprocals of these moments, fundamental constants and quantum numbers describing the rotational states. The fact that real molecules are not rigid, their moments of inertia changing slightly for different states on account of centrifugal distortion, is interpreted in refined theory by the evaluation of several further constants characteristic of the molecule, but the dominant factors in deciding the frequencies at which a molecule absorbs microwave radiation are I_A, I_B and I_C, coupled with the vector components of the molecular dipole moment in the A, B, C axis system. These vector components decide the selection rules for rotational transitions. Both the moments of inertia and the dipole components are highly specific to molecular structure.

Usually, a dipolar molecule will give rise to a considerable number of absorptions, more or less randomly distributed through the microwave region. The sensitivity of the frequency of a typical line to changes in molecular moment of inertia is very high; molecules of which the moments of inertia differed by as little as 1 in 10^5 would produce microwave absorptions at measurably different frequencies, even if their dipole orientation and other spectral features were identical. The sensitivity of microwave frequencies to molecular constitution is higher than for any other type of spectroscopy, essentially because microwave spectroscopy is the only type of radio spectroscopy which responds directly to the properties of the whole molecular structure. There is, moreover, virtually no absorption in the frequency space between the lines, and hence overlapping of spectra is much rarer than in other techniques of optical spectroscopy. Many microwave spectra, however, are rich enough to occupy an appreciable fraction of the whole frequency space, and overlap between spectra of two or more such substances can easily occur, although it does not normally prevent absolute characterisation or the availability of parts of the range where analyses can be made without problems due to overlap (Section 8.6).

Sensitivity to molecular mass distribution is such that spectra of different isotopic forms of a substance can usually be completely resolved from one another and analyses can, therefore, be made for isotopic composition and location of isotopes in molecules (Section 8.6).

8.4.2   Fine structures and Stark effects

Closely spaced groups of lines, frequently not fully resolvable, are found in many microwave spectra. Some are due to small changes in mean moments of inertia of molecules in different vibrational states, when several such states are populated, but, more commonly, they are caused by the coupling of spins of certain nuclei to the molecular rotation through the nuclear quadrupole moment. The latter effect, for example, produces weak splittings in spectra of nitrogen-14 compounds, while progressively larger splittings are found in compounds of chlorine, bromine and iodine, as well as of some less commonly encountered nuclei.

Electric fields, which are applied during molecular modulation for sensitive detection of spectra (Section 8.5) themselves produce fine structure (the Stark splittings) which are normally presented in addition to those lines which occur in the absence of field.

Although fine structures and Stark effects may serve to characterise transitions of particular molecules, and have frequently been used in this way in the assignment of spectra, their effects upon microwave specificity and potential for quantitative analysis are in general disadvantageous. While quadrupole or Stark signatures of lines will sometimes aid a qualitative detection of a substance, the distortions of line shape and reduction in resolution which accompany such fine structures must be carefully considered in any quantitative procedure.

8.4.3   Classes of substances which can be detected and measured

Any dipolar substance which can exist as a gas at a pressure of some 10^{-3} Torr or more, at the temperature of the absorption cell, will give a microwave spectrum. Since it is easy to use cell temperatures up to c. 200 °C, and not difficult to use considerably higher cell temperatures, the techniques may be used to analyse for a very wide range of substances, from the most volatile to those with a very low volatility. Less volatile substances naturally involve greater difficulty in sample introduction and maintenance of proper cell temperature. Even when a total sample cannot be vaporised, significant progress can be made by examination of volatile fractions from a sample, or volatile material obtained from it by chemical reaction such as pyrolysis, oxidation, etc.

Molecules that can exist in several conformations, or that possess several easily excited vibrations, are less suitable, since the variable molecular geometry spreads the spectra into more diffuse, weaker, absorptions, and the same is true of substances that contain several quadrupolar nuclei in the molecule. So-called 'low-resolution microwave spectroscopy', in which large frequency ranges are presented on a compressed scale, can, however, reveal spectra for such substances that would not be easily seen in high-resolution instruments, and sensitivity to molecular structure is still remarkable, even with this reduced resolution (Section 8.6).

Spectra will not be observed for substances with no dipole moment, but these are, in any case, exceptional. Quite small dipole moments, such as those of a number of hydrocarbons, are sufficient to give clear microwave spectra, although a large polarity is naturally advantageous from the point of view of detectability of a substance.

8.5 SENSITIVITY TO AMOUNT OF SUBSTANCE

8.5.1 Absorption coefficients

The intensity of microwave absorption at a given frequency is expressed by the absorption coefficient, γ, defined as the fractional fall in radiation power, P, due to molecular absorption, per unit of path length x. Hence:

$$\gamma = -(1/P)\mathrm{d}P/\mathrm{d}x$$

If the power entering the cell is P_0, and that leaving after traversing a cell of length l is P_l, the integrated form is:

$$P_l = P_0\,\mathrm{e}^{-\gamma l}$$

When, as is always the case here, γ is much less than unity, this becomes:

$$P_0 - P_l = P_0 \gamma l$$

The expression of γ in terms of molecular properties, quantum numbers, and physical conditions, and the variation of γ with frequency in the region of a line are derived in standard works. For lines of which the width is predominantly due to gas-phase collisions, the factors in γ consist of the following, in addition to fundamental constants:

(a) the square of the dipole moment matrix element of the transition, which is a function of the quantum numbers and the dipole moment of the molecule;

(b) the square of the frequency concerned (v);

(c) the factor Nf/T, where N is the total number of absorbing molecules per unit volume, f the fraction of these in the initial energy state, and T the absolute temperature;

(d) a line-shape term, which, for a narrow line centred on frequency v_0 is:

$$\Delta v / [(v - v_0)^2 + \Delta v^2]$$

where Δv and v have their earlier significance.

Factors (a) and (b) indicate the greater intensities in general, of lines due to highly dipolar substances, and of lines of high frequencies. In factor (c), f is determined in complex ways according to the nature of the molecule, but is constant at a given temperature. Hence, Nf/T depends only on the total number of absorbing molecules at constant temperature. The value of this term is normally such that the value of γ increases markedly as the temperature is reduced, and therefore microwave absorptions are more intense at low cell temperatures. Factor (d) produces the so-called Lorentz line shape, seen, for example, in the lines in Figure 8.4.

An important assumption in the above is that γ is independent of the power P, and that the thermodynamic equilibrium among molecular states is not appreciably disturbed by the microwave power. The nett absorption depends on the difference in populations per unit statistical weight (here called simply the populations) of the upper and lower states. The population of the lower state, at thermodynamic equilibrium, exceeds that of the upper state by only a very small factor when the energy separation of the levels is as small as a microwave quantum, and this, in addition to the smallness of the factor f, contributes to the smallness of microwave absorption coefficients. The population difference, moreover, is easily displaced from its equilibrium value by the high powers of monochromatic radiation which are available, especially at low sample pressures, when the collisions maintaining thermodynamic equilibrium are reduced in number. Too much power at the line frequency, therefore, tends to equalise the populations of the upper and lower levels which are at equilibrium in the presence of the radiation. This effect is known as power saturation, and is accompanied by a fall in the apparent absorption coefficient as the power is increased. The line shape also becomes a function of power. Although power saturation underlies an ingenious procedure for quantitative analysis of mixtures (Section 8.7), it will be assumed that it has been avoided in what follows, unless specific mention is made to the contrary. This implies low power levels, and use of suitable tests for the absence of any influence of changes in these levels on the absorption coefficient.

Values of the absorption coefficient at the peak of a line, γ_{max}, are, under the usual cell condition, typically from 10^{-5} cm^{-1}, down to values below the limit of detection near 10^{-10} cm^{-1}. A value of γ_{max} of 10^{-6} cm^{-1} corresponds to a strong line in a microwave spectrum. Hence, very high sensitivities are required in a microwave spectrometer to detect typical absorptions, which cause loss of perhaps only 0.001% of the power in 1 m of path. Essentially standard radio-techniques, coupled with molecular modulation, allow very remarkable sensitivities to be attained.

8.5.2   Stark-effect modulation

Almost all microwave spectrometers depend critically on the use of Stark effects to modulate the power changes which are specifically due to the molecules of the sample. An electric field, variable in intensity from zero to perhaps 2000 V cm^{-1} is applied to the gas by means of an insulated electrode in the cell, the walls of which are grounded. This field is switched on and off, uniformly, to give so-called 'square wave modulation' at a frequency between a few kHz and c. 100 kHz. If the Stark splittings of the lines are fully, or partly, resolvable under the experimental conditions, the power variations at the detector which are coherent with the modulation can be selectively amplified to give purely molecular signals. By use of the methods of phase-sensitive detection, the sensitivity can be raised to a limit which is set only by the inconvenience of the slow sweep speeds which must be employed when the band pass of the amplifiers is made extremely narrow to reject noise. With conveniently narrow pass bands, many good spectrometers can detect lines with absorption coefficients as small as 10^{-9} cm^{-1} at wavelengths near 1 cm, and sensitivities of this order are, for example, specified for the Hewlett Packard 8460A spectrometer[13].

As a rough generalisation, therefore, spectrometer sensitivity is typically sufficient to detect a substance with an average microwave spectrum in concentrations of as little as 1 %, and often less; certain substances with very intense absorptions, such as ammonia, linear molecules or molecules of the class known as symmetric-tops, are detectable in concentrations of 0.1 or even 0.01 %.

The design of cells to permit Stark effect modulation, while retaining other desirable properties, is of some importance. The Hewlett Packard 8460A spectrometer[13] uses an X-band cell of conventional geometry, in which the Stark septum is centrally supported parallel to the broad faces of the waveguide, particular attention being paid to precise dimensioning, and support of the electrode with the minimum of insulator. Other spectrometers, including the 'Camspek' model, made by Cambridge Scientific Instruments Limited[14], employ other designs in which the cell is split, and the modulation applied between two insulated sections[15]. Such cells are more suitable than the conventional type for high temperature work, and can be more easily dismantled, but the Stark modulation effects in them can be more complicated than in a cell containing a central insulated septum. We may expect considerable further development work on the design of cells for analytical use.

The instruments for providing the modulation voltages, and the amplification systems, are available in a number of commercial forms. It is desirable that the quality of the square wave used for modulation should be very high.

Stark effect modulation can easily produce distortion of line shapes, particularly when it is not possible to apply sufficient modulating field to remove all of the Stark spectrum from the frequency at which absorption occurs in the absence of an electric field. Although phase-sensitive detection allows the Stark spectrum to be distinguished as displacements of the spectral recording in the opposite sense to those due to the zero-field line, incomplete Stark modulation is very common, and leads in an obvious way to dependence of both peak intensity and line shape on the amplitude of the modulating voltage (Section 8.7).

8.6 CHARACTERISATION AND IDENTIFICATION OF SUBSTANCES

8.6.1 Low-resolution microwave spectra

The use of sweepers as sources and of wide-banded components, has permitted microwave spectra to be presented at much reduced resolution, some thousands of MHz, perhaps even an entire radar band, being presented on a page-size pen recording. These 'band microwave spectra' have attracted publicity, doubtless because of their formal resemblance to the spectral traces of optical spectroscopy[13, 14]. Figure 8.5 shows an example of such

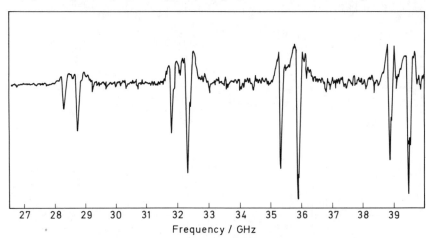

Figure 8.5 Low-resolution microwave spectrum of crotonic acid. Two families of peaks (pointing downwards) are observed, members of each family being equidistantly spaced in frequency, the spacing being about 3.59 GHz for one family, and about 3.53 GHz for the second. The families of lines are due to the geometrical isomers

respectively
(Reproduced with permission by courtesy of the Hewlett Packard Company)

spectra. Even at a resolution only one-thousandth of that customarily used, sensitivity to molecular constitution is obvious, as in this case where the rotation of a carboxyl group by 180 degrees with respect to the rest of the molecule produces a second set of completely resolved absorptions. Many similar cases are on record[13, 14, 16]. Since it is always easy (Section 8.6.2) to expand the resolution and to characterise line frequencies with high accuracy, it would seldom be necessary to make direct comparisons of such spectral traces for analytical purposes, although in many cases it would be quite a discriminating method. The spectra, for example, of the three isomers 1-bromoethylbenzene, 2-bromoethylbenzene and p-bromoethylbenzene[13] show sharp peaks which are nearly all completely resolved, even when all three substances are present together and with such low resolution. It must be remembered, however, that the form of such traces is a function of the Stark-modulation voltage, and that this form of presentation, while of surprisingly wide applicability, works best for the specific, although frequently encountered, type of molecle in which an axis containing a large

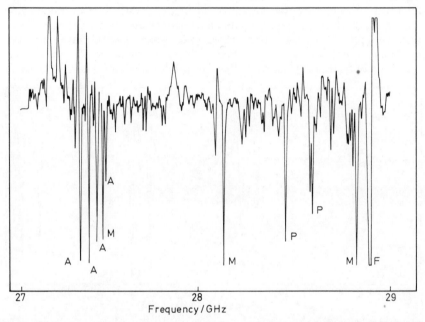

Figure 8.6 Low-resolution spectrum of a mixture containing formaldehyde (line marked F), acrolein (lines marked A), propionaldehyde (lines marked P) and methanol (lines marked M). With resolution at only one-thousandth of its usual value, completely resolved lines are recognisable for each component

(Reproduced with permission by courtesy of the Hewlett Packard Company)

component of the dipole moment is an axis of easily the least moment of inertia (the so-called near-prolate molecules).

Since, unlike the optical spectroscopist, the operator of a microwave spectrometer can instantly expand the resolving power by many orders of magnitude beyond that of low-resolution work, it is normal for qualitative analysis to be made at much higher resolutions, where the specificity is clearly increased correspondingly. Exceptional specificity is already reached with the most common line-width conditions, under which lines 1 MHz apart can be commonly resolved, although, for special purposes, it is easy to extend the resolution to resolve separations well under 1 MHz. The high specificity under such conditions is well emphasised by the fact that many totally isolated lines from different components in mixtures can be distinguished long before a resolution reaches its normal value, as is seen in the example of Figure 8.6.

For the discussion of the attainable specificity, it is not normally necessary to consider spectral traces, but to consider the frequencies, line widths and any other properties of absorptions, as recorded numerically.

8.6.2 Attainable specificity

The evaluation of this depends upon the circumstances considered. Since the spectra in general contain many lines which can be reduced in width to a fraction of 1 MHz in a spectral space which will accommodate tens of thousands of such lines, specificity of identification of a single substance alone is measured by how few such lines need to be measured accurately in frequency for identification to be accepted as unequivocal. This must be viewed against the background of what is known about the substance already, and what the chances may be that another substance could be responsible for a line very close in frequency to a particular line seen. In the more general case, where we have a mixture of substances, specificity is defined in the same way, with the additional background factor that now the possibility of line coincidences in the spectra of components which are known to be potentially present in the mixture must be taken into account. Under all circumstances, specificity will be dependent upon the availability of a large data-bank of spectra of particular molecules, to which data must be added for any components of the mixture for which previous logging of data is insufficient. A large body of such data is available in continuing publications from the United States National Bureau of Standards[17], and numerous lines which may not have been specifically measured may be calculated in frequency from the molecular spectroscopic constants, where these are included. It must be emphasised, however, that, for analytical purposes, it is not essential to assign the lines to specific quantum transitions, and many analytically useful lines are assigned only in the sense of belonging to a particular molecule. The volume of new microwave spectroscopic data published annually is now large, and increasing. An analytical spectrometer would normally make reference to the data-bank of known frequencies by use of a computer-search procedure.

8.6.2.1 Single substances

While, at a particular frequency, a large number of substances might potentially give rise to an absorption, it is clear that the chance of a correct characterisation from the frequency of a single line is quite high when, as normally, there are sharp limitations on the possible nature of the substance. By measurement of a small number of lines, characterisation becomes virtually unequivocal.

8.6.2.2 Mixtures

Clearly, the specificity in this case is smaller the fewer conditions restricting the range of compounds that might be present. It is also smaller, the more dense the spectra of the components. Any attempt to analyse the specificity quantitatively must do so in terms of these factors.

In the limiting case, where all substances present have strong, sparsely distributed lines, powers of diagnosis are obviously very high. Thus, if the lines of any one component occupy only one-thousandth of the frequency range, overlapping of lines is clearly highly improbable, and some tens of such substances could be separately identified in a mixture through the characterisation of a small number of lines of each, provided these are intense enough to be observed under such conditions. Most substances of interest, however, possess quite dense spectra, with lines occupying a considerable proportion of the frequency range. A single molecular species might produce measurable absorption over some 10% of this range, although some of it would be relatively weak. The number of such substances that could be separately identified might typically then be no more than ten, and their characterisation would require measurement of numerous lines and checking of frequencies against a data-bank. This procedure has been described in detail for the Camspek spectrometer[14].

Jones and Beers[18] have recently made a computer analysis of 10 000 lines of 33 substances, between 26 and 40 GHz, to give a critical appraisal of their potential for qualitative analysis. The spectra of many of the substances are dense, several having enough lines, not counting those unrecorded, to fill between 5 and 10% of the frequency space with line widths of 1 MHz. For these 33 substances, there was about one chance in two that a given line would be overlapped by one from the other 32 specta, within a frequency interval of 0.2 MHz. Use of two or more lines brought rapid increase in the chances of unequivocal identification. On the whole, the chances of positive identification proved even better than might have been anticipated for such a complex mixture, probably because a good proportion of the substances were associated with relatively few reported lines. With comprehensive specta, it is likely that the results would have called for rather a large number of lines to be measured for the same degree of certainty to be reached. As expected, the use of the sparser ladders of lines above a certain intensity, which are those likely to be used in an analysis, reduced the effects of overlap. This work confirms the powers of the method, and suggests that in many cases quite restricted frequency ranges could be used, particularly if frequency measurements are made with the high accuracy which is now possible as a routine procedure.

A further favourable aspect is the ability of the techniques to deal with very small samples, perhaps as little as a few micrograms for qualitative work. The method has also had much success in dealing with unstable substances. At a more empirical level, spectra have been published[14] characterising compounds in the pyrolysis products of such materials as tobacco, cannabis resin and motor oils.

In the limiting case where it may be very important to detect, unequivocally, substances in mixtures showing very dense spectra, an effective extension of the techniques is already well established for research purposes. This is microwave double resonance, in which a line is observed in the presence of intense microwave radiation at a different frequency (the pump radiation), this frequency being chosen such that it affects only one of the molecules which might be responsible for the line. If an effect, the 'double-resonance signal', is observed, the molecule responsible must be involved in two transitions of known frequencies with a common energy level, and such double resonances are clearly very much more specific to molecular constitution even than conventional microwave spectra. The pump radiation may be used to replace the Stark field as a means of molecular modulation, and high sensitivities attained. Use of double resonance for analysis has been reviewed[19], but at present it would be considered only in difficult special cases.

8.6.2.3 Examples

Refs. 20–26 summarise some publications on the use of the method for qualitative analysis. They are not typical of the present situation, since many more examples are now being studied in the increasing number of laboratories using commercial spectrometers, and we may expect much work of this nature to become available in the immediate future.

8.7 QUANTITATIVE ANALYSIS

For most other methods of quantitative spectroscopic analysis, difficulties have been concerned mainly with overlapped absorptions. They are discussed in various standard works[27].

Typically, the quantitative absorption curve for a mixture of two components may be used to derive the concentration of each, if the corresponding curves for the separate components are available. Simultaneous equations are set up, which become trivial if the overlap of the lines is very small. Difficulties can be marked for mixtures of numbers of components, although progress can be made with sufficient computational aids[28]. On the other hand, optical spectra are usually relatively free from effects due to the different environments of molecules in mixtures, as compared with those in pure substances. Wide use is normally made of calibration procedures, etc., to eliminate instrumental parameters, and to simplify the conversion of absorption measurements to concentrations. It is not always easy to find statements as to what is to be regarded as an acceptable accuracy in such determinations, but the ability to gauge concentrations to within a few per cent of their value seems generally to be accepted as a satisfactory performance.

In microwave spectroscopy, the situation is essentially the reverse. The overlap of lines, as has been seen, can be avoided, and the arithmetic accordingly greatly simplified. On account of the sensitivity of line widths to molecular collisions, however, the shapes of lines depend strongly on the nature of the collision partners in mixtures, and, therefore, on the composition of the mixture itself. Thus, any method which seeks to compare peak absorption coefficients must take full account of this effect, which is virtually absent in spectra for which the quantum jumps are too large for variation in collision conditions to affect them appreciably.

8.7.1 Methods using low power levels

A large proportion of the work on intensities of lines has concentrated on conditions under which power saturation (Section 8.5.1) is avoided by use of power levels too low to disturb appreciably the Boltzmann distribution of energy level populations. The methods fall into two types, those designed to measure peak intensities, and those suitable for measuring integrated intensities or line areas. The former methods are easier, but potentially less useful, while the latter are of increasing importance at the present time.

8.7.1.1 *Measurement of peak intensities*

A considerable amount of experimental effort has been devoted to the evolution of satisfactory measuring procedures for peak intensities. Development of such procedures has been rendered difficult by a problem connected with the use of coherent radiation. Such use can easily cause the overall response of an instrument to vary quite markedly over small frequency ranges, through incomplete elimination of mismatch and other instrumental effects. Even with the refined procedures that are now possible, it is still true that, for intensity measurement by any means whatsoever, a microwave spectrometer must be designed and constructed with a high degree of precision. With this proviso, however, methods now exist that allow a practised operator to make measurements with confidence of the relative peak intensities of absorptio   over reasonably broad ranges of frequency.

Methods that merit mention are:

(a) The direct comparison method for closely-spaced lines. While such comparisons have been much used, with many variants of procedure, the best known standard routine is that developed by Wilson and his colleagues[29–31], which is used in many research studies as a quick method of comparing peak intensities with an accuracy of perhaps 10%, or sometimes better. The heights of lines are compared directly at the same voltage displacements of the spectral trace by including a calibrated attenuator in the lead from the detector in order to reduce the height of the stronger line by a known amount, thus eliminating errors due to variable amplifier response. Inclusion of ferrite isolators at both ends of the cell is found to improve the reliability of such methods by minimising standing wave effects.

(b) The 'antimodulation' method[32, 33]. This uses electronic devices, the antimodulators, which introduce a small variation in microwave power, this attenuation being variable in intensity with a periodicity corresponding to a frequency of up to several kHz. The signal obtained from this artificial power variation is combined in opposition with the absorption signals obtained from a Stark modulation system with a modulation frequency equal to that of the antimodulation. The depths of antimodulation required to cancel different lines are measures of their relative intensities. The method is probably somewhat more precise, but also more complicated than the first mentioned.

(c) Use of PIN-diode 'signal calibrators'. This essentially resembles the antimodulation method, with the advantage of using the recently developed PIN-diodes, which, when suitably biased, can produce 100% attenuation of microwave power, an attenuation which can be applied electronically at high frequencies, with very short rise- and fall-times. In the Hewlett Packard 8460A instrument[13], the signal calibrator imposes 100% square-wave attenuation on a sample of the radiation, at the same frequency as the Stark modulation, 33.3 kHz. The power modulated by the PIN-diode is varied with calibrated attenuators and adjusted in phase, and, after traversing the spectrometer cell, forms a standard signal that can be compared with any absorption line or, with advantage, combined with it in opposite phase, and adjusted to give a null signal. The fact that this 'electronic molecular signal' can be varied in a calibrated way over wide ranges of microwave frequency gives this method advantages for the comparison of lines well separated in spectra. With such devices, while it is not possible to eliminate entirely the variation of instrumental performance with frequency, it is possible to form an accurate and reproducible calibration of instrumental performance as a function of frequency, and hence go a long way towards minimising the problems raised by the use of coherent radiation.

8.7.1.2   Analysis using peak intensities

As already indicated, a proper procedure must allow for the different collision efficiencies of all the molecules in the sample, the concentrations of which may vary. The half-width of a line due to substance A can reasonably be expressed as a series of terms, each due to collisions of A with a particular species of molecule in the mixture, i.e.

$$\Delta v_A = k_{AA}p_A + k_{AB}p_B + k_{AC}p_C + \dots k_{AX}p_X + \dots$$

where the coefficient of type k_{AX} is a measure of the efficiency of a collision with molecule X in broadening the line due to A, and the p_X-terms represent the partial pressures of the various components. For a given common collision partner, the constants k_{AX} roughly increase as the dipole moment of X increases, since a large collision radius is associated with a large dipole moment. Accordingly, at a given molecular concentration of a substance, its lines are wide and of low peak intensity in the presence of polar partners, and narrower and of higher peak intensity in the presence of less polar partners. An extreme case of the effects of polarity of partners on line widths is seen in Figure 8.7.

In principle, analyses can be made through peak intensity studies, if supporting measurements are made to determine the appropriate k_{AX}-terms, e.g. by studies of widths of lines due to A in the presence of an excess of each of the other components in turn. Use of such a method is dependent on the assumption that the line shape is good, which is by no means always the case. Studies would be restricted to relatively simple mixtures in which the number and nature of all possible species present is known.

Only in the case where the k_{AX}-terms are all essentially equal is a favourable situation found. The line width is then independent of changes in composition of the mixture, and peak intensity data can serve to compare concentrations accurately. These conditions are met for mixtures of isotopic forms of the

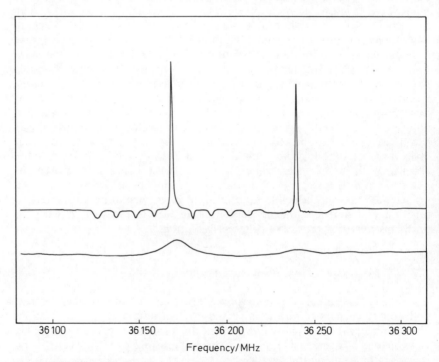

Figure 8.7 Effect of polar diluents on line-shape. The upper curve shows two lines due to methanol in a sample of methanol vapour alone. The lower curve shows the same lines, for the same partial pressure of methanol, diluted with a 49-fold excess of water vapour. Note the great reduction in peak intensity, and in resolution of the Stark components (pointing downwards), although the line areas remain roughly constant
(Reproduced with permission from data provided by Cambridge Scientific Instruments Limited)

same substance, and when samples contain varying proportions of molecules in different vibrational states, each state giving a separate spectrum. Isotopic substitution, or vibrational excitation, have, except in a few predictable special cases, negligible effects on the collisional properties of molecules, and peak intensity studies are commonly used to establish low-lying vibrational energies of molecules and, in analytical work, for isotopic estimations. In such studies, moreover, the effects of incompletely resolved Stark spectra are largely self-cancelling, provided the same transitions are compared for each form of the molecule, since the substances' dipolar properties are essentially the same in all the states concerned. The Stark voltage must, of course, be maintained constant during comparison.

Early isotopic analysis of considerable precision was reported for carbon-13 and nitrogen-15 by Southern, Morgan, Keilholtz and Smith[34]. Other isotopic studies have been concerned largely with complex mixtures met in kinetic investigations, and include work on mixtures of deuterated forms of cyclo-butyl chloride[35], propene[36, 49] and butene[37, 38]. As the isotopic substitutions are exactly located in the molecule, the method has some unique advantages, and will be further applied.

It seems probable that rather similar methods have been used to study mixtures of chemically different substances, including mixtures of geometrical and rotational isomeric forms of molecules. Neglect of variations in the k_{AX}-terms must lead to the ascribing of higher probable errors to any deductions, although these can still be chemically significant. An example is the comparison of the amounts of SO, SO_2, S_2O and H_2CO produced in the pyrolysis of ethylene episulphoxide[39].

8.7.1.3 Methods using integrated absorption

These were little used in early work, largely because spectrometers were not well-suited to the measurement of shapes of weak lines. In principle, simplified methods, depending on measurements of Δv, or its calculation from k_{AX} coefficients, also give the integrated absorption, with an assumed line-shape.

Digital integration instruments have now facilitated line-area and line-shape studies, and allow tests to be made for the proper conditions for analysis by integrated absorption. Such a system is used in the Camspek spectrometer[14]. The integration can be carried out over selected fractions of the line width, symmetrical about the peak, each such measurement being converted to a 'normalised' line area, on the assumption of a Lorentzian shape. If the normalised area varies with the portion of the line integrated, it is known that the shape is not Lorentzian, a circumstance usually caused, in the absence of fine structure, by the incomplete removal of Stark spectra from the line itself. If adjustments, such as increasing the Stark voltage, do not lead to a consistent area, this must be obtained over a narrower frequency range near the peak of the line. The findings can be checked by adding a measured extra amount of the substance concerned, and repeating the procedure (the 'spiking' method).

A striking illustration of the close proportionality of line areas to molecular concentration, over a range of conditions under which the peak intensities varied strongly from proportionality to concentration is given by measurements made in the above way on methanol, and methanol–water mixtures (Table 8.1). It is seen that a tenfold increase in amount of methanol, whether alone, or in presence of a large excess of water molecules, is accompanied by an accurate parallel increase of tenfold in the line area. The peak heights, meanwhile, become constant after relatively small increases in sample size, and, as indicated, are less in the presence of dominant collisions with the highly polar water molecules.

Many lines will have to be used for which the influence of the Stark spectrum on the line shape is unavoidable. They will require addition of a calibration procedure, in terms of Stark voltage, carried out with known concentrations.

We may expect extensive further testing of integrated absorption methods

Table 8.1   Quantitative analysis for methanol

Volume of liquid methanol vaporised/ μl	As methanol alone		10% solution (of methanol) in water	
	Peak intensity (arbitrary units)	Peak area (arbitrary units)	Peak intensity (arbitrary units)	Peak area (arbitrary units)
0.1	52.5	78	27.2	72
0.2	106.1	149	51.3	151
0.3	136.2	225	73.9	228
0.4	156.7	307	91.0	300
0.5	170.5	377	115.0	369
0.6	177.2	452	124.9	450
0.7	182.8	527	133.8	522
0.8	186.0	602	149.0	597
0.9	188.2	688	156.1	682
1.0	191.0	756	167.2	750

in the immediate future. They have the advantage of using low source-powers, which are readily obtainable over wide frequency ranges. As with all procedures, success is dependent on uniform and calibratable spectrometer performance, over the frequency range concerned.

8.7.2   Methods using power saturation

The study of absorption coefficients as functions of microwave power has become important, both as a scientific means to knowledge of molecular collision processes, and as a basis for procedures which allow the varied molecular collisional effects on peak intensities to be automatically taken into account, at the expense of some instrumental elaboration, without the measurement of line widths. These developments are due largely to Harrington[40, 41], and have been applied in the design of the Hewlett Packard 8460A spectrometer.

Since the collision processes which, through relaxation, oppose the onset of power saturation, are very similar to those that are responsible for line broadening, it is readily seen qualitatively that analytical advantage can be gained through correlation of the two phenomena. Basically, lines that are much broadened through the efficiency of collisions made by the absorbing molecules, are, through essentially the same efficiency, more reluctant to lose intensity as the power level is raised, since the collisions delay the onset of power saturation. Conversely, lines that are narrow on account of inefficient collision processes are easily saturated and lose intensity more readily with increasing power. Power saturation thus counteracts the effects of variable collisional broadening, and it is not surprising to find that correct power levels can be chosen at which peak intensities of lines truly represent the number of absorbing molecules.

In the quantitative expression of the effects, the absorption coefficient at power density P, γ_P, is written as

$$\gamma_P = \gamma_0/(1+KP)$$

where K is the saturation coefficient for the transition; KP is obviously dimensionless. If a new 'intensity coefficient', Γ, is now written as the product of the absorption coefficient and the square root of the power density, i.e.

$$\Gamma = \gamma_P P^{\frac{1}{2}}$$

it is obvious that if P is constant, Γ is the product of two terms, η and ϕ, such that

$$\eta = \gamma_0/K^{\frac{1}{2}}$$

and

$$\phi = (KP)^{\frac{1}{2}}/(1+KP)$$

Clearly, ϕ is dimensionless, and is the only part of Γ containing P. In practice, P is not constant, but is distributed in space in a manner that is a predetermined instrumental property, and it remains possible to write

$$\Gamma = \eta\phi'$$

where ϕ' is still a dimensionless instrumental function of KP_0, P_0 being the input power. The value of ϕ', as a function of KP_0, rises sharply as P_0 is increased from zero, and then passes through a broad maximum. Provided that a sufficient range of power is available, and that K is not too small (that is, that collision broadening is not so efficient as to make the line very difficult to saturate), KP_0 can be chosen to give ϕ' any value from zero to past the maximum.

Furthermore, it can be shown[40] that the line signal S in a Stark-modulated spectrometer is proportional to Γ, provided the detector crystal current is kept constant. Hence, if S is measured while ϕ' is maintained constant, S is a measure of η. The value of this formulation now emerges, since the common collisional basis of line-broadening and saturation causes η to be independent of the collisions of the absorbing molecule, and proportional simply to the number of such molecules, N. More specifically, if we look at the two factors of η, for the peak of a line, γ_0 is seen to be proportional to N and to the reciprocal of Δv; for collision-broadened lines, this reciprocal is in turn proportional to the mean time between collisions, τ (Section 8.3.2). For such lines, however, the denominator in η, $K^{\frac{1}{2}}$, also turns out to be proportional to τ, and is independent of N. Accordingly, η and hence S, is proportional to N and independent of τ[40-44].

The easiest value of ϕ' to recognise is its maximum, since this is established when P_0 is adjusted to give maximum signal. Comparison of such signals for a given line from various samples gives accurate comparison of numbers of absorbing molecules, whatever the collision conditions. Comparison of powers needed to give maximum signals allows changes in K, and hence of τ, to be compared, and leads to information on molecular rotational relaxation.

The Hewlett Packard 8460A system (Figures 8.2 and 8.3) contains a microwave bridge, for measurements of the above type. This configuration is necessary in order that, at high power levels in the cell, most of the power can be cancelled before the radiation meets the detector, so ensuring that detector performance is not degraded, and that detector current is maintained constant.

For lines with considerable pressure broadening, it may not be possible to make P_0 great enough to raise ϕ' to its maximum. The formulation may be generalised, however, in ways which allow such cases to be treated[41]. Inspection shows that variations in $\log S$ with power for a given line (η constant) will be the same as variations in $\log \phi'$ with power, and accordingly a plot of $\log S$ $v.$ $\log P_0$ will have the same shape, though not the same origin, as a plot of $\log \phi'$ $v.$ $\log KP_0$. The latter plot is a defined and purely instrumental property, referred to by Harrington as the 'intensity law'. The origin displacements of the $\log S$ $v.$ $\log P_0$ plots for different lines are seen to be interpreted as follows: displacement in the $\log S$ axis reflects changes in $\log \Gamma$ not due to changes in ϕ' and, therefore, reflects changes in $\log \eta$, while displacement in the $\log P_0$ direction reflects changes in $\log K$. The $\log S$ $v.$ $\log P_0$ plots can be superimposed, even if they do not contain the maximum of S, and the origin shifts required are interpreted in the above terms.

These relationships have been thoroughly tested for simple cases, and the influence upon them of instrumental imperfections has been examined. Good construction, especially of the cell, remains essential, but factory-adjusted instruments can meet the necessary specifications. At present, not many results are available, showing the application of these methods to the analysis of mixtures, although some progress has been made[45, 48–50]. More data of this type will be available for evaluation in the near future, from laboratories where these techniques can now be operated.

Power saturation methods, like the others listed, are adversely affected by poor line shape, due, for example, to incompletely resolved fine structure or, particularly, to incomplete Stark modulation. As with the other methods, tests must be made for such a situation, which must then be eliminated or allowed for in calibration procedures.

8.7.3   The status of quantitative methods

The high resolution and specificity, coupled with the fact that the behaviour of line shapes can be expressed in terms which allow general methods to be applied to the measurement of concentrations, favours high applicability of the microwave method. There is a good chance that, in many cases, sufficiently well-behaved absorptions will be available for the method to be more independent of calibration procedures than some other methods of spectroscopic analysis. Where additional calibrations prove necessary, they will usually have comparable counterparts in optical spectrometry. A complication peculiar to the microwave method is the presence of Stark spectra, which may sometimes be moved over considerable frequencies by the modulation field, and particular care will be necessary to avoid distortion through their effects in the results of quantitative work on mixtures of any complexity. In a field of rapidly developing techniques, however, we may expect ways of dealing with such effects to be found. Thus, an obvious way of reducing Stark effect interferences is to blur the Stark spectra by applying an inhomogeneous field in the on-period of an accurately maintained 50% modulation duty-cycle.

In general, there are few factors which do not favour the applicability of the method, except the present high cost of commercial spectrometers designed for general operation over wide frequency ranges. In an estimation of the information available from microwave spectroscopy[14, 46], the method is found to compare well with all other forms of spectrometric analysis, and with mass 'spectrometry'. There can be no doubt that the method will now take its place as an accepted regular procedure, to be used perhaps initially in special cases, and ultimately more generally for the solution of analytical problems. We may also expect microwave spectrometers to be coupled with other analytical devices, such as chromatographs, in the analysis of complex mixtures.

8.8 MORE SPECIALISED INSTRUMENTS AND FUTURE DEVELOPMENTS

An important area for immediate development is the use of spectrometers with restricted frequency ranges for specific analytical problems. Such instruments would be much cheaper than the commercial models at present available, and the use of rugged, solid-state sources would reduce maintenance costs. A natural extension of this development would be that of process control instruments, perhaps using one or more fixed frequencies, with instrument geometry exactly fitted to them. For monitoring of continuous industrial processes, new sampling methods would be used, involving streaming of samples continuously at low pressure through the cell, a method of sampling that has proved valuable in reducing decomposition effects, and in ensuring equilibration of the sample with the cell. While the fundamental laws of physics prevent on-line monitoring at greater pressures, the need for low-pressure sampling is by no means confined to microwave spectroscopy. For example, tuneable lasers in the infrared, which have been proposed for monitoring of atmospheric pollutants[47], use sample pressures of only a few Torr to achieve resolution in any way comparable with that available throughout the huge spectral space of the microwave range. With their fast electronic response, specialised microwave spectrometers and control systems are due for development.

Acknowledgements

I thank the Hewlett Packard Company and Cambridge Scientific Instruments Limited for permission to make use of data from their publications. Dr. Howard Harrington, of Hewlett Packard, and Dr. John Cuthbert, of Cambridge Scientific Instruments, have kindly discussed many details.

References

1. Dailey, B. P. (1949). *Anal. Chem.*, **21**, 540
2. Hughes, R. H. (1952). *Ann. New York Acad. Sci.*, **55**, 872
3. Dailey, B. P. (1956). *Physical Methods in Chemical Analysis,* (W. G. Berl, editor), Vol. III, 281 (New York: Academic Press)
4. Gordy, W. (1956). *Technique of Organic Chemistry Vol. IX* (W. West, editor), 71 (New York: Interscience)
5. Sheridan, J. (1957). *Proceedings of Conference on Automatic Measurement of Quality in Process Plants,* 185 (London: Butterworths)
6. Lide, D. R. (1966). *Advan. Anal. Chem. Instrum.*, **5**, 235
7. Millen, D. J. (1968). *Chem. in Britain,* **4**, 202
8. Gordy, W., Smith, W. V. and Trambarulo, R. F. (1953). *Microwave Spectroscopy* (New York: Wiley)
9. Townes, C. H. and Schawlow, A. L. (1955). *Microwave Spectroscopy* (New York: McGraw-Hill)
10. Sugden, T. M. and Kenney, C. N. (1965). *Microwave Spectroscopy of Gases* (London: Van Nostrand)
11. Wollrab, J. E. (1967). *Rotational Spectra and Molecular Structure* (New York: Academic Press)
12. Gordy, W. and Cook, R. L. (1970). *Microwave Molecular Spectra* Vol. 9 in *Technique of Organic Chemistry (A. Weissberger, editor) 2nd Edition* (New York: Interscience)
13. Hewlett Packard Company, Palo Alto, California (1964–72). 'Application Notes— Molecules and Microwaves'; *Hewlett Packard Journal,* June 1971, 2
14. Cuthbert, J., Denney, E. J., Silk, C., Stratford, R., Farren, J., Jones, T. L., Pooley, D., Webster, R. K. and Wells, F. H. (1971). *The Design of an Analytical Microwave Spectrometer* (Cambridge Scientific Instruments Ltd: Cambridge, U.K.)
15. Lide, D. R. (1965). *J. Chem. Phys.*, **42**, 1013
16. Harrington, H. W. (1971). Personal Communication
17. National Bureau of Standards, Washington, D.C., U.S.A. (1954–72). *Microwave Spectral Tables, Monograph No. 70,* Vols. I–V
18. Jones, G. E. and Beers, E. T. (1971). *Anal. Chem.*, **43**, 656
19. Volpicelli, R. J., Stiefvater, O. L. and Flynn, G. W. (1967). NASA Contract Report, *Chem. Abst.* 1968, **69**, 243879; *Chem. Eng. News,* 6th Feb., 1967, 70
20. Loomis, C. C. and Strandberg, M. W. P. (1951). *Phys. Rev.*, **81**, 798
21. Trambarulo, R., Ghosh, S. N., Burrus, C. A. and Gordy, W. (1953). *J. Chem. Phys.*, **21**, 851
22. Hughes, R. H. (1956). *J. Chem. Phys.*, **24**, 131
23. Sanders, T. M., Schawlow, A. L., Dousmanis, G. C. and Townes, C. H. (1954). *J. Chem. Phys.*, **22**, 245
24. Richardson, W. S. and Wilson, E. B. (1950). *J. Chem. Phys.*, **18**, 694
25. Amble, E. and Dailey, B. P. (1950). *J. Chem. Phys.*, **18**, 1422
26. Ghosh, S. N., Trambarulo, R. F. and Gordy, W. (1952). *J. Chem. Phys.*, **20**, 605
27. e.g. Willard, H. H., Merritt, L. L. and Dean, J. A. (1965). *Instrumental Methods of Analysis* (New York: Van Nostrand); Bauman, R. P. (1962). *Absorption Spectroscopy* (New York: Wiley)
28. e.g. Hugus, Z. Z. and El-Awady, A. A. (1971). *J. Phys. Chem.*, **75**, 2954
29. Verdier, P. H. and Wilson, E. B. (1958). *J. Chem. Phys.*, **29**, 340
30. Esbitt, A. S. and Wilson, E. B. (1963). *Rev. Sci. Instrum.*, **34**, 901
31. Stiefvater, O. L. and Wilson, E. B. (1969). *J. Chem. Phys.*, **50**, 5385
32. Dynamus, A. (1959). *Physica,* **25**, 859
33. Dynamus, A., Dijkerman, H. A. and Zijderveld (1960). *J. Chem. Phys.*, **32**, 717
34. Southern, A. L., Morgan, H. W., Keilholtz, G. W. and Smith, W. V. (1951). *Anal. Chem.*, **23**, 1,000
35. Kim, H. and Gwinn, W. D. (1964). *Tetrahedron Letters,* **37**, 2535
36. Hirota, K. and Hironaka, Y. (1966). *Bull. Chem. Soc. Japan,* **39**, 2638
37. Sakurai, Y., Onishi, T. and Tamaru, K. (1971). *Trans. Faraday Soc.*, **67**, 3094
38. Sakurai, Y., Kaneda, Y., Kondo, S., Hirota, E., Onishi, T. and Tamaru, K. (1971). *Trans. Faraday Soc.*, **67**, 3275
39. Saito, S. (1969). *Bull. Chem. Soc. Japan,* **42**, 667

40. Harrington, H. W. (1967). *J. Chem. Phys.,* **46,** 3698
41. Harrington, H. W. (1968). *J. Chem. Phys.,* **49,** 3023
42. Crable, G. F. and Wahr, J. C. (1969). *J. Chem. Phys.,* **51,** 5181
43. Quade, C. R. (1970). *J. Chem. Phys.,* **52,** 1588
44. Curl, R. F. (1969). *J. Molec. Spectrosc.,* **29,** 375
45. Funkhouser, J. T., Armstrong, S. and Harrington, H. W. (1968). *Anal. Chem.,* **40,** 11, 22A
46. Kaiser, H. (1970). *Anal. Chem.,* **42,** 24A
47. Hinkley, E. D. (1971). International I.E.E.E. convention, New York; Second High Resolution Spectroscopy Conference, Dijon, France
48. Scharpen, L. H. and Laurie, V. W. (1972). *Anal. Chem.,* **44,** 378R
49. Scharpen, L. H., Rauskolb, R. F. and Tolman, C. A. (1972). *Anal. Chem.,* in the press
50. Scharpen, L. H. (1972). 'Axial–Equitorial Energy Difference in Cyclohexyl Fluoride from Rotational Transition Intensity Measurements', in preparation

9
Liquid–Liquid Distribution (Solvent Extraction)

JIŘÍ STARÝ
Technical University of Prague

9.1 INTRODUCTION

Liquid–liquid distribution LLD (a term preferable to 'solvent extraction' because of its generality and freedom from the restrictions implied by the latter term) is one of the most important and versatile of all analytical techniques because of its simplicity, speed and wide scope. This technique can be used for the purpose of preparation, purification, enrichment and analysis, on all scales of working, from micro-analysis to production processes. LLD is often very selective and the isolation and concentration of the species involved into a small' volume of non-aqueous phase can substantially increase the sensitivity of its determination.

LLD has also found increasing use in the study of kinetics as well as thermodynamic aspects of chemical reactions, particularly metal-complex formation, in both aqueous and non-aqueous phases.

The importance of LLD can be judged from the steadily increasing number of papers covering both theory and application. A complete bibliography of inorganic compounds[1] contains 4509 references for the period 1945–1962, whereas 4466 papers were published in the following 5 year period (1963–1967). The current literature is reviewed by Freiser[2] every 2 years in the journal *Analytical Chemistry*. Among the monographs in this field mention should be made of those by Morrison and Freiser[3], Starý[4], Zolotov[5, 6], Marcus and Kertes[7], De, Khopkar and Chalmers[8] and Shmidt[9]. The *Proceedings of the International Conferences on Solvent Extraction Chemistry* held in Göteborg (1966)[10], Jerusalem and Haifa (1968)[11] and in The Hague (1971) provide valuable records of the latest developments and trends.

9.2 THEORY

9.2.1 General

For the quantitative description of LLD processes and for prediction of the behaviour of untried systems, it is necessary to determine the chemical composition of distributed species, equilibrium constants of chemical reactions occurring in both phases and the corresponding thermodynamic data. Measurement of the dependence of the distribution ratio, D, of species investigated on the pertinent concentration variables is usually sufficient for the determination of all the above-mentioned data for relatively simple systems such as metal chelates. These investigations can be automated using, e.g. AKUFVE[12] which allows the determination of c. 600–800 experimental points in 10 h [13, 14].

The general treatment of LLD of ion-association compounds is usually much more complicated because of the extensive interaction between two phases both with each other and with distributed species. In these cases, in addition to distribution studies, direct measurement of the chemical and physical properties of the organic phase (visible, ultraviolet, infrared and Raman spectra, paramagnetic resonance, viscosity, etc.) at macro-concentrations of the distributed species has to be carried out to ascertain its chemical composition.

9.2.2   Elements and simple covalent compounds

Only a few elements can be extracted into non-polar solvents in the elementary state (S, Se, Br, I, At, noble gases) or as oxides (RuO_4, OsO_4) and halogenides ($GeCl_4$, SnI_4, $AsBr_3$, AsI_3, $SbBr_3$, SbI_3, $SeBr_4$, SeI_4, $TeBr_4$, TeI_4, $HgCl_2$, $HgBr_2$ and HgI_2). In these cases a quantitative treatment of LLD is relatively simple[7, 15].

9.2.3   Metal chelates

The theory of LLD of metal ions M^{n+} using a chelating agent, HA, is fairly well evaluated[3–8, 10, 11]. The species extracted into non-polar solvents are in most cases of the type MA_n. When a chelate is coordinatively unsaturated, it can add on HA molecules to give $MA_n \cdot HA$ or $MA_n \cdot 2HA$. The formation of these adducts is sometimes rather unexpected, for example, lanthanum is extracted by N-benzoyl-N-phenylhydroxylamine in chloroform as $LaA_3 \cdot HA$ [16], whereas other lanthanides extract as MA_3. Similarly, uranium-(VI) is extracted by 2-thenoyltrifluoroacetone as UO_2A_2 [17]. With benzoylacetone or dibenzoylmethane, however, the formation of species such as $UO_2A_2 \cdot HA$, occurs[4].

Extraction into polar solvents is more complicated because of the coordinating effect of the solvent on the neutral chelate (see Section 9.2.4). A study of the LLD of the zinc chelate of 2-thenoyltrifluoroacetone into 36 oxygen-containing solvents has shown that the distributed species is $ZnA_2 \cdot S$, where S can be alcohol, ether, esther or ketone[18].

When the reagent is present in the organic phase predominantly as a dimer $(HA)_2$ (e.g. dialkylphosphoric acids), the distributed species is in most cases $M(HA_2)_n$[19, 20]. However, other species have also been determined ($LiA \cdot 3HA$, $CsA \cdot 5HA$ [21], $SrA_2 \cdot 3HA$, $SrA_2 \cdot 4HA$, $SrA_2 \cdot 5HA$ [22], $ZrA_4 \cdot 2HA$ [23], $UA_2(HA_2)_2$[24], etc.) depending on the reagent concentration and solvent used.

Formally, the same composition extends to the extractable metal salts of carboxylic acids, for example, lead can be extracted by capric acid as $PbA_2 \cdot 2HA$ or $PbA_2 \cdot 4HA$ [25]; however, the theoretical treatment of LLD is complicated by the dimerisation of the carboxylic acids and aggregation of their metal salts (e.g. $[AlA_2(OH)]_6$ [26]).

The extraction of mixed complexes of the type $MA_{n-p}X_p$ (where X is $[ClO_4]^-$, Cl^-, NO_3^-, etc.) takes place usually from solutions of high ionic strength (e.g. GaA_2Br, and $GaABr_2$ using 2-thenoyltrifluoroacetone[27] or AuA_2Cl and $AuACl_2$ using dithizone[28]).

The overall equilibrium constant of LLD processes (the extraction constant) can be usually determined from distribution studies at various pH values and reagent concentrations[4, 11]. Other methods include: the investigation of exchange reactions in two-phase equilibria[29] or in the organic phase[30]. Using all the above methods, the best values for the extraction constants of metal dithizonates and diethyldithiocarbamates have been evaluated[29].

The distribution constants of metal chelates $K_D(MA_n)$ can be directly determined only for those values which are lower than 10^3 and higher than 10^{-3} with accuracy. However, even in these cases, considerable errors can be found in published data because hydrolysis and formation of other non-extractable species has been neglected. Very high $K_D(MA_n)$ values can be best determined from the plot log D v. $f(pA)$ for various solvents[14] or from the solubility data of metal chelates in both organic and aqueous phases[31, 32].

It was found for closely related reagents (e.g. alkyl-substituted β-diketones) that the distribution constants for both the reagent $K_D(HA)$ and the metal chelate $K_D(MA_n)$ were linear functions of the number of carbon atoms in the molecule[33].

The influence of solvents has been investigated in detail. Good correlation was found between $K_D(HA)$ and the 'solubility parameter' of the solvent used and the relation log $K_D(MA_n) = n$ log $K_D(HA)+$ const. has been confirmed[14, 34].

It follows from theory[4] as well as from published data that the separation factor of two metals in question does not change to a great extent when various chelating agents of the same type (e.g. alkyl-substituted β-diketones[35]) are used. The search for new derivatives of known chelating agents is important only for agents used in the separation of chemically very similar metals such as lanthanides or actinides. Recently, it was found that 1-methylheptylphenylphosphonic acid is twice as effective for this purpose as the well known di-(2-ethylhexyl)phosphoric or 2-ethylhexylphosphonic acids[36].

A large number of equilibrium constants in LLD reactions of acidic organophosphorus reagents was recently compiled[37]; the compilation of these data for other chelating agents is now in progress (see Table 9.1).

9.2.4 Ion-association compounds

The quantitative treatment of LLD of simple ion-association compounds formed by heavy organic cations such as tetraphenylarsonium or tetraphenylphosphonium with oxyanions is relatively simple[7]. It can be treated simply as the distribution of compounds formed by a cation of basic dye T^+ and an anionic metal complex. Recently, a great number of basic dyes (e.g. Methyl

Table 9.1 Decadic logarithms of extraction constants K ($K = [MA_n]_{org}[H^+]^p[M^{n+}]^{-1}[HA]_{org}^{-p}$) of metal chelates with acetylacetone (HAA), benzoylacetone (HBA), dibenzoylmethane (HDBM), 2-thenoyltrifluoroacetone (HTTA), 8-hydroxyquinoline (HOx), N-nitrosophenylhydroxylamine (Cupferron, HCup), N-benzoyl-N-phenylhydroxylamine (HBPHA), 8-mercaptoquinoline (thio-oxine, HTOx), diphenylthiocarbazone (dithizone, H_2Dz) and diethyldithiocarbamic acid (HDDC) (20–25°C, $I = 0.1$).

Metal chelate	HAA C_6H_6	HBA C_6H_6	HDBM C_6H_6	HTTA C_6H_6	HOx CHCl₃	HCup CHCl₃	HBPHA CHCl₃	HTOx CHCl₃	H_2Dz CCl₄	HDDC CCl₄
LiA	N	N	N	−10.16	N	N	N	N	N	N
NaA	N	N	N	−11.16	N	N	N	N	N	N
KA	N	N	N	−11.16	N	N	N	N	N	N
CsA	N	N	N	−10.2	N	N	N	N	N	N
BeA₂	−2.79	−3.88	−3.46	−3.2	−9.62	−1.54	E	N	N	N
MgA₂	P	−16.65	−14.72	−10	−15.13	N		N	N	N
CaA₂	N	−18.28	−18.0	−12	−17.89*	N	N	N	N	N
SrA₂	N	−20.0	−20.9	−12	−19.7†	N	N	N	N	N
BaA₂	N	N	P	−14.4	−20.9†	N	N	N	N	P
AlA₃	−6.48	−7.60	−8.92	−5.25	−5.22	−3.5	−7?	−6.5?	N	
GaA₃	−5.51	−6.34	−5.76		3.72	4.92	E	8.5	−1.3	P
InA₃	−7.2	−9.3	−7.61	−4.34	0.89	2.42	−1.74	−4.4	7.2	12
TlA	N			−5.1	−9.4	P	−7.58	−4.4	−3.8	−0.5
TlA₃	E			E	5	3			N	E
PbA₂	−13.3	−9.61	−9.45	−5.24	−8.04	−1.53	−8.2	−0.5	1.0	8.0
AsA₃	N				N	N			N	E
SbA₃	N				N	E		E	N	E
BiA₃	N	P	P	−1.9	−1.2	5.08	5.3	12.6	10.8	16.8
CuA₂	−3.93	−4.17	−3.80	−0.53	1.77	2.66	−0.66	E	10.4	14.0
AgA	N	−7.81	−8.58		−4.5*	P	P		8.9	11.9
ZnA₂	−11	−10.79	−10.67	−8.0	{ −5.2* / −2.4† }	P	−9.94	3.8	2.6	2.8
CdA₂	N	−14.11	−13.98		−5.29†	P	−12.06		2.0	5.8
HgA₂	P	P	E		P	0.91	P	25.3	26.8	30
ScA₃	−5.83	−5.99	−6.04	−0.3	−13.0	3.34		N	N	N
YA₃	P	−16.95	−6.8	−6.8	−15.66	−4.74		N	N	N
LaA₃	P	−20.6	−19.46	−10.51	−15.66	−6.22	−13.59*	N	N	N

	1	2	3	4	5	6	7	8	9	10
CeA_3	P			−9.43			−13.12	N	N	N
PrA_3	P			−9.0				N	N	N
NdA_3	P			−8.76	−14.7			N	N	N
PmA_3	P			−7.82	−15			N	N	N
SmA_3	P			−7.68		−5.8		N	N	N
EuA_3	−18.9			−7.66				N	N	N
GdA_3	P			−7.57			−12.95	N	N	N
TbA_3	P			−7.51				N	N	N
DyA_3	P			−7.03				N	N	N
HoA_3	P			−7.25	−14			N	N	N
TmA_3				−6.96				N	N	N
YbA_3	P			−6.72				N	N	N
LuA_3	P			−6.77				N	N	N
TiA_4	P	−15.2	−15.2	9.2	2			N	N	N
ZrA_4	P	E		7.9		>10	12.6	N	N	N
HfA_4	P	3?		1.0		10	−0.65	N	N	N
ThA_4	−12.16	−7.68	−6.38	1.0	−7.12 / 1.67 / 4.4*	4.4		N	N	N
VO_2A	P							N	N	N
VOA_2	P	P	P	5.3	9.88	E	1.5	1.7	N	N
Mo_2A_2	P	P	P	−2.6				8.3	N	N
UA_4	−5.3							N	N	N
UO_2A_2	P	−4.68*	−4.12*		−1.60* / −9.32		−3.14	N	P	−4
MnA_2	P	−14.63	−13.71		4.11	9.85		1.3	P	
FeA_3	−1.39	−0.50	−1.93	−6.7			6	6.8		
CoA_2	P	−11.11	−10.78		−2.16† / −2.18 / −0.1*	−3.5	−0.5*	3.9	1.6	2.3
NiA_2	P	−12.12	−11.02	−6.6	15	P	−9	3.4	−0.6	E
PdA_2	P	1.2		5.6	P			N	>26	>26
NpA_4	−4.97			6.85				N	N	N
PuA_4	−1.8						2.95	N	N	N
AmA_3				−7.48				N	N	N

*Complexes of the type MA_xHA are formed

†Complexes of the type $MA_x \cdot 2HA$ are formed; N, no extraction occurs; P, only partial extraction occurs; E, chelate is completely extracted

Violet, Methylene Blue, Brilliant Green, etc.) have found widespread application in analytical chemistry because of their high sensitivity[38]. The composition of the distributed species can be represented as TX (where X is Cl^-, $[ClO_4]^-$, Br^-, I^-, $[ReO_4]^-$), or as TMX_{n+1} (where MX_{n+1} is $[BF_4]^-$, $[NbF_6]^-$, $[TaF_6]^-$, $[GaCl_4]^-$, $[TlCl_4]^-$, $[SnCl_5]^-$(?), $[SbCl_4]^-$, $[SbCl_6]^-$, $[AuCl_4]^-$, $[HgCl_3]^-$, $[InBr_4]^-$, $[CdI_3]^-$, $[HgI_3]^-$ and $[Ag(CN)_2]^-)$[38]. The extraction of species of another type (e.g. T_2MX_{n+2}) has not been proved[38]. Only a few theoretical studies have been published; the investigation of LLD of Crystal Violet showed that a change in entropy is the governing force in the distribution[39] and a strong influence from the nature of solvent used can be expected. The best solvent or mixture of solvents has to be found experimentally to ensure the maximum extraction of TMX_{n+1} and minimum extraction of TX (the maxima of absorption and molar absorptivity of both species are identical or very similar). From the large amount of experimental data available it seems that two types of dyes — triphenylmethane and rhodamine dyes — are sufficient to ensure the distribution of all the species mentioned above. It is not to be expected that some specific dyes could be prepared. The best way to increase the selectivity of isolation lies in a study of various anionic metal complexes or chelates.

LLD using long-chain amines has become increasingly popular in recent years because of its selectivity in the reprocessing of nuclear fuel. However, the quantitative treatment of these systems is very difficult. The main reason is that co-extraction of water and supporting mineral acids leads to the formation of mixed complexes and their aggregation in the organic phase (e.g. the formation of 6-membered cyclic polymers[40]). This factor is strongly dependent on the conditions used. Water does not seem to play any role in the aggregation using, e.g. trilaurylmethylammonium thiocyanate in carbon tetrachloride. On the other hand, trilaurylmethylammonium chloride in the same solvent was found to be in the form of several polyhydrated species[41].

The study of the system: iron(III)–hydrochloric acid–trilaurylamine (TLA)–benzene shows that the metal:amine ratio varies between 1 and 2, whereas the absorption spectra invariably correspond to the presence of only the $[FeCl_4]^-$ anion[42]. Thus in the organic phase the species $TLA \cdot HFeCl_4$ and $TLA \cdot HCl \cdot TLA \cdot HFeCl_4$ are present in addition to the various oligomers of both $TLA \cdot HCl$ and $TLA \cdot HFeCl_4$ [42]. An analogous study with zinc also proved the formation of polymers in the organic phase[43].

It is necessary to mention that the simple reaction suggested on the basis of slope analysis and the corresponding equilibrium constants (compilation see[37]) is only an expression of experimental dependence and does not reflect the complicated reactions that occur in the organic phase[44] where even the aggregated amine salt may be mono-ionised[7].

Although several empirical parameters have been suggested to explain the extraction efficiency in terms of the type or structure of the amine[45–47] and the solvent[48], our ability to predict behaviour of untried system is only improving at a slow rate.

The selectivity of the separation is in general determined by the tendency of metals to form anionic complexes irrespective of the class of the amine. At present, most attention has been given to chloride, nitrate and sulphate complexes[7]. It is to be expected that the utilisation of other types of anionic complex, such as cyanides[49], acetates[50], citrates[51], EDTA[52] etc. will offer new possibilities from the theoretical as well as practical point of view.

The LLD of metal halides and halometallic complexes using oxygenated solvents, such as ethers, ketones and alcohols, is also very complicated (formation of extractable species of the type $H(H_2O)_x S_y BiX_4(H_2O)_2$, $H_2(H_2O)_x S_y BiX_5(H_2O)$, $H_3(H_2O)_x S_y BiX_6$ and $H(H_2O)_x S_y PbX_3(H_2O)$, where X is Cl^-, Br^- or I^-)[53]. These solvents are at the present time replaced by neutral organophosphorus reagents such as tri-n-butyl phosphate (TBP), tri-n-butyl phosphine oxide (TBPO) and tri-n-octyl phosphine oxide (TOPO)[7]. Other neutral reagents are similar in their LLD behaviour and the type of complexes they form. Nitrate, chloride and sulphate complexes have received most attention whereas other complexes have not been investigated in detail. Nitrates are usually extracted as solvates, e.g. $Zr(NO_3)_4 \cdot 2TBP$[54] or $Np(NO_3)_4 \cdot 2TBP$[55], chlorides and sulphates can be extracted at higher acidities also by the oxonium mechanism (e.g. $H_2(TBPO)_y U(SO_4)_3$)[56]. Quantitative treatment is complicated by the mutual interaction of water, mineral acid and reagent[57-59].

Recently, organic sulphides or sulphoxides have been suggested as extractants[60, 61] and their characteristics were compared with TBP and TOPO[62]. The extracted species are of the type $UO_2(NO_3)_2 \cdot 2R_2SO$[63] or $H(R_2SO)_y FeCl_4$[64].

9.2.5 Synergic systems

Synergic enhancement of the distribution of metals using certain combinations of two extractants can be described as a change in extractive power when in the presence of each other and/or by a change of the composition of the distributed species. The highest enhancement is observed on using hydroxy-group-containing chelating agents (e.g. 2-thenoyltrifluoroacetone[65, 66], 1-phenyl-3-methyl-4-benzoyl-5-pyrazolone[67], salicylic acid[68]) and neutral organophosphorus reagents (tri-n-butylphosphate, tri-n-octylphosphine oxide). In these cases, an extractable adduct MA_nS_y is formed and the distribution constants (see Table 9.2) can easily be determined by slope-analysis. It has been found[69] that the extraction constants of zinc and copper benzoyltrifluoroacetonates increase in the presence of donor-active substances in the following order: hexyl alcohol, tri-n-butylphosphate, tri-n-octylphosphine oxide, acridine, quinoline, lepidine and isoquinoline.

Using the SH group-containing chelating agents the synergic effect is usually absent[70]. On the other hand, manganese(II) dithizonate was found to form an adduct with pyridine of the type MnA_2S_2[71].

When two chelating agents, HA and HA', are used, the formation of a mixed chelate $MA_nA'_{m-n}$ or an adduct $MA_n \cdot HA'$ may occur[72]. The formation of adducts usually leads to greater synergic enhancement than those observed with mixed chelates[73, 74].

9.2.6 Heteropoly acids

Molybdophosphoric acid was found to be distributed into butyl alcohol (BuOH) as a solvate $H_3PMO_{12}O_{40} \cdot 6\,BuOH \cdot xH_2O$[75]. A study of the LLD of

12-heteropoly acids $H_3PMo_{12}O_{40}$, $H_3PW_{12}O_{40}$ and $H_3SiMo_{12}O_{40}$ into tri-n-butylphosphate showed the presence of ion-association compounds of the type $H^+3TBP \cdot H_2PMo_{12}O_{40}^-$ [76, 77].

Table 9.2 Decadic logarithms of the extraction constants K ($K = [MA_nS_y]_{org} \cdot [H^+]^n \cdot [M^{n+}]^{-1}[HA]_{org}^{-n}[S]_{org}^{-y}$) of the adducts of metal 2-thenoyltrifluoroacetonate and tri-n-butylphosphate (TBP) or tri-n-octylphosphine oxide (TOPO) (20–25 °C, $I = 0.1$–1.0)[10, 11].

Species extracted	TBP			TOPO		
	CCl_4	C_6H_6	C_6H_{12}	CCl_4	C_6H_6	C_6H_{12}
$LiAS_2$		−4.20			−2.20	−5.43
$NaAS_2$		−6.90			−5.0	
KAS_2		−8.16			−6.36	
$CsAS_2$		−8.42			−6.90	
CaA_2S	−9.29			−7.76		
CaA_2S_2	−5.18	−5.27	−3.27	−2.72	−3.27	−1.99
SrA_2S	−11.27			−9.64		
SrA_2S_2	−7.51			−5.25		
BaA_2S				−9.3		
BaA_2S_2				−5		
CuA_2S	1.19		0.48	2.20		1.14
ZnA_2S	−4.26	−4.15	−3.56			
PmA_3S_2		−1.04			2.43	5.00
EuA_3S	−2.70			−0.67		
EuA_3S_2	0.3			4.10		8.1
TmA_3S			−0.42		−0.01	1.24
TmA_3S_2		−0.34	2.62			5.72
AcA_3S	−8.7					
AcA_3S_2	−4.0					
ThA_4S	6.4	5.7	7.95		7.70	10.58
UO_2A_2S		2.48	3.20		3.1	3.7
$UO_2A_2S_3$					12.9	
CoA_2S			−3.78			
CoA_2S_2			−1.11			
$NpA_3(NO_3)S$			7.75			
$NpO_2A(HA)S$			5.3			
PuA_3S_2			5.30			
$PuA_3(NO_3)S$			8.67			
PuO_2A_2S			3.13			
AmA_3S	−3.82					
AmA_3S_2	0.0	−0.96	2.40		2.51	5.14
CmA_3S_2		−0.70	2.50		2.83	5.44

9.2.7 Molten-phase systems

Recently the LLD of the lanthanides and actinides has been studied using molten $Mn(NO_3)_2 \cdot 6H_2O$ and TBP in dodecane[78]. The distribution of promethium, curium and californium between molten lithium chloride and liquid lithium–bismuth indicates that californium(II) is present in the salt phase (640 °C) [79]. When molten lithium fluoride–beryllium fluoride was used, the valencies of the lanthanides and actinides were as expected with the possible exception of samarium(II) and americium(II) [80].

9.3 APPLICATION

9.3.1 General

From the previous section it follows that the search for new types of chelating agent (for metal chelate systems), or new types of metal complex (for ion-association systems) may be a better way to improve selective separation procedures. Many new reagents have been proposed for this purpose. However, only in a few cases has a systematic study of their distribution characteristics been carried out. Thus, selenazone – the selenium analogue of dithizone – was found to form brilliantly coloured chelates with lead, bismuth, copper, silver, zinc, cadmium, mercury, manganese, cobalt and nickel, extractable into organic solvents. The diselenide is more selective, for it reacts only with lead, silver, mercury and nickel[81]. The extraction of 20 metal cations with a series of aliphatic carboxylic acids (C_3–C_{12}) in chloroform was examined over the pH range 2–10. With lower members of the series few metals are extracted, but the number of metals increases with increase of the number of carbon atoms in the chain[82]. A systematic study of the LLD of many metals has been carried out using di-isoamylmethyl phosphonate in chloroform[83], sulphoxides[60], etc.[4, 7, 8].

Recently, about 20 new reagents have been suggested for isolation of copper and palladium. Unfortunately, from data published it is not evident that these reagents are better in their selectivity and sensitivity than those which are well known[4, 7, 8]. On the other hand, only a few papers critically investigate published methods in order to find the best procedures for the isolation of individual elements.

Selectivity of separation can be increased in many cases relatively simply using new masking agents or a mixture which can be chosen on the basis of stability constant data. For example a systematic study of the LLD of 44 metals by 1 % diethylammonium diethyldithiocarbamate in chloroform and their re-distribution into acid, base, sulphide, fluoride or EDTA solutions shows new possibilities for increasing the selectivity of this reagent[84]. In an analogous study, 0.1 M N-benzoyl-N-phenylhydroxylamine in chloroform has been used[85].

A preliminary isolation of the element in question or of the interfering elements can often be used to make known separation procedures almost specific. There are also many possibilities unexplored in the utilisation of synergic systems or organic-phase reactions.

For the separation of chemically very similar elements the Craig method or extraction chromatography can be used advantageously. When chelating agents are applied, it is even possible to calculate the optimum conditions for the separation from known distribution constants[86].

Lastly, applications in radioanalytical methods should be mentioned. At present very sensitive and selective LLD procedures have been developed for the substoicheiometric determination of 33 elements using isotope dilution analysis and/or activation analysis[87]. Isotope exchange reactions in the organic extract can be used for highly selective determinations of many elements[88, 89].

In the following section the best of the recently developed procedures are reviewed and compared with older methods[4, 8]. When the complex extracted is suitable for absorption spectrophotometry the value of the molarabsorptivity ε (l mol^{-1} cm^{-1}) is given.

9.3.2 Group I (main group)

9.3.2.1 Potassium

Potassium can be quantitatively distributed from alkaline solution into chloroform in the presence of dibenzo-18-crown-6 (2,3,11,12-dibenzo-1,4,7,10,13,16-hexaoxycyclo-octadeca-2,11-diene) and dipicrylaminate or tetraphenyl borate anions[90]. Lithium, sodium and multivalent elements remain in the aqueous phase, whereas caesium is also transferred into the organic phase.

9.3.2.2 Rubidium

Rubidium and caesium can be completely extracted from acidic media into nitrobenzene in the presence of the complex acid H[As(catechol)$_3$] prepared by mixing arsenic acid with catechol in a mole ratio of 1:3 [91].

9.3.2.3 Caesium

The quantitative extraction of caesium with dibenzo-18-crown-6 [90], arsenic acid–catechol complex[91] and alkyl or aryl phenols[92, 93] has been reported. Using 2-(α-methylbenzyl)-4-chlorophenol in kerosene, the distribution ratio decreased in the order: caesium (4.4), rubidium (0.21) and potassium (0.015) (pH.13)[93].

9.3.3 Group II (main group)

9.3.3.1 Beryllium

The acetylacetone method remains the most selective method for the isolation of this element (EDTA used as masking agent)[4, 8].

9.3.3.2 Magnesium

Magnesium forms a complex with Eriochrome Black T that can be extracted at pH 11–12 into tetradecyldimethylbenzylammonium chloride in 1,2-dichloroethane[94]. By measuring the difference in absorbance at 690 nm between the blank and test solutions, the highest sensitivity can be reached[94]. Even a tenfold excess of calcium does not interfere in the determination.

9.3.3.3 Calcium

The extraction of calcium as a complex with 5,7-dibromo-8-hydroxyquinoline and rhodamine S into benzene can be used for the sensitive (ε_{550nm} = 3.6×10^4), but not selective (magnesium and strontium interfere) determination of this element[95]. Calcium can also be distributed from 0.05 M NaOH by 0.2% solutions of Azo-azoxy BN into a 20% solution of TBP in carbon tetrachloride[96]. By repeating the extraction procedure (10–15% of strontium is extracted in these conditions) calcium can be separated from strontium and barium. The preliminary extraction of calcium from 2.4 M NaSCN by TBP (EDTA used a masking agent) increases the selectivity of the determination[97].

9.3.3.4 Strontium

Quantitative distribution of strontium from an aqueous solution using 0.2% Azo-azoxy BN in a 20% TBP solution in carbon tetrachloride occurs at a higher pH (1–2 M NaOH) than for calcium[96].

9.3.3.5 Barium

Barium can be extracted from alkaline media by a 1 M solution of hexafluoro-acetylacetone in isoamyl acetate[98].

9.3.4 Group III (main group)

9.3.4.1 Boron

The extraction of the ion-association compound of tetrafluoroborate and methylene blue into 1,2-dichloroethane has been applied in steel analysis[99]. According to Blyum and Oparina[38] the extraction of the borosalicylate complex is more selective.

9.3.4.2 Aluminium

The only selective method for the isolation of aluminium is based on its extraction as 8-hydroxyquinolinate[4, 8] in the presence of CN^- and o-phenanthroline as masking agents[100]. Iron(III) has to be removed by preliminary extraction with amyl acetate.

9.3.4.3 Gallium

The selectivity of the 8-hydroxyquinoline method[4] (in the presence of potassium thiocyanate, ascorbic and citric acids, pH = 2.5)[101] can be greatly increased by a preliminary extraction of gallium from 6 M HCl by isopropyl-ether, in the presence of titanium trichloride to reduce iron(III) which would otherwise interfere[101].

Recently, several sensitive methods were developed based on the extraction of tetrachlorogallate with Basic Blue K, Xylenol Orange, Acridine Orange, Methylene Blue, Victoria Blue R etc. The LLD of gallium from 1.5–2.5 M H_2SO_4 which is 3.5 M in NaCl in the presence of Victoria Blue B into benzene–acetone mixture (10:1) is perhaps the most sensitive absorptiometric procedure for its detemination ($\varepsilon_{610nm} = 1.1 \times 10^5$)[102]. In the presence of titanium trichloride and thiourea only tungsten(VI), iodide, thiocyanate and perchlorate anions interfere.

9.3.4.4 Indium

Indium can be distributed from 2.25 M H_2SO_4 which is 2 M in KBr in the presence of Rhodamine B into a mixture of benzene–di-isopropyl ether–acetylacetone (6:3:1). This method is more sensitive ($\varepsilon_{557nm} = 1.1 \times 10^5$)[103] than analogous methods using methylene blue or Pyronine G. Zinc, cadmium, copper, iron(II), aluminium, gallium and thallium(I) (iron(III) and thallium(III) can be reduced by ascorbic acid) do not interfere in the determination, whereas tin(II) and tin(IV) do.

9.3.4.5 Thallium

Many basic dyes (Methyl Violet, Brilliant Green, Pyrazolone Green, Victoria Blue B, Methylene Blue, Toluidine Blue, etc.) has been proposed for the extraction of the ion-association complexes of tetrachloro or tetrabromo thallium(III). The most sensitive method is that based on the extraction of thallium(III) from 0.2–0.5 M HCl in the presence of Toluidine Blue into a 1:1 mixture of 1,2-dichloroethane and trichloroethylene[104] ($\varepsilon_{660nm} = 1.17 \times 10^5$). Only gold(III), antimony(V) and large amounts of mercury(II) interfere.

In the determination of thallium by activation analysis, thallium(I) is distributed into dithizone in chloroform from KCN–NaOH solution followed by its displacement from the organic extract using a sub-stoicheiometric amount of mercury(II)[105].

The extraction of thallium(I) diethyldithiocarbamate from an alkaline solution of ammonium tartrate and potassium cyanide into carbon tetrachloride, followed by an isotope exchange reaction with $^{204}Tl^{III}$ diethyldithiocarbamate in the organic extract furnishes one of the most selective and sensitive methods available for the determination of this element[89]. 100-Fold excesses of 36 metals reacting with diethyldithiocarbamic acid do not interfere in the determination.

9.3.5 Group IV (main group)

9.3.5.1 Silicon

Silicic acid can be extracted into a 1:5 mixture of n-amyl alcohol and diethyl ether from 0.4 M $HClO_4$ which is 0.01 M in molybdate[106]. Phosphorus and arsenic interfere.

9.3.5.2 Germanium

Germanium can be extracted into $CHCl_3$ as a complex with 6,7-dihydroxy-2, 4-diphenylbenzopyranol in the presence of perchlorate ($\varepsilon_{530nm} = 1.2 \times 10^4$)[107] or with 9-phenyl-2,3,7-trihydroxy-6-fluorone in the presence of bromide and antipyrine (ε_{515nm} 1.14×10^5)[108]. The preliminary extraction of germanium from 9 MHCl into chloroform or carbon tetrachloride followed by its back extraction into water considerably increases the selectivity of the above methods[8].

9.3.5.3 Tin

The extraction of tin(IV) into benzene from M $HClO_4$ which is 6.5 M in NaI is perhaps the most selective method for the isolation of this element[109, 110]. At higher iodide concentrations only antimony(III) and germanium(IV) are extracted into the organic phase.

9.3.5.4 Lead

Dithizone extraction of lead from slightly ammoniacal cyanide solution[4, 8] remains the most selective and sensitive method for the isolation and absorptiometric determination of this element.

9.3.6 Group V (main group)
9.3.6.1 Phosphorus

Propyl and butyl acetates were found to be the most effective and selective solvents for the extraction of molybdophosphoric acid (pH 0.2–0.5)[111, 112]. Silicic acid and arsenic are not extracted in these conditions. For the isolation of phosphorus from zirconium, niobium, titanium and tungsten, the extraction of molybdovanadatophosphoric acid into isobutylmethylketone has been recommended[113]. Fluoride and citrate can be used as suitable masking agents.

Phosphorus can be very sensitively determined in saturated silicate solution by the extraction of molybdophosphoric acid into butyl or propyl acetates in the presence of crystal violet[114]. A ten-fold excess of arsenic does not interfere in the determination.

9.3.6.2 Arsenic

Trivalent arsenic may be rather selectively extracted from 3.5 M H_2SO_4 made 0.8 M in KI into carbon tetrachloride[115]. Under these conditions only tin(IV) and germanium(IV) are also extracted.

Two radioanalytical methods have been evaluated for the determination of traces of arsenic: extraction with a substoicheiometric amount of zinc diethyldithiocarbamate[116] or extraction of arsenic(III) diethyldithiocarbamate followed by an isotope exchange reaction with labelled arsenic(III) iodide in carbon tetrachloride[117].

9.3.6.3 Antimony

Antimony(III) can be extracted from 2 M HCl made M in $MgCl_2$ into a 20%

solution of TBP in toluene. By shaking the organic extract with aqueous 0.05% Brilliant Green solution, antimony can be determined by measuring the absorbance of the complex formed directly in the organic phase at 640 nm ($\varepsilon = 7.3 \times 10^4$). Thallium(I), gold(III), chromium(VI) and bromide strongly interfere in the determination[118].

Hexachloro-complexes of antimony(V) can be extracted as ion-association compounds with various basic dyes: Basic Blue K ($\varepsilon_{638nm} = 6.58 \times 10^4$ in toluene; thallium(I) interferes)[119], Butyl rhodamine S ($\varepsilon_{560nm} = 1.18 \times 10^5$ in carbon tetrachloride–cyclohexanone mixture; gallium interferes)[120] and especially 1,4-dimethyl-1,2,4-triazolinium-(3-azo-4)-N,N-diethylaniline ($\varepsilon_{540nm} = 6.7 \times 10^4$ in chlorobenzene)[121] are most selective.

9.3.6.4 Bismuth

Bismuth can be extracted from 0.05–0.15 M sulphuric acid in the presence of potassium iodide and methylene green into a 1:1 mixture of benzene and nitrobenzene ($\varepsilon_{650nm} = 4.3 \times 10^4$)[122]. This method is more selective (only equal amounts of mercury(II), gold(III) and tellurium(IV) interfere) than the extraction of bismuth iodide complexes into oxygen-containing solvents.

A highly selective and sensitive (up to 0.01 µg/10 cm^3) radioanalytical method is based on the homogeneous isotope exchange reaction between bismuth diethyldithiocarbamate extracted from a KCN–EDTA mixture and a standard bismuth-210 iodide complex in the organic extract[88]. Amongst all the metals extractable as diethyldithiocarbamates only antimony(III) (its interference can be eliminated by preliminary oxidation with potassium permanganate) and greater amounts of thallium(I) interfere.

9.3.7 Group VI (main group)

9.3.7.1 Selenium

3,3'-Diaminobenzidine forms with selenium(IV) at pH 1.5–3 a yellow complex which can be extracted at pH 6–7 into benzene, toluene or xylene. EDTA and fluoride can be used as masking agents, however the formation of the complex is strongly influenced by the conditions used (temperature of heating, concentration of masking agents and salts, etc.)[123].

Recently, an interesting new method has been proposed based on the extraction of selenium(IV) into a 0.03 M 2-thenoyltrifluoroacetone solution in xylene (pH 0.5–4.5). Selenium can be determined directly in the extract after the addition of 3,3'-diaminobenzidine[124]. Similarly, selenium(IV) can be determined in the organic extract after its extraction from a 4 M HCl–2 M MgCl$_2$ mixture into a 60% TBP solution in toluene[125].

1,2-Diaminobenzene[126] and especially 2,3-diaminonaphthalene[127] seem to be more selective reagents for selenium, because the extraction of selenium occurs at a lower pH than with 3,3'-diaminobenzidine.

The extraction of selenium(IV) from 6 M HCl into a 5×10^{-5} M dithizone solution in carbon tetrachloride is probably the most sensitive absorptiometric method for the determination of the element ($\varepsilon_{420nm} = 7-8 \times 10^4$)[128, 129]. All interfering metals can be removed by a preliminary dithizone extraction at a higher pH.

9.3.7.2 Tellurium

The selectivity of the diethyldithiocarbamate method[4] can be increased by using triethanolamine as a masking agent[123]. The tellurium(IV) complex with Bismutiol II can be extracted into chloroform ($\varepsilon_{330nm} = 3.5 \times 10^4$)[130]. The selectivity of the method can be increased by a preliminary extraction with isobutyl methyl ketone from hydrochloric acid. Under these conditions, tellurium(VI) remains in the acid phase, whereas tellurium(IV) is quantitatively extracted.

A very sensitive method for the determination of tellurium(IV) is based on its extraction from 5 M H_2SO_4 made 2 M in KBr in the presence of Victoria Blue R into a 2:1 mixture of benzene and nitrobenzene ($\varepsilon_{602nm} = 8 \times 10^4$)[131]. Indium and mercury(II) interfere in the determination.

9.3.7.3 Polonium

A dithizone method[4, 8] can be used for the separation of polonium (RaF) from lead (RaD) and bismuth (RaE)[133]. These three elements can be easily separated using the same reagent on an extraction chromatography column[86].

9.3.8 Group VII (main group)

9.3.8.1 Fluorine

Fluoride can be sensitively determined by measuring absorbance at 570 nm after the extraction of the mixed cerium(III)–fluoride–Alizarin Complexan complex into a 5% triethylamine solution in n-amyl alcohol[132].

9.3.8.2 Chlorine

Chloride ion may be extracted at pH 5.1–6.1 into chloroform as the chloro-pyridine mercury(II) complex which may then be reacted with dithizone to give mercury(II) dithizonate. This reaction has been applied to the determination of traces of chloride ion[134]. Ions that usually accompany chloride do not interefere in the determination except bromide, iodide, thiocyanate and cyanide.

A sensitive method for the determination of perchlorate ion is based on its extraction from a citrate buffer (pH 1.9) as an ion-association compound with Neutral Red into nitrobenzene ($\varepsilon_{552nm} = 9.39 \times 10^4$)[135]. Iodide, nitrate and thiocyanate interfere.

A very sensitive radio-analytical method involves extraction of the perchlorate ion as an ion-association compound with zinc-65–*ortho*phenanthroline complex into nitrobenzene[136].

9.3.8.3 Bromine

Bromide and also iodide, ions can be rather selectively extracted as halogenosulphinates ($BrSO_2^-$ or ISO_2^-) into 1,2-dichloroethane using triphenylbenzylphosphonium ions and sulphite ($\varepsilon_{359nm} = 3.29 \times 10^3$ and $\varepsilon_{321.5nm} = 2.35 \times 10^4$ for bromide ions; $\varepsilon_{381nm} = 1.35 \times 10^4$ for iodide ions)[137].

9.3.8.4 Iodine

Iodide ion forms, at pH 3, an ion-association compound with the Neutral Red cation extractable into nitrobenzene ($\varepsilon_{552nm} = 3.25 \times 10^4$)[138]. Chloride, bromide and cyanide ions do not interfere, whereas perchlorate, iodate, nitrate and thiocyanate do.

Submicrogram amounts of iodide ions can also be determined by extractive titration using mercury-203[139].

9.3.9 Group I (transition)

9.3.9.1 Copper

Although about 20 new procedures have been suggested for the extraction of copper, the dithizone[4], diethyldithiocarbamate[4] and *ortho*phenanthroline[8] methods seem to be amongst the most selective and sensitive for the determination of this metal.

9.3.9.2 Silver

Only a few new methods have been proposed for the extractive isolation of silver, however, their selectivity is substantially lower than those of the dithizone[4] or *ortho*phenanthroline methods[8].

9.3.9.3 Gold

Gold can be separated from large amounts of rhodium(III), palladium(II), iridium(IV) and platinum(IV) by extraction with 4-dimethylaminophenyl-4-methylbenzylaminophenyl-antipyrinylcarbinol (Chrompyrazole I) ($\varepsilon_{580nm} = 6 \times 10^4$ in toluene, pH 0.5–2)[140] or with pyridine-*N*-oxide in methylene chloride[141].

Activation analysis and isotope dilution analysis using extraction with a substoicheiometric amount of these reagents are amongst the most sensitive and selective methods for the determination of gold (see reference 87).

9.3.10 Group II (transition)
9.3.10.1 Zinc

From many new organic reagents proposed for the extraction of zinc from aqueous media, 1-[(5-chloro-2-pyridyl)-azo]-2-naphthol (5-chloro-2-PAN) is perhaps the most sensitive ($\varepsilon_{565nm} = 8.4 \times 10^4$ in diethyl ether)[142]. In the presence of cyanide (pH 8–11) only large amounts of cadmium interfere.

The selectivity of the dithizone method[4] (diethanolamine dithiocarbamate is the best masking agent) can be further increased by preliminary extraction of zinc from 2 M HCl into a mixture of tri-n-octylamine and trichloroethylene[143].

9.3.10.2 Cadmium

The preliminary extraction of cadmium from M HCl into 0.1 M tri-n-octylamine in cyclohexane[144] or from 3 M H_2SO_4 made 0.2 M in KI, into iso-butyl methyl ketone[145] can be used to increase the selectivity of the dithizone method[4].

9.3.10.3 Mercury

The extraction of mercury(II) into 2-thiothenoyltrifluoroacetone in xylene followed by removal of the excess reagent by washing the extract with an alkaline solution, is less sensitive ($\varepsilon_{370nm} = 3 \times 10^4$)[146] but more selective than the dithizone method[4]. Only tin(II), bromide, iodide and thiocyanate anions strongly interfere (the interference of gold has not been studied).

Other new methods are based on the extraction of $HgCl_3^-$ or $HgBr_3^-$ anions with various basic dye cations, e.g. with Crystal Violet, Methylene Green, Methylene Blue, Variamine Blue B, Brilliant Green and Bindschedler's Green. The most sensitive reagent for the isolation of mercury(II) (as $HgBr_3^-$) is perhaps Bindschedler's Green ($\varepsilon_{740nm} = 1.7 \times 10^5$ in 1,2-dichloroethane, citrate buffer pH $= 2$)[147]. Tin(II),(IV), iodide, cyanide and thiocyanate interfere[147].

9.3.11 Group III (transition)

9.3.11.1 Scandium

The extraction of scandium in the presence of chlorphosphonazo III and diphenylguanidine (or tetraphenylarsonium chloride) into n-butyl alcohol is a rather sensitive method for its determination[148]. Citric or tartaric acids can be used for masking a tenfold excess of aluminium, zirconium and lanthanides. Thorium strongly interferes however.

9.3.11.2 Lanthanides

The separation of individual lanthanides can only be achieved using the Craig method[149] or extractive chromatography[19].

Several methods have been published for the extractive-photometric determination of individual lanthanides. The extraction as a mixed complex of 5,7-dibromo-8-hydroxyquinoline and Rhodamine B into benzene is the most sensitive ($\varepsilon_{540nm} = 3.7 \times 10^4$, 4.2×10^4, 6.3×10^4 and 8.7×10^4 for lanthanum, neodymium, erbium and lutetium, respectively)[150].

Cerium(III) can be separated from other fission products utilising the high synergic enhancement of its extraction by a mixture of 2-thenoyltrifluoroacetone and tri-n-octylphosphine oxide[151]. Uranium and zirconium are also extracted, however and they remain in the extract, whereas cerium may be selectively back-extracted into a mineral acid solution.

A highly sensitive and selective method for the determination (but not for the isolation) of cerium(III) is based on its extraction by a 5% solution of di-(2-ethylhexyl)phosphoric acid in toluene (pH ≈ 2.5) followed by an isotope exchange reaction with ^{144}Ce^{IV}–di-(2-ethylhexyl)phosphate in the organic extract[152]. Large amounts of the 36 other metals studied (including other lanthanides, scandium, zirconium, uranium, etc.) do not interfere.

9.3.12 Group IV (transition)

9.3.12.1 Titanium

The selectivity of the 8-hydroxyquinoline method[4] can be further increased using an excess of hydrogen peroxide as masking agent and by removing the interfering metals from the organic extract with 0.1 M sulphuric acid[153].

The extraction of titanium from 0.5 M HCl into chloroform in the presence of 4,4′-diantipyrinylmethane and stannous chloride is probably the most sensitive method for the determination of this element ($\varepsilon_{390nm} = 6.8 \times 10^4$)[154].

9.3.12.2 Zirconium

A highly selective method for the separation of zirconium is based on its extraction into N-benzoyl-N-phenylhydroxylamine in benzene from an ammonium carbonate solution containing several masking agents[155]. After its back extraction into a HBF$_4$–HCl mixture, zirconium can be determined as its chelate with 8-hydroxyquinoline ($\varepsilon_{390nm} = 1.2 \times 10^4$ in xylene)[155]. Of 60 metals tested, only hafnium interferes.

A more sensitive, but not so selective, method is based on the extraction of zirconium from 7 M nitric acid into 0.05 M TOPO in toluene[156]. After washing the organic extract with 7 M nitric acid, zirconium can be determined directly in the extract by the addition of aniline and Pyrocatechol Violet ($\varepsilon_{655nm} = 5 \times 10^4$)[156].

9.3.12.3 Hafnium

The chemical similarity of zirconium and hafnium makes it difficult to separate them and, in general, any system that will extract one will extract both. The best method for their mutual separation remains the extraction of hafnium into cyclohexanone from a solution that is 3 M in NH$_4$CNS, 2 M in (NH$_4$)$_2$SO$_4$ and 1.5 M in HCl. At 30 °C the distribution ratio D is 15 and 0.14 for hafnium and zirconium, respectively[8].

9.3.12.4 Thorium

The extraction of the mixed thorium–5,7-dibromo-8-hydroxyquinoline–Rhodamine S complex into benzene (pH ≈ 5) can be used for the sensitive determination of thorium in the presence of the lanthanides ($\varepsilon_{551nm} = 8.8 \times 10^4$)[157]. Thorium can be selectively extracted from 1.5–2 M sulphuric acid into a 7% solution of N-benzylaniline in chloroform[158]. In the presence of ascorbic acid and EDTA, only uranium(VI) is extracted simultaneously with thorium.

9.3.13 Group V (transition)

9.3.13.1 Vanadium

8-Hydroxyquinoline[4, 8] and N-benzoyl-N-phenylhydroxylamine[4, 8] have been established as selective reagents for the extractive-photometric determination of vanadium(V). The extraction of vanadium from a mixture of 2 M sulphuric acid–2 M hydrofluoric acid into 0.1% N-benzoyl-N-phenylhydroxylamine in chloroform is rather selective[159]; the only interferences, due to cerium(IV) and chromium(VI), can be eliminated by reduction with iron(II) ($\varepsilon_{475nm} = 4.28 \times 10^3$).

Recently[160], over 150 N-substituted hydroxamic acids have been studied as reagents for vanadium. It was found that N-phenyl-3-styrylacrylohydroxamic acid was the most selective and sensitive reagent for the determination of this element ($\varepsilon_{555nm} = 7.8 \times 10^3$ in chloroform)[160].

9.3.13.2 Niobium

Niobium(V) can be extracted from 5–7 M sulphuric acid as butyl dithiocarbamate into a 3:1 mixture of chloroform and isoamyl alcohol. By measuring the absorbance at 425 nm, even a 4000-fold excess of titanium and zirconium and a 60-fold excess of tantalum did not interfere[161]. Microgram amounts of niobium in vanadium can be determined by the extraction of the niobium–pyrocatechol complex into tribenzylamine in chloroform[162]. For the separation of niobium from titanium, extraction from 8.4–8.6 M hydrochloric acid into 0.1–0.2 M TOA in chloroform can be used[163].

9.3.13.3 Tantalum

The extraction of ion-association compounds of hexafluorotantalate with various basic dyes (e.g. Methylene Blue, Methylene Green, Brilliant Green, Methyl Violet, Toluidine Blue) can be used for sensitive determination of this element. The highest sensitivity was obtained in the extraction of tantalum from 1 M sulphuric acid made 0.5% in hydrofluoric acid in the presence of methylene blue into 1,2-dichloroethane ($\varepsilon_{660nm} = 10^5$)[164]. Only boron, niobium and perchlorate interfere. However, microgram amounts of tantalum can be separated from milligram amounts of niobium by preliminary extraction from 1 M hydrofluoric acid made 2 M in hydrochloric acid using isobutylmethyl ketone as solvent[164].

Triphenylguanidine in nitrobenzene can be used for substoicheiometric determination of tantalum[165].

9.3.13.4 Protactinium

Protactinium can be separated from zirconium, niobium and tantalum by extraction from 6 M hydrochloric acid made 0.3 M in oxalic acid, using 0.05 M tetraphenylarsonium chloride in chloroform followed by re-distribution into 5 M hydrofluoric acid[166].

Another method involves the extraction of protactinium from 2.5 M sulphuric acid made 0.005 M in hydrofluoric acid into 0.1 M 1-phenyl-3-methyl-4-benzoyl-5-pyrazolone in benzene followed by its back extraction into 0.1 M hydrofluoric acid[167].

9.3.14   Group VI (transition)

9.3.14.1   Chromium

The acetylacetone method[4] seems to be a more selective method for the isolation of chromium(III) than extraction of the chromium(III) thiocyanate complex into TBP or cyclohexanone[168].

Chromium(VI) can be extracted from 1 M hydrochloric acid made 1 M in ammonium chloride into 30% TBP in xylene[169] or from 1 M hydrochloric acid made 2 M in potassium chloride into mesityl oxide[170]. Long chain amines can also be utilised for the isolation of chromium(VI)[171].

9.3.14.2   Molybdenum

The dithiol method[4] seems to be the most selective method for the determination of molybdenum[172]. Ascorbic acid and thiourea are recommended as masking agents for steel analysis[173], whereas tartaric and hydrofluoric acid can be used for the determination of molybdenum in tantalum[174].

The most sensitive method is based on the extraction of a mixed molybdenum–antipyrine–salicylfluorone–bromide complex into chloroform ($\varepsilon_{515nm} = 1.3 \times 10^5$)[175]. Preliminary extraction of molybdenum as its diethyldithiocarbamate from acid medium followed by back-extraction using an excess of copper(II) ions considerably increases the selectivity of the determination.

9.3.14.3   Tungsten

The extraction of the tungsten thiocyanate complex (tin(II) chloride and titanium(III) chloride are used to reduce tungsten (VI)) as an ion-association compound with tetraphenylarsonium cation into chloroform is rather selective (even a 25-fold excess of molybdenum does not interfere). The procedure is also very sensitive ($\varepsilon_{402nm} = 1.47 \times 10^4$)[176, 177].

9.3.14.4   Uranium

The uranyl benzoinate complex can be selectively extracted in the presence of Methylene Blue and EDTA into a 1 : 1 mixture of 1,2-dichloroethane and trichloroethane[178]. Another sensitive method is based on the extraction of the mixed uranium(VI)–3-pyridinecarboxylic acid–Rhodamine S complex into a 1:1 mixture of benzene and acetone ($\varepsilon_{556nm} = 1.2 \times 10^5$)[179].

The extraction of uranium(VI) by TOPO into cyclohexane followed by direct colour development in the organic extract with 2-(5-bromo-2-pyridyl-azo)-diethylaminophenol provides a highly selective method for its determination[180].

9.3.15 Group VII (transition)

9.3.15.1 Manganese

Manganese(II) can be extracted from aqueous media at pH ≈ 9.8 into dithizone in carbon tetrachloride in the presence of pyridine (0.6 M), hydroxyl-amine and potassium cyanide ($\varepsilon_{510nm} = 5.7 \times 10^4$)[71]. The interfering metals (with the exception of thallium(I)) can be removed by a preliminary dithizone extraction at pH c. 7.

The isolation of permanganate as an ion-association compound with the tetraphenylarsonium cation can be improved by using 1,2-dichloroethane as the organic solvent[181].

9.3.15.2 Technetium

Pertechnetate can be separated from fission products by extraction from 4 M sodium hydroxide using a freshly prepared mixture of tetrapropylam-monium hydroxide and bromoform (1:10), followed by back extraction into 1 M hydrochloric acid[182]. The extraction of ^{99}Tc by isobutyl methyl ketone in the pH range 0.5–13 can be used for its separation from ^{99m}Mo [183].

9.3.15.3 Rhenium

The reduction of rhenium in 1 M sulphuric acid with liquid zinc amalgam and extraction by isoamyl alcohol from 3 M sulphuric acid, separates rhenium from almost all interfering elements. Small amounts of molybdenum, uranium, iron and ruthenium still accompanying the rhenium are removed by a thiocyanate–amyl acetate or a 8-hydroxyquinoline–chloroform extraction[184].

The extraction of perrhenate with Victoria Blue R into benzene from 0.5 M sulphuric acid or from an oxalate medium (pH ≈ 7) provides a very sensitive procedure for the determination of rhenium ($\varepsilon_{610nm} = 8 \times 10^4$)[185].

9.3.16 Group VIII

9.3.16.1 Iron

The extraction of the iron(II)– o-phenanthroline–azide mixed complex into nitrobenzene (pH 3.5–6) can be used for the absorptiometric determination of this element[186]. Interferences caused by copper(II) and tungsten(VI) can be eliminated by using EDTA as a masking agent.

The preliminary extraction of iron(III) from 6 M hydrochloric acid into a 3:2 mixture of methyl ethyl ketone and carbon tetrachloride[187] considerably increases the selectivity of the acetylacetone or 2-thenoyltrifluoroacetone methods[4, 8].

9.3.16.2   Cobalt

An interesting new method for the sensitive determination of cobalt is based on extraction of the cobalt complex of nitroso-R-salt into 5% trioctyl-methylammonium chloride in chloroform[188]. Ammonium phosphate, fluoride and citrate mask large amounts of iron, nickel and copper[188]. Other new proposed methods using 8-mercaptoquinoline, 2,2'-pyridylmono-oxime, 3-hydroxy-1,3-diphenyltriazine, etc. are less selective than the 2-nitroso-1-naphthol method[4] in the presence of citrate as a masking agent[123].

9.3.16.3   Nickel

Several new methods have been proposed for nickel. The extraction of nickel from an acetate buffer by the zinc complex of p-hydroxydithiobenzoic acid in isoamyl alcohol is perhaps the most sensitive absorptiometric method available for the determination of this metal $(\varepsilon_{530nm} = 4.3 \times 10^4)$[189]. Treatment of the organic extract with 0.1 M sodium hydroxide, followed by 1 M hydrochloric acid, increases the selectivity of the method. Large amounts of iron(III), manganese(II), chromium(III), uranium(VI), 50-fold excesses of silver, cadmium, cobalt, mercury, molybdenum(VI), vanadium(V) and tungsten(VI) and 25-fold excesses of antimony(III), lead and bismuth do not interfere. Large amounts of copper can be masked by thiourea.

9.3.16.4   Ruthenium

Ruthenium(III) reacts on heating with tropolone (pH 4.5–6.0) to form a chelate that may be extracted into chloroform $(\varepsilon_{415nm} = 1.87 \times 10^4)$[190]. Preliminary extraction of the tropolone chelates of interfering metals or measurement of absorbance at 480 nm increases the selectivity considerably.

Ruthenium can also be selectively extracted as its tetroxide (after oxidation with a mixture of potassium persulphate and sodium periodate) from sodium hydroxide solution into carbon tetrachloride or other non-polar solvents[191, 192]. Ruthenium can be back-extracated from the extract into concentrated hydrochloric acid (only traces of iodine are also back extracted under these conditions).

9.3.16.5   Rhodium

Rhodium can be extracted as its tin chloride complex $[Rh_2Cl_2(SnCl_3)_4]^{4-}$ from 6 M hydrochloric acid made 0.06 M in stannous chloride, into 10% TBP in benzene[193]. After diluting the extract with benzene, rhodium can be back extracted into 0.25–0.5 M hydrochloric acid. The analogeous tin bromide complex is extracted from 2 M hydrobromic acid made 0.05 M in stannous bromide into 0.1 M diantipyrinylmethane (or diphenylguanidine or tribenzylamine) in chloroform ($\varepsilon_{430nm} = 4.9 \times 10^4$)[194].

9.3.16.6 Palladium

Although numerous reagents have been proposed for the isolation of palladium, the dimethylglyoxime and nitrosonaphthol methods still seem to be the most selective[4, 8]. Dithizone has also been suggested for the substoicheiometric determination[195].

9.3.16.7 Osmium

The extraction of osmium tetroxide is perhaps the most selective method for the isolation of osmium[3].

9.3.16.8 Iridium

Iridium (similarly to rhodium) can be extracted as its tin bromide complex from 2 M hydrobromic acid, made 0.1 M in stannous bromide, into 0.1 M diphenylguanidine in chloroform ($\varepsilon_{410nm} = 6.0 \times 10^4$) [194].

Iridium can be selectively determined as a complex with N,N'-di-(2-naphthyl)-p-phenylenediamine which can be extracted into chloroform[196]. By measuring the absorbance at 488 nm, even a 100-fold excess of rhodium(III), platinum(IV), osmium(IV), ruthenium(III) and palladium caused no interference; gold interferes seriously however.

9.3.16.9 Platinum

Platinum(II) can be extracted from acidic solutions into organic solvents containing dithizone[4]. All interfering metals can be removed by a preliminary dithizone extraction if the platinum is present in the inactive tervalent state. Isobutyl methyl ketone has been recommended[197] for the determination of platinum in rocks.

9.3.17 Transuranic elements
9.3.17.1 Neptunium

Neptunium(IV) can be quantitatively extracted by 0.5 M 2-thenoyltrifluoroacetone in xylene[4] from 1 M hydrochloric acid or by 0.5 M mono-(2-ethylhexyl)phosphoric acid in toluene from 12 M hydrochloric acid[8]. Under suitable reducing conditions, a solution may contain neptunium(IV), plutonium(III) and uranium(VI); only neptunium is extracted.

9.3.17.2 Plutonium

Plutonium(IV) can be separated from neptunium(V), uranium(VI) and higher transuranium elements by extraction with 2-thenoyltrifluoroacetone or di-(2-ethylhexyl)phosphoric acid in various organic solvents[4].

9.3.17.3 Americium

Americium(III) can best be separated from the higher transuranium elements by extractive chromatography using di-(2-ethylhexyl)phosphoric, 2-ethylhexylphosphonic or 1-methylheptylphenylphosphonic acids[36, 198].

References

1. Bagreev, V. V., Zolotov, Yu. A., Kurilina, N. A. and Kalinina, G. F. (1971). *Ekstraktsiya Neorganicheskikh Soedinenii*, Part I 328, Part II 336, (Moscow: Nauka)
2. Freiser, H. (1968). *Anal. Chem.*, **40**, 552R
3. Morrison, G. H. and Freiser, H. (1957). *Solvent Extraction in Analytical Chemistry*, 269 (New York: J. Wiley)
4. Starý, J. (1964). *The Solvent Extraction of Metal Chelates* (Oxford: Pergamon Press; (1966) Moscow: Mir)
5. Zolotov, Yu. A. (1968). *Ekstraktsiya Vnutrokompleksnykh Soedinenii*, 313 (Moscow: Nauka; (1970) Ann Arbor: Humprey Sci. Publ.)
6. Zolotov, Yu. A. and Kuz'min, N. M. (1971). *Ekstraktsionnoe Kontsentrirovanie*, 272 (Moscow: Khimiya)
7. Marcus, Y. and Kertes, A. S. (1969). *Ion Exchange and Solvent Extraction of Metal Complexes*, 1137 (New York: J. Wiley)
8. De, A. K., Khopkar, S. M. and Chalmers, R. A. (1970). *Solvent Extraction of Metals*, 259 (London: Van Nostrand Reinhold Co.)
9. Shmidt, V. S. (1970). *Ekstraktsiya Aminami*, 312 (Moscow: Atomizdat)
10. Dyrssen, D., Liljenzin, J. O. and Rydberg, J. (1967). *Solvent Extraction Chemistry*, 682 (Amsterdam: North-Holland)
11. Kertes, A. S. and Marcus, Y. (1970). *Solvent Extraction Research*, 434 (New York: J. Wiley)
12. Reinhardt, H. and Rydberg, J. (1969). *Acta Chem. Scand.*, **23**, 2773
13. Starý, J. and Liljenzin, J. O. (1969). *Radiochem. Radioanalyt. Letters*, **1**, 273
14. Liljenzin, J. O. and Starý, J. (1970). *J. Inorg. Nucl. Chem.*, **32**, 1357
15. Sekine, T. and Ishii, T. (1970). *Bull. Chem. Soc. Japan*, **43**, 2422
16. Riedl, A. (1970). *J. Radioanalyt. Chem.*, **6**, 75
17. Akiba, K., Suzuki, N., Asano, H. and Kanno, T. (1971). *J. Radioanalyt. Chem.*, **7**, 203
18. Suzuki, N. and Akiba, K. (1971). *J. Inorg. Nucl. Chem.*, **33**, 1897
19. Michelsen, O. B. and Smutz, M. (1971). *J. Inorg. Nucl. Chem.*, **33**, 265
20. Levin, I. S., Rodina, T. F. and Vorsina, I. A. (1970). *Zh. Neorg. Khim.*, **15**, 496
21. McDowell, W. J. (1971). *J. Inorg. Nucl. Chem.*, **33**, 1067
22. Šístková, N. V., Kolařík, Z. and Chotívka, V. (1970). *J. Inorg. Nucl. Chem.*, **32**, 637
23. Sato, T. and Nakamura, T. (1971). *J. Inorg. Nucl. Chem.*, **33**, 1081
24. Kiwan, A. M. and Amin, R. S. (1971). *J. Inorg. Nucl. Chem.*, **33**, 2221
25. Nakasuka, N., Nakai, M. and Tanaka, M. (1970). *J. Inorg. Nucl. Chem.*, **32**, 3667
26. Tanaka, M., Nakasuka, N. and Yamada, H. (1970). *J. Inorg. Nucl. Chem.*, **32**, 2791
27. Lobel, E., Zangen, M. and Kertes, A. S. (1970). *J. Inorg. Nucl. Chem.*, **32**, 483
28. Zolotov, Yu. A., Demina, L. A. and Petrukhin, O. M. (1970). *Zhur. Analit. Khim.*, **25**, 2315
29. Starý, J. and Burcl, R. (1971). *Radiochem. Radioanalyt. Letters*, **7**, 235
30. Starý, J. (1969). *Talanta*, **16**, 359
31. Usatenko, Yu. I., Barkalov, V. S. and Tulyupa, F. M. (1970). *Zhur. Analit. Khim.*, **25**, 1458
32. Chera, L. M. and Bankovskis, Yu. A. (1970). *Zhur. Analit. Khim.*, **25**, 1661
33. Hasegawa, Y. (1969). *Bull. Chem. Soc. Japan*, **42**, 3425
34. Suzuki, N. and Akiba, K. (1971). *J. Inorg. Nucl. Chem.*, **33**, 1169
35. Koshimura, H. and Okubo, T. (1970). *Anal. Chim. Acta*, **49**, 67
36. Weaver, B. and Shoun, R. R. (1971). *J. Inorg. Nucl. Chem.*, **33**, 1909
37. Marcus, Y. and Kertes, A. S. (1972). *Equilibrium Constants of Liquid-Liquid Distribution Reactions*, in the press
38. Blyum, I. A. and Oparina, L. I. (1970). *Zavod. Lab.*, **36**, 897

39. Krasnov, K. S. and Garceva, L. A. (1971). *Radiokhimiya*, **13**, 545, 551
40. Mrnka, M., Čeleda, J. and Stankov, V. (1971). *Chem. Prum.*, **21**, 55
41. Desreux, J. F. (1971). *Anal. Chim. Acta*, **53**, 117
42. Levy, O., Markovits, G. and Kertes, A. S. (1971). *J. Inorg. Nucl. Chem.*, **33**, 551
43. Danesi, P. R., Chiarizia, R. and Scibona, G. (1970). *J. Inorg. Nucl. Chem.*, **32**, 2349
44. Komarov, E. V. and Komarov, V. N. (1970). *Radiokhimiya*, **12**, 302
45. Makarov, V. M. (1970). *Radiokhimiya*, **12**, 584
46. Frolov, Yu. G., Sergievskii, V. V. and Ryabov, V. P. (1971). *Radiokhimiya*, **13**, 634
47. Shmidt, V. S., Shesternikov, V. N. and Mezhov, E. A. (1970). *Radiokhimiya*, **12**, 399
48. Shmidt, V. S., Mezhov, E. A. and Shesteruikov, V. N. (1970). *Radiokhimiya*, **12**, 590
49. Irving, H. M. N. H. and Damodaran, A. D. (1971). *Anal. Chim. Acta*, **53**, 267, 277
50. Vieux, A. S. (1969). *Bull. Soc. Chim. Fr.*, **9**, 3364
51. Pyatnitskii, I. V. and Tabenskaya, T. V. (1970). *Zhur. Analit. Khim.*, **25**, 943
52. Irving, H. M. N. H. and Al-Jarrah, R. H. (1971). *Anal. Chim. Acta*, **55**, 135
53. Iofa, B. Z., Yuscenko, A. S. and Kireev, G. I. (1971). *Radiokhimiya*, **13**, 391
54. Solovkin, A. S. (1971). *Zhur. Neorg. Khim.*, **16**, 865
55. Moskvin, A. I. (1971). *Zhur. Neorg. Khim.*, **16**, 759
56. Ellert, G. V., Bolotova, G. T. and Krasovskaya, T. I. (1971). *Zhur. Neorg. Khim.*, **16**, 789
57. Davis, V. J., Mrochek, J. and Judkins, R. R. (1970). *J. Inorg. Nucl. Chem.*, **32**, 1689
58. Lodhi, M. A., Danesi, P. R. and Scibona, G. (1971). *J. Inorg. Nucl. Chem.*, **33**, 1889
59. Levin, V. I., Novoselov, V. S. and Kozlova, M. D. (1971). *Radiokhimiya*, **13**, 538
60. Kennedy, D. C. and Fritz, J. S. (1970). *Talanta*, **17**, 823
61. Shanker, R. and Venkateswarlu, K. S. (1970). *J. Inorg. Nucl. Chem.*, **32**, 229
62. Rozen, A. M., Murinov, Yu. I. and Nikitin, Yu. E. (1970). *Radiokhimiya*, **12**, 355
63. Laurence, G. (1970). *J. Inorg. Nucl. Chem.*, **32**, 3065
64. Shanker, R. and Venkateswarlu, K. S. (1970). *J. Inorg. Nucl. Chem.*, **32**, 2369
65. Subramanian, M. S. and Pai, S. A. (1970). *J. Inorg. Nucl. Chem.*, **32**, 3677
66. Jacobs, E. B. and Walker, W. R. (1970). *Aust. J. Chem.*, **23**, 2413
67. Chmutova, M. K. and Kochetkova, N. E. (1970). *Zhur. Analit. Khim.*, **25**, 710
68. Irving, H. M. N. H. and Sinha, S. P. (1970). *Anal. Chim. Acta*, **51**, 39
69. Sigematsu, T., Honjyo, T., Tabushi, M. and Matsui, M. (1970). *Bull. Chem. Soc. Japan*, **43**, 796
70. Zolotov, Yu. A., Petrukhin, O. M. and Gavrilova, L. G. (1970). *J. Inorg. Nucl. Chem.*, **32**, 1679
71. Marczenko, Z. and Mojski, M. (1971). *Anal. Chim. Acta*, **54**, 469
72. Kolařik, Z. (1971). *J. Inorg. Nucl. Chem.*, **33**, 1135
73. Shigematsu, T. and Honjyo, T. (1970). *Bull. Chem. Soc. Japan*, **43**, 796
74. Woo, C., Wagner, W. F. and Sands, D. E. (1971). *J. Inorg. Nucl. Chem.*, **33**, 2661
75. Murata, K. and Kiba, T. (1970). *J. Inorg. Nucl. Chem.*, **32**, 1667
76. Lakshmanan, V. I. and Haldar, B. C. (1970). *J. Indian Chem. Soc.*, **47**, 231
77. Levin, I. S., Yukhín, Yu. M. and Vorsina, I. A. (1970). *Izv. Sib. Otd. Akad. Nauk. SSSR, Ser. Khim. Nauk.*, **(1)**, 61
78. Akatsu, E. and Asano, M. (1971). *Anal. Chim. Acta*, **55**, 333
79. Mailen, J. C. and Ferris, L. M. (1971). *Inorg. Nucl. Chem. Lett.*, **7**, 431
80. Ferris, L. M., Mailen, J. C., Lawrence, J. J., Shmith, F. J. and Nogueira, E. D. (1970). *J. Inorg. Nucl. Chem.*, **32**, 2019
81. Ramakrishna, R. S. and Irving, H. M. N. H. (1969). *Anal. Chim. Acta*, **48**, 251
82. Pietsch, R. and Sinic, H. (1970). *Anal. Chim. Acta*, **49**, 51
83. Perricos, D. C., Tsolis, A. K. and Belkas, E. P. (1970). *Talanta*, **17**, 551
84. Förster, H. (1970). *J. Radioanalyt. Chem.*, **4**, 1
85. Förster, H. (1970). *J. Radioanalyt. Chem.*, **6**, 11
86. Šebesta, F. (1971). *J. Radioanalyt. Chem.*, **7**, 41
87. Starý, J. and Ružička, J. (1971). *Talanta*, **18**, 1
88. Starý, J., Kratzer, K. and Zeman, A. (1970). *J. Radioanalyt. Chem.*, **5**, 71
89. Starý, J., Kratzer, K. and Zeman, A. (1971). *Radiochem. Radioanalyt. Letters*, **6**, 1
90. Rais, J. and Selucký, P. (1971). *Radiochem., Radioanalyt. Letters*, **6**, 257
91. Rais, J., Selucký, P. and Dražanová, S. (1970). *Radiochem. Radioanalyt. Letters*, **4**, 195
92. Everett, R. J. and Mottola, H. A. (1971). *Anal. Chim. Acta*, **54**, 309
93. Radionova, G. S., Alekseeva, V. V. and Starostin, V. V. (1970). *Zhur. Neorg. Khim.*, **15**, 176

94. Fukamachi, K., Kohara, H. and Ishibashi, N. (1970). *Bunseki Kagaku*, **19**, 1529
95. Bel'tyukova, S. V. and Poluektov, N. S. (1970). *Zhur. Analit. Khim.*, **25**, 1714
96. Gorbenko, F. P., Kuchkina, E. D. and Olevinskii, M. I. (1970). *Radiokhimiya*, **12**, 661
97. Gorbenko, F. P., Sachko, V. V., Lapitskaya, E. V. and Nodezhda, A. A. (1969). *Zhur. Prikl. Khim. (Leningrad)*, **42**, 2212
98. Edelbeck, L. and West, P. W. (1970). *Anal. Chim. Acta*, **52**, 447
99. Vernon, F. and Williams, J. M. (1970). *Anal. Chim. Acta*, **51**, 533
100. Dagnall, R. M., West, T. S. and Young, P. (1965). *Analyst*, **90**, 13
101. Keil, R. (1970). *Fresenius' Z. Anal. Chem.*, **249**, 172
102. Kish, P. P. and Bukovich, A. M. (1969). *Ukr. Khim. Zh.*, **35**, 1290
103. Garčic, A. and Sommer, L. (1970). *Coll. Czech. Chem. Commun.*, **35**, 1047
104. Tarayan, V. M., Ovsepyan, E. N. and Artsruni, V. Zh. (1970). *Zhur. Analit. Khim.*, **25**, 691
105. Adámek, A. and Obrusnik, I. (1971). *Radiochem. Radioanalyt. Letters*, **7**, 147
106. Halász, A., Polyák, K. and Pungor, E. (1971). *Talanta*, **18**, 691
     Nazarenko, V. A. and Makrinich, N. I. (1970). *Zhur. Analit. Khim.*, **25**, 719
108. Nazarenko, V. A., Makrinich, N. I. and Shustova, M. B. (1970). *Zhur. Analit. Khim.*, **25**, 1595
109. Shirodker, R. and Schibilla, E. (1969). *Fresenius' Z. Anal. Chem.*, **248**, 173
110. Asmus, E. and Weinert, H. (1970). *Fresenius' Z. Anal. Chem.*, **249**, 179
111. Babko, A. K., Shkaravsky, Yu. F. and Ivashkovich, E. M. (1971). *Zhur. Analit. Khim.*, **26**, 854
112. Fujinuma, H. and Yoshida, I. (1970). *Bunseki Kagaku*, **19**, 1273
113. Pakalns, P. (1970). *Anal. Chim. Acta*, **50**, 103
114. Shkaravskii, Yu. F., Lynchak, K. A. and Chernogorenko, V. B. (1970). *Zavod. Lab.*, **36**, 524
115. Stará, V. and Starý, J. (1970). *Talanta*, **17**, 341
116. Arnold, A., Davis, S. and Jordan, A. L. (1969). *Analyst*, **94**, 664
117. Zeman, A., Starý, J. and Kratzer, K. (1970). *Radiochem. Radioanal. Letters*, **4**, 1
118. Yadav, A. A. and Khopkar, S. M. (1971). *Bull. Chem. Soc. Japan*, **44**, 693
119. Narushkyavichus, L. R., Kazlauskas, R. M. and Shkadauskas, Yu. S. (1971). *Zhur. Analit. Khim.*, **26**, 922
120. Kish, P. P. and Onishenko, Yu. K. (1971). *Zhur. Analit. Khim.*, **26**, 514
121. Kish, P. P. and Onishenko, Yu. K. (1970). *Zhur. Analit. Khim.*, **25**, 500
122. Shestidesyatnaya, N. L., Kish, P. P. and Merenich, A. V. (1970). *Zhur. Analit. Khim.*, **25**, 1547
123. Patrovský, V. (1970). *Chem. Listy*, **64**, 715
124. Akki, S. B. and Khopkar, S. M. (1971). *Separ. Sci.*, **6**, 455
125. Yadar, A. A. and Khopkar, S. M. (1971). *Chem. Anal. (Warsaw)*, **16**, 299
126. Kunte, N. S. and Ranade, S. N. (1970). *Indian J. Chem.*, **8**, 370
127. Nazarenko, I. I., Kislov, A. M., Kislova, I. V. and Maslevskii, A. Yu. (1970). *Zhur. Analit. Khim.*, **25**, 1135
128. Kasterka, B. and Dobrovolski, J. (1970). *Chem. Anal. (Warsaw)*, **15**, 303
129. Starý, J., Marek, J., Kratzer, K. and Šebesta, F. (1972). *Anal. Chim. Acta*, **57**, 393
130. Kawamura, K., Ito, H. and Tanabe, T. (1970). *Bunseki Kagaku*, **19**, 824
131. Kish, P. P. and Kremeneva, S. G. (1970). *Zhur. Analit. Khim.*, **25**, 2200
132. Haarsma, J. P. S. and Agterdenbos, J. (1971). *Talanta*, **18**, 747
133. Bishop, C. T. (1970). *U.S. Atom. Energy Commun. MLM-1721*
134. Einaga, H. and Ishii, H. (1969). *Bunseki Kagaku*, **18**, 1211
135. Tsubouchi, M. and Yamamoto, Yu. (1970). *Bunseki Kagaku*, **19**, 966
136. Shigematsu, T., Matsui, M., Aoki, T. and Ito, M. (1970). *Bunseki Kagaku*, **19**, 412
137. Behrends, K. and Klein, H. (1970). *Fresenius' Z. Anal. Chem.*, **249**, 165
138. Tsubouchi, M. (1971). *Bull. Chem. Soc. Japan*, **44**, 554
139. Jutte, B. A. H. G., Agterdenbos, J. and Elberse, P. A. (1970). *Talanta*, **17**, 1130
140. Busev, A. I., Gorbunova, N. N. and Ivanov, V. M. (1971). *Zavod. Lab.*, **37**, 26
141. Ziegler, M. and Stephan, G. (1970). *Mikrochim. Acta*, **(3)**, 628
142. Shibata, S., Furukawa, M. and Sasaki, S. (1970). *Anal. Chim. Acta*, **51**, 271
143. Uny, G., Mathien, C., Tardif, J. P. and Tran Van Danh (1971). *Anal. Chim. Acta*, **53**, 109
144. Ghersini, G. and Mariottini, S. (1971). *Talanta*, **18**, 442
145. Kono, T. and Kobayashi, S. (1970). *Bunseki Kagaku*, **19**, 1491

146. Hashitani, H. and Katsuyama, K. (1970). *Bunseki Kagaku*, **19**, 355
147. Tsubouchi, M. (1970). *Anal. Chem.*, **42**, 1087
148. Eremin, Yu. G. and Katochkina, V. S. (1969). *Zavod. Lab.*, **35**, 1425
149. Rossmanith, K. (1970). *Monatsh. Chem.*, **101**, 1665
150. Poluektov, N. S. and Mishchenko, V. T. (1969). *Zhur. Analit. Khim.*, **24**, 1434
151. Pfeifer, V. (1970). *Mikrochim. Acta*, **(6)**, 1232
152. Kratzer, K., Starý, J., Zeman, A. and Majer, V. *Collect. Czech. Chem. Commun.* (in press).
153. Titkov, Yu. B. (1971). *Ukr. Khim. Zh.*, **37**, 57
154. Hofer, A. and Heidinger, R. (1970). *Fresenius' Z. Analyt. Chem.*, **249**, 177
155. Villarreal, R., Young, J. O. and Krsul, J. R. (1970). *Anal. Chem.*, **42**, 1419
156. Kajiyama, R. and Senuma, K. (1970). *Bunseki Kagaku*, **19**, 1163
157. Mishchenko, V. T. and Zavarina, T. V. (1970). *Zhur. Analit. Khim.*, **25**, 1533
158. Khosla, M. M. L. and Rao, S. (1961). *Anal. Chim. Acta*, **54**, 315
159. Donaldson, E. M. (1970). *Talanta*, **17**, 583
160. Bhura, D. C. and Tandon, S. G. (1971). *Anal. Chim. Acta*, **53**, 379
161. Uvarova, K. A., Usatenko, Yu. I. and Klopova, Zh. G. (1970). *Zavod. Lab.*, **36**, 909
162. Yagnyatinskaya, G. Ya. (1970). *Zavod. Lab.*, **36**, 158
163. Alimarin, I. P., Ivanov, N. A. and Gibalo, I. M. (1969). *Zhur. Analit. Khim.*, **24**, 1521
164. Onishi, H. and Nagai, H. (1971). *Bunseki Kagaku*, **20**, 86
165. Bilimovich, G. N., Alimarin, I. P. and Tikhonova, T. V. (1971). *Zhur. Analit. Khim.*, **26**, 122
166. Davydov, A. V. and Palshin, E. S. (1970). *Zhur. Analit. Khim.*, **25**, 1558
167. Myasoedov, B. F., Molochnikova, N. P. and Palci, P. N. (1970). *Radiokhimiya*, **12**, 829
168. Sukhanovskaya, A. I., Solov'ev, E. A., Tikhonov, G. P., Golubev, V. Yu. and Bezhevol'-nov, E. A. (1970). *Zhur. Analit. Khim.*, **25**, 1563
169. Yadav, A. A. and Khopkar, S. M. (1970). *Indian J. Chem.*, **8**, 290
170. Shinde, V. M. and Khopkar, S. M. (1970). *Fresenius' Z. Anal. Chem.*, **249**, 239
171. Adam, J. and Přibil, R. (1971). *Talanta*, **18**, 91
172. Busev, A. I. and Rodionova, T. V. (1971). *Zhur. Analit. Khim.*, **26**, 578
173. Tanaka, K. and Takagi, N. (1970). *Bunseki Kagaku*, **19**, 790
174. Yamane, T., Iida, K., Mukoyama, T. and Fukasawa, T. (1970). *Bunseki Kagaku*, **19**, 808
175. Nazarenko, V. A., Shustova, M. B. and Shelikhina, E. I. (1970). *Zhur. Analit. Khim.*, **25**, 2139
176. Fogg, A. G., Marriott, D. R. and Burns, D. T. (1970). *Analyst*, **95**, 854
177. Kajiyama, R., Ichihashi, K. and Ichikawa, K. (1969). *Bunseki Kagaku*, **18**, 1500
178. Tarayan, V. M., Ovsepyan, E. N. and Petrosyan, A. A. (1971). *Zhur. Analit. Khim.*, **26**, 322
179. Poluektov, N. S. and Bel'tyukova, S. V. (1971). *Zhur. Analit. Khim.*, **26**, 541
180. Johanson, D. A. and Florence, T. M. (1971). *Anal. Chim. Acta*, **53**, 73
181. Goto, H. and Kakita, Y. (1971). *Fresenius' Z. Anal. Chem.*, **254**, 18
182. Swindle, D. L. and Kuroda, P. K. (1971). *Radiochem. Radioanalyt. Letters*, **7**, 229
183. Baishya, N. K., Heslop, R. B. and Ramsey, A. C. (1970). *Radiochem. Radioanalyt. Letters*, **4**, 15
184. Yatirajan, V. and Kakkar, L. R. (1970). *Talanta*, **17**, 759
185. Pilipenko, A. T., Kish, P. P. and Zheltvaj, I. I. (1971). *Ukr. Khim. Zh.*, **37**, 477
186. Rao, V. P. R. and Sarma, P. V. R. B. (1970). *Mikrochim. Acta*, **(4)**, 783
187. Gagliardi, E. and Wöss, H. P. (1969). *Fresenius' Z. Anal. Chem.*, **248**, 302
188. Adam, J. and Přibil, R. (1971). *Talanta*, **18**, 733
189. Rudzit, G. P., Pastare, S. Ya. and Yansons, E. Yu. (1970). *Zh. Analit. Khim.*, **25**, 2407
190. Rizvi, G. H., Gupta, B. P. and Singh, R. P. (1971). *Anal. Chim. Acta*, **54**, 295
191. Krtil, J., Mencl, J. and Bulovič, V. (1970). *Radiochem. Radioanalyt. Letters*, **4**, 355
192. Smulek, W. (1969). *Radiochem. Radioanalyt. Letters*, **2**, 265
193. Kalinin, S. K., Katikhin, G. S., Nikitin, M. K. and Yakovleva, G. A. (1970). *Zhur. Analit. Khim.*, **25**, 535
194. Pilipenko, A. T., Danilova, V. N. and Lisichenok, S. L. (1970). *Zhur. Analit. Khim.*, **25**, 1154
195. Briscoe, G. B. and Humphries, S. (1971). *Talanta*, **18**, 39
196. Nasouri, F. G. and Witwit, A. S. (1970). *Anal. Chim. Acta*, **50**, 163
197. Simonsen, A. (1970). *Anal. Chim. Acta*, **49**, 368
198. Starý, J. (1966). *Talanta*, **13**, 421